Engineering Graphics
with SOLIDWORKS 2019

A Step-by-Step Project Based Approach
Utilizing 3D Solid Modeling

David C. Planchard
CSWP & SOLIDWORKS Accredited Educator

SDC Publications
P.O. Box 1334
Mission, KS 66222
913-262-2664
www.SDCpublications.com
Publisher: Stephen Schroff

INTRODUCTION

Engineering Graphics with SOLIDWORKS® 2019 is written to assist students, designers, engineers and professionals who are new to SOLIDWORKS.

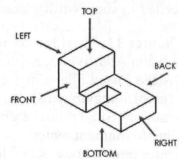

The book combines the fundamentals of engineering graphics and dimensioning practices with a step-by-step project based approach to learning SOLIDWORKS. The book is divided into four sections with 11 Chapters.

Chapters 1 - 3: Explore the history of engineering graphics, manual sketching techniques, orthographic projection, Third vs. First angle projection, multiview drawings, dimensioning practices (ASME Y14.5-2009 standard), line type, fit type, tolerance, fasteners in general, general thread notes and the history of CAD leading to the development of SOLIDWORKS.

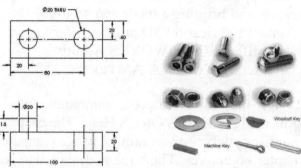

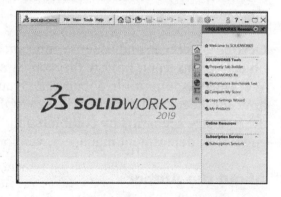

Chapters 4 - 9: Comprehend the SOLIDWORKS User Interface and CommandManager, Document and System properties, simple machine parts, simple and complex assemblies, proper design intent, design tables, configurations, multi-sheet, multi-view drawings, BOMs, and Revision tables using basic and advanced features.

Follow the step-by-step instructions in over 80 activities to develop eight parts, four sub-assemblies, three drawings and six document templates.

Chapter 10: Prepare for the Certified SOLIDWORKS Associate (CSWA) exam. Understand the curriculum and categories of the CSWA exam and the required model knowledge needed to successfully take the exam.

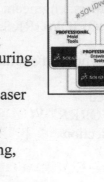

Chapter 11: Provide a basic understanding between Additive vs. Subtractive manufacturing. Discuss Fused Filament Fabrication (FFF), STereoLithography (SLA), and Selective Laser Sintering (SLS) printer technology. Select suitable filament material. Comprehend 3D printer terminology. Knowledge of preparing, saving, and printing a model on a Fused Filament Fabrication 3D printer. Information on the Certified SOLIDWORKS Additive Manufacturing (CSWA-AM) exam.

Review individual features, commands, and tools using SOLIDWORKS Help. The chapter exercises analyze and examine usage competencies based on the chapter objectives. The book is designed to complement the SOLIDWORKS Tutorials located in the SOLIDWORKS Help menu.

Desired outcomes and usage competencies are listed for each project. Know your objectives up front. Follow the step-by step procedures to achieve your design goals. Work between multiple documents, features, commands, and properties that represent how engineers and designers utilize SOLIDWORKS in industry. The author developed the industry scenarios by combining his own industry experience with the knowledge of engineers, department managers, vendors and manufacturers.

About the Author

David Planchard is the founder of D&M Education LLC. Before starting D&M Education, he spent over 27 years in industry and academia holding various engineering, marketing, and teaching positions. He holds five U.S. patents. He has published and authored numerous papers on Machine Design, Product Design, Mechanics of Materials, and Solid Modeling. He is an active member of the SOLIDWORKS Users Group and the American Society of Engineering Education (ASEE). David holds a BSME, MSM with the following professional certifications: CCAI, CCNP, CSWA-SD, CSWSA-FEA, CSWA-AM, CSWP, CSWP-DRWT and SOLIDWORKS Accredited Educator. David is a SOLIDWORKS Solution Partner, an Adjunct Faculty member and the SAE advisor at Worcester Polytechnic Institute in the Mechanical Engineering department. In 2012, David's senior Major Qualifying Project team (senior capstone) won first place in the Mechanical Engineering department at WPI. In 2014, 2015 and 2016, David's senior Major Qualifying Project teams won the Provost award in the Mechanical Engineering department for design excellence. In 2018, David's senior Major Qualifying Project team (Co-advisor) won the Provost award in the Electrical and Computer Engineering department. Subject area: Electrical System Implementation of Formula SAE Racing Platform.

David Planchard is the author of the following books:

- **SOLIDWORKS® 2019 Reference Guide**, 2018, 2017, 2016, 2015, 2014, 2013, 2012, 2011, 2010, and 2009

- **Engineering Design with SOLIDWORKS® 2019**, 2018, 2017, 2016, 2015, 2014, 2013, 2012, 2011, 2010, 2009, 2008, 2007, 2006, 2005, 2004, and 2003

- **Engineering Graphics with SOLIDWORKS® 2019**, 2018, 2017, 2016, 2015, 2014, 2013, 2012, and 2011

- **SOLIDWORKS® 2019 Quick Start**, 2018

- **SOLIDWORKS® 2017 in 5 Hours with video instruction**, 2016, 2015, and 2014

- **SOLIDWORKS® 2019 Tutorial**, 2018, 2017, 2016, 2015, 2014, 2013, 2012, 2011, 2010, 2009, 2008, 2007, 2006, 2005, 2004, and 2003

- **Drawing and Detailing with SOLIDWORKS® 2014**, 2012, 2010, 2009, 2008, 2007, 2006, 2005, 2004, 2003, and 2002

- **Official Certified SOLIDWORKS® Professional (CSWP) Certification Guide with video instruction, Version 4: 2015 - 2017**, Version 3: 2012 - 2014, Version 2: 2012 - 2013, Version 1: 2010 - 2010

- **Official Guide to Certified SOLIDWORKS® Associate Exams: CSWA, CSWA-SD, CSWSA-FEA, CSWA-AM Version 4: 2017 - 2019**, Version 3: 2015 - 2017, Version 2: 2012 - 2015, Version 1: 2012 -2013

- **Assembly Modeling with SOLIDWORKS® 2012**, 2010, 2008, 2006, 2005-2004, 2003, and 2001Plus

- **Applications in Sheet Metal Using Pro/SHEETMETAL & Pro/ENGINEER**

Acknowledgements

Writing this book was a substantial effort that would not have been possible without the help and support of my loving family and of my professional colleagues. I would like to thank Professor John M. Sullivan Jr., Professor Jack Hall and the community of scholars at Worcester Polytechnic Institute who have enhanced my life, my knowledge and helped to shape the approach and content to this text.

The author is greatly indebted to my colleagues from Dassault Systèmes SOLIDWORKS Corporation for their help and continuous support: Avelino Rochino and Mike Puckett.

Thanks also to Professor Richard L. Roberts of Wentworth Institute of Technology, Professor Dennis Hance of Wright State University, Professor Jason Durfess of Eastern Washington University and Professor Aaron Schellenberg of Brigham Young University - Idaho who provided vision and invaluable suggestions.

SOLIDWORKS certification has enhanced my skills and knowledge and that of my students. Thank you to Ian Matthew Jutras (CSWE) who is a technical contributor and Stephanie Planchard, technical procedure consultant.

Contact the Author

We realize that keeping software application books current is imperative to our customers. We value the hundreds of professors, students, designers, and engineers that have provided us input to enhance the book. Please contact me directly with any comments, questions or suggestions on this book or any of our other SOLIDWORKS books at dplanchard@msn.com or planchard@wpi.edu.

Note to Instructors

Please contact the publisher www.SDCpublications.com for additional classroom support materials: PowerPoint presentations, Adobe files along with avi files, additional design projects, quizzes with initial and final SOLIDWORKS models and tips that support the usage of this text in a classroom environment.

Trademarks, Disclaimer, and Copyrighted Material

SOLIDWORKS®, eDrawings®, SOLIDWORKS Simulation, and SOLIDWORKS Flow are a registered trademark of Dassault Systèmes SOLIDWORKS Corporation in the United States and other countries; certain images of the models in this publication courtesy of Dassault Systèmes SOLIDWORKS Corporation.

Microsoft Windows®, Microsoft Office® and its family of products are registered trademarks of the Microsoft Corporation. Other software applications and parts described in this book are trademarks or registered trademarks of their respective owners.

The publisher and the author make no representations or warranties with respect to the accuracy or completeness of the contents of this work and specifically disclaim all warranties, including without limitation warranties of fitness for a particular purpose.

No warranty may be created or extended by sales or promotional materials. Dimensions of parts are modified for illustration purposes. Every effort is made to provide an accurate text. The author and the manufacturers shall not be held liable for any parts, components, assemblies or drawings developed or designed with this book or any responsibility for inaccuracies that appear in the book. Web and company information was valid at the time of this printing.

The Y14 ASME Engineering Drawing and Related Documentation Publications utilized in this text are as follows: ASME Y14.1 1995, ASME Y14.2M-1992 (R1998), ASME Y14.3M-1994 (R1999), ASME Y14.41-2003, ASME Y14.5-1982, ASME Y14.5-2009, and ASME B4.2. Note: By permission of The American Society of Mechanical Engineers, Codes and Standards, New York, NY, USA. All rights reserved.

Additional information references the American Welding Society, AWS 2.4:1997 Standard Symbols for Welding, Braising, and Non-Destructive Examinations, Miami, Florida, USA.

References

- SOLIDWORKS Users Guide, SOLIDWORKS Corporation, 2019

- ASME Y14 Engineering Drawing and Related Documentation Practices

- Beers & Johnson, <u>Vector Mechanics for Engineers</u>, 6th ed. McGraw Hill, Boston, MA

- Lockhart & Johnson, <u>Engineering Design Communications</u>, Addison Wesley, 1999

- Olivo C., Payne, Olivo, T, <u>Basic Blueprint Reading and Sketching</u>, Delmar, 1988

- Planchard & Planchard, <u>Drawing and Detailing with SOLIDWORKS</u>, SDC Pub., Mission, KS 2014

- Simpson Strong Tie Product Manual, Simpson Strong Tie, CA, 2008

- Ticona Designing with Plastics - The Fundamentals, Summit, NJ, 2008

- Emhart - A Black and Decker Company, On-line catalog, Hartford, CT, 2012

During the initial SOLIDWORKS installation, you are requested to select either the ISO or ANSI drafting standard. ISO is typically a European drafting standard and uses First Angle Projection. The book is written using the ANSI (US) overall drafting standard and Third Angle Projection for drawings.

View the provided models to enhance the user experience.

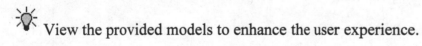

TABLE OF CONTENTS

View the provided models to enhance the user experience. Additional projects are included in the exercise section.

Overview of Chapters

Chapter 1: History of Engineering Graphics

Chapter 1 provides a broad discussion of the history of Engineering Graphics and the evolution of manual drawing/drafting.

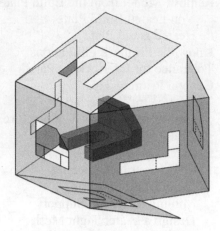

The chapter provides a general understanding of the following: 2D hand sketching techniques, Alphabet of lines, Precedence of line types, Global and Local Coordinate system, 2D and 3D Cartesian Coordinate system, Terminology and divisions of Projection, Orthographic Projection, Glass Box, Principle views and First and Third Angle Projection type.

Chapter 2: Isometric Projection and Multi View Drawings

Chapter 2 provides a general introduction into Isometric Projection (six principle views) using Third Angle projection and sketching along with additional projections and arrangement of views.

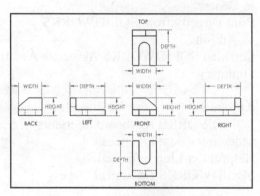

It covers freehand engineering sketching and drawing techniques, the three main projection divisions (Axonometric, Obliques and Perspective), along with Boolean operation (Union, Difference and Intersection), proper design intent, advanced drawing views and an introduction to the evolution from manual drawing/drafting to early CAD systems and finally to SOLIDWORKS.

Chapter 3: Dimensioning Practices, Scales, Tolerancing and Fasteners

Chapter 3 provides an introduction into dimensioning systems, dimensioning units, and scales along with knowledge of the ASME Y14.5-2009 standard.

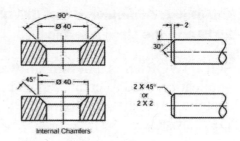

- Understand and apply the ASME Y14.5-2009 dimensioning standard.

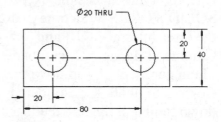

- Awareness of measurement units:

  o Metric system (MMGS).

  o English system (IPS).

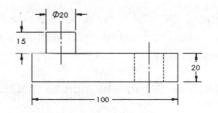

- Familiarity of dual dimensioning:

  o (Primary vs. Secondary).

- Understand Scale type:

  o Engineer's scale, Architect's scale, Linear scale, Vernier scale and Linear encoder.

- Ability to correctly dimension the following features, objects and shapes: rectangle, Cone, Sphere, hole, cylinder, angle, point or center, arc, chamfer and more.

- Understand and apply part and drawing Tolerance.

- Read and understand Fastener notation.

- Recognize single, double and triple thread.

- Distinguish between Right-handed and Left-handed thread.

- Recognize annotations for a simple hole, Counterbore and Countersink in a drawing.

- Identify Fit type.

Chapter 4: Overview of SOLIDWORKS 2019 and the User Interface

SOLIDWORKS is a design software application used to create 2D and 3D sketches, 3D parts and assemblies and 2D drawings.

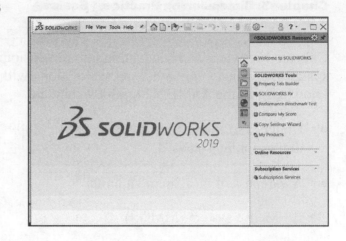

Chapter 4 introduces the user to the SOLIDWORKS Welcome dialog box, the User Interface (UI) and CommandManager: Menu bar toolbar, Menu bar menu, Drop-down menus, Context toolbars, Consolidated drop-down toolbars, System feedback icons, Confirmation Corner, Heads-up View toolbar, Document Properties and more.

Start a new SOLIDWORKS Session. Create a new part. Open an existing part and view the created features and sketches using the Rollback bar. Design the part using proper design intent.

Chapter 5: Introduction to SOLIDWORKS Part Modeling

Chapter 5 provides a comprehensive understanding of System Options, Document Properties, Part templates, File management and more.

Create two Part Templates utilized for the parts in this chapter:

- PART-IN-ANSI
- PART-MM-ISO

A Template is the foundation for a SOLIDWORKS document. Templates are part, drawing, and assembly documents that include user-defined parameters and are the basis for new documents.

Create two parts for the FLASHLIGHT assembly:

- BATTERY
- BATTERYPLATE

Chapter 6: Revolved Boss/Base Features

In Chapter 6, create two parts for the FLASHLIGHT assembly:

- LENS

- BULB

You apply the following features: Extruded Base, Extruded Boss, Extruded Cut, Revolved Base, Revolved Boss Thin, Revolved Thin Cut, Dome, Shell, Hole Wizard, and Circular Pattern along with the following Geometric relations: Equal, Coincident, Symmetric, Intersection and Perpendicular.

Chapter 7: Swept, Lofted, Rib, Mirror and Additional Features

In Chapter 7, you create four parts: O-RING, SWITCH, LENSCAP and HOUSING.

Chapter 7 covers the development of the Swept Base, Swept Boss, Lofted Base, Lofted Boss, Mirror, Draft, Shape, Rib, and Linear Pattern features and strengthens the use of the previously applied features and sketch tool along with proper design intent.

Chapter 8: Assembly Modeling - Bottom-up method

In Chapter 8, you learn about the Bottom-up assembly technique and create four assemblies: LENSANDBULB, CAPANDLENS, BATTERYANDPLATE and the FLASHLIGHT assembly.

You insert the following Standard mate types: Coincident, Concentric, and Distance and use the following tools: Insert Component, Hide/Show, Suppress/Unsuppress, Mate, Move Component, Rotate Component, Exploded View and Interference Detection.

Chapter 9: Drawing Fundamentals

Chapter 9 covers the development of a customized drawing template with a Company logo and Custom Properties.

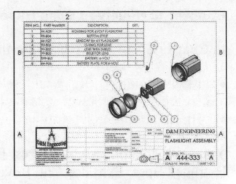

Create a BATTERY drawing with five views: Front, Top, Right, Detail and Isometric. Insert all needed dimensions and annotations in the correct drawing views.

Create an Exploded Isometric FLASHLIGHT assembly drawing.

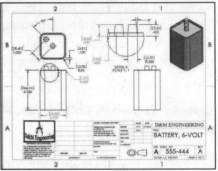

Insert a Bill of Materials (BOM) with Balloons and a Revision table along with Custom Properties.

Create an O-RING TABLE drawing.

Chapter 9 introduces Design Tables and configurations in the drawing.

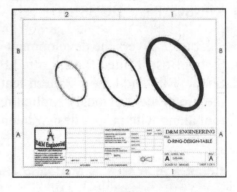

Create three configurations of the O-RING part.

Chapter 10: Introduction to the Certified SOLIDWORKS Associate (CSWA) Exam

Chapter 10 provides a basic introduction into the curriculum and five exam categories for the Certified SOLIDWORKS Associate (CSWA) exam.

Review the exam procedure, process and required model knowledge needed to take and pass the exam.

- Review the five exam categories: *Drafting Competencies, Basic Part Creation and Modification, Intermediate Part Creation and Modification, Advanced Part Creation and Modification,* and *Assembly Creation and Modification.*

Chapter 11: Additive Manufacturing - 3D Printing

Provide a basic understanding between Additive vs. Subtractive manufacturing. Discuss Fused Filament Fabrication (FFF), STereoLithography (SLA), and Selective Laser Sintering (SLS) printer technology. Select suitable filament material. Comprehend 3D printer terminology. Knowledge of preparing, saving, and printing a model on a Fused Filament Fabrication 3D printer. Information on the Certified SOLIDWORKS Associate Additive Manufacturing (CSWA-AM) exam.

On the completion of this chapter, you will be able to:

- Discuss Additive vs Subtractive manufacturing.

- Review 3D printer technology: Fused Filament Fabrication (FFF), STereoLithography (SLA), and Selective Laser Sintering (SLS).

- Select the correct filament material:

  o PLA (Polylactic acid), FPLA (Flexible Polylactic acid), ABS (Acrylonitrile butadiene styrene), PVA (Polyvinyl alcohol), Nylon 618, and Nylon 645.

- Create an STL (*.stl) file, an Additive Manufacturing (*.amf) file and a 3D Manufacturing format (*.3mf) file.

- Prepare G-code.

- Comprehend general 3D printer terminology.

- Understand optimum build orientation.

- Enter slicer parameters:

  o Raft, brim, skirt, layer height, percent infill, infill pattern, wall thickness, fan speed, print speed, bed temperature, and extruder (hot end) temperature.

- Address fit tolerance for interlocking parts.

- Define general 3D Printing tips.

- Print directly from SOLIDWORKS.

- Knowledge of the Certified SOLIDWORKS Associate Additive Manufacturing exam.

About the Book

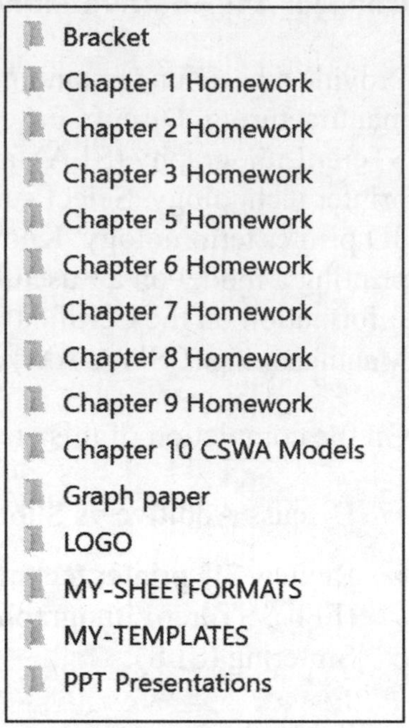

The following conventions are used throughout this book:

▓	Bracket
▓	Chapter 1 Homework
▓	Chapter 2 Homework
▓	Chapter 3 Homework
▓	Chapter 5 Homework
▓	Chapter 6 Homework
▓	Chapter 7 Homework
▓	Chapter 8 Homework
▓	Chapter 9 Homework
▓	Chapter 10 CSWA Models
▓	Graph paper
▓	LOGO
▓	MY-SHEETFORMATS
▓	MY-TEMPLATES
▓	PPT Presentations

- The term document is used to refer a SOLIDWORKS part, drawing, or assembly file.

- The list of items across the top of the SOLIDWORKS interface is the Menu bar menu or the Menu bar toolbar. Each item in the Menu bar has a pull-down menu. When you need to select a series of commands from these menus, the following format is used: Click **View**, **Hide/Show**, check **Origins** from the Menu bar. The Origins are displayed in the Graphics window.

- The ANSI overall drafting standard and Third Angle projection is used as the default setting in this text. IPS (inch, pound, second) and MMGS (millimeter, gram, second) unit systems are used.

- The book is organized into various chapters. Each chapter is focused on a specific subject or feature.

- View the provided models to enhance the user experience. All needed templates, logos and model documents for this book are available.

- Screen shots in the book were made using SOLIDWORKS 2019 SP0 running Windows® 10.

The following command syntax is used throughout the text. Commands that require you to perform an action are displayed in **Bold** text.

Format:	Convention:	Example:
Bold	<ul><li>All commands actions.</li><li>Selected icon button.</li><li>Selected geometry: line, circle.</li><li>Value entries.</li></ul>	<ul><li>Click **Options** ⚙ from the Menu bar toolbar.</li><li>Click **Corner Rectangle** ⬜ from the Sketch toolbar.</li><li>Click **Sketch** ⌐ from the Context toolbar.</li><li>Select the **centerpoint**.</li><li>Enter **3.0** for Radius.</li></ul>
Capitalized	<ul><li>Filenames.</li><li>First letter in a feature name.</li></ul>	<ul><li>Save the **FLATBAR** assembly.</li><li>Click the **Fillet** ▱ feature.</li></ul>

Windows Terminology in SOLIDWORKS

The mouse buttons provide an integral role in executing SOLIDWORKS commands. The mouse buttons execute commands, select geometry, display Shortcut menus and provide information feedback.

Item:	Description:
Click	Press and release the left mouse button.
Double-click	Double press and release the left mouse button.
Click inside	Press the left mouse button. Wait a second, and then press the left mouse button inside the text box. Use this technique to modify Feature names in the FeatureManager design tree.
Drag	Point to an object, press and hold the left mouse button down. Move the mouse pointer to a new location. Release the left mouse button.
Right-click	Press and release the right mouse button. A Shortcut menu is displayed. Use the left mouse button to select a menu command.
Tool Tip	Position the mouse pointer over an Icon (button). The tool name is displayed below the mouse pointer.
Large Tool Tip	Position the mouse pointer over an Icon (button). The tool name and a description of its functionality are displayed below the mouse pointer.
Mouse pointer feedback	Position the mouse pointer over various areas of the sketch, part, assembly or drawing. The cursor provides feedback depending on the geometry.

A mouse with a center wheel provides additional functionality in SOLIDWORKS. Roll the center wheel downward to enlarge the model in the Graphics window. Hold the center wheel down. Drag the mouse in the Graphics window to rotate the model.

Visit SOLIDWORKS website: http://www.SOLIDWORKS.com/sw/support/hardware.html to view their supported operating systems and hardware requirements.

Hardware & System Requirements

Research graphics cards hardware, system requirements, and other related topics.

SolidWorks System Requirements
Hardware and system requirements for SolidWorks 3D CAD products.

Data Management System Requirements
Hardware and system requirements for SolidWorks Product Data Management (PDM) products.

SolidWorks Composer System Requirements
Hardware and system requirements for SolidWorks Composer and other 3DVIA related products.

SolidWorks Electrical System Requirements
Hardware and system requirements for SolidWorks Electrical products.

Graphics Card Drivers
Find graphics card drivers for your system to ensure system performance and stability.

Anti-Virus
The following Anti-Virus applications have been tested with SolidWorks 3D CAD products.

Hardware Benchmarks
Applications and references that can help determine hardware performance.

The book does not cover starting a SOLIDWORKS session in detail for the first time. A default SOLIDWORKS installation presents you with several options. For additional information for an Education Edition, visit the following site: http://www.SOLIDWORKS.com/sw/engineering-education-software.htm.

The Instructor's information contains over 45 classroom presentations, along with helpful hints, What's new, sample quizzes, avi files of assemblies, projects, and all initial and final SOLIDWORKS model files.

The CSWA certification indicates a foundation in and apprentice knowledge of 3D CAD design and engineering practices and principles.

The CSWA Academic exam is provided either in a single 3 hour segment, or 2 - 90 minute segments. The CSWA exam for industry is only provided in a single 3 hour segment. All exams cover the same material.

To obtain additional CSWA exam information, visit the SOLIDWORKS VirtualTester Certification site at https://SOLIDWORKS.virtualtester.com/.

Chapter 1

History of Engineering Graphics

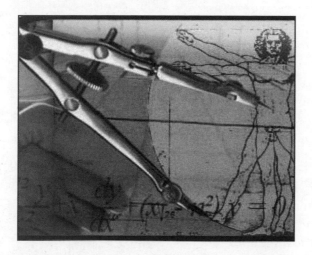
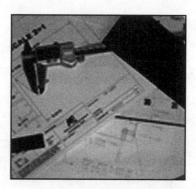

Below are the desired outcomes and usage competencies based on the completion of Chapter 1.

Desired Outcomes:	Usage Competencies:
• Appreciate the history of Engineering Graphics.	• Identify categories and disciplines related to Engineering Graphics.
• Knowledge of the Cartesian Coordinate system.	• Apply 2D and 3D Cartesian Coordinate system: Absolute, Relative, Polar, Cylindrical, and Spherical.
• Understand Geometric entities. • Comprehend Free Hand Sketches.	• Points, Circles, Arcs, Planes, etc. • Solid Primitives. • Generate basic 2D shapes and objects. • Create 2D and 3D freehand sketches.
• Recognize Alphabet of Lines and Precedence of Line types.	• Create and understand correct line precedence.
• Grasp the concept of Multi-view drawings. • Comprehend Orthographic Projection/Glass Box.	• Select the proper Front view. • Explain First and Third Angle projection type. • Identify the six principal views.

Notes:

Chapter 1 - History of Engineering Graphics

Chapter Overview

Chapter 1 provides a broad discussion of the history of Engineering Graphics and the evolution from manual drawing/drafting along with an understanding of the Cartesian Coordinate system, Geometric entities, general sketching techniques, alphabet of lines, precedence of line types and Orthographic projection.

On the completion of this chapter, you will be able to:

- Appreciate the history of Engineering Graphics.

- Comprehend Global and Local Coordinate system.

- Understand 2D and 3D Cartesian Coordinate system:

  o   Right-handed vs. Left-handed.

  o   Absolute, Relative, Polar, Cylindrical, and Spherical.

- Understand Geometric entities:

  o   Point, Circle, Arc, Plane, etc.

- Recognize Alphabet of Lines and Precedence of Line types.

- Grasp the concept of Multi-view drawings:

  o   Select the proper Front view.

- Understand the general terminology and divisions of Projection.

- Comprehend Orthographic Projection and the Glass Box method.

- Identify the six Principal views.

- Explain First and Third Angle Projection type.

History of Engineering Graphics

Engineering Graphics is the academic discipline of creating standardized technical drawings by architects, interior designers, drafters, design engineers and related professionals.

Standards and conventions for layout, sheet size, line thickness, text size, symbols, view projections, descriptive geometry, dimensioning, tolerancing, abbreviations and notation are used to create drawings that are ideally interpreted in only one way.

A technical drawing differs from a common drawing by how it is interpreted. A common drawing can hold many purposes and meanings, while a technical drawing is intended to concisely and clearly communicate all needed specifications to transform an idea into physical form for manufacturing, inspection or purchasing.

We are all aware of the amazing drawings and inventions of Leonardo da Vinci (1453-1528). It is assumed that he was the father of mechanical drafting. Leonardo was probably the greatest engineer the world has ever seen. Below are a few freehand sketches from his notebooks.

Example 1:

The first freehand sketch is of a crossbow. Note the detail and notes with the freehand sketch.

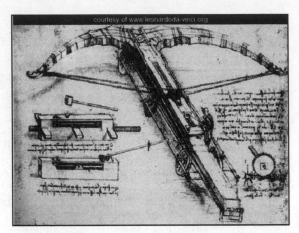

Example 2:

The second freehand sketch is of an early example of an exploded assembly view.

The only source for the detailed history of Leonardo's work is his own careful representations. His drawings were of an artist who was an inventor and a modern-day engineer. His drawings were three-dimensional (3D) and they generally were without dimensional notations.

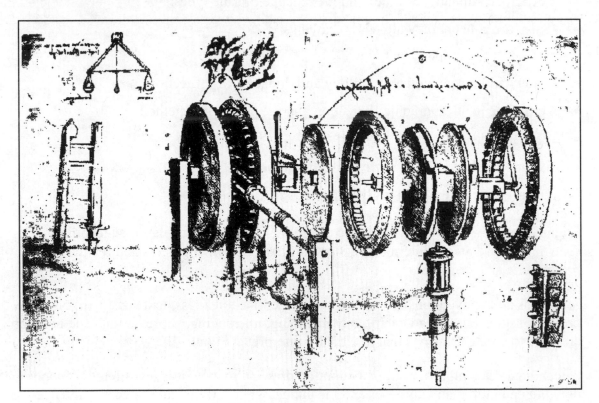

Craftsmen created objects from his drawings, and each machine or device was a one-of-a-kind creation. Assembly line manufacturing and interchangeable parts were not a concern.

Engineering graphics is a visual means to develop ideas and convey designs in a technical format for construction and manufacturing. Drafting is the systematic representation and dimensional specification and annotation of a design.

The basic mechanics of drafting is to place a piece of paper (or other material) on a smooth surface with right-angle corners and straight sides - typically a drafting table. A sliding straightedge known as a T-square is then placed on one of the sides, allowing it to be slid across the side of the table and over the surface of the paper.

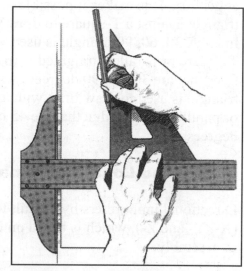

"Parallel lines" can be drawn simply by moving the T-square and running a pencil or technical pen along the T-square's edge, but more typically the T-square is used as a tool to hold other devices such as set squares or triangles. In this case, the drafter places one or more triangles of known angles on the T-square, which is itself at right angles to the edge of the table and can then draw lines at any chosen angle to others on the page.

Modern drafting tables (which have by now largely been replaced by CAD workstations) come equipped with a parallel rule that is supported on both sides of the table to slide over a large piece of paper. Because it is secured on both sides, lines drawn along the edge are guaranteed to be parallel.

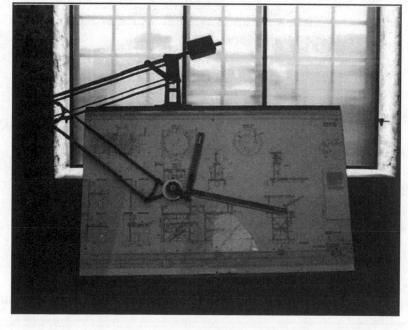

In addition, the drafter uses several tools to draw curves and circles. Primary among these are the compasses, used for drawing simple arcs and circles; the French curve, typically made out of plastic, metal, or wood composed of many different curves; and a spline, which is a rubber coated articulated metal that can be manually bent to most curves.

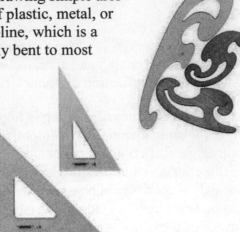

💡 A drafting triangle always has one right angle 90°. This makes it possible to put a triangle against a T-square to draw vertical lines. A 30, 60, 90 triangle is used with a T-square or parallel straightedge to draw lines that are 30, 60, 90 degrees. A 45, 90 triangle is used to draw lines with a T-square or parallel straightedge that are 45 or 90 degrees.

Global and Local Coordinate System

Directional input refers by default to the Global coordinate system (X-, Y- and Z-), which is based on Plane1 with its origin located at the origin of the part or assembly.

The figure below illustrates the relationship between the Global coordinate system and Plane 1 (Front), Plane 2 (Top) and Plane 3 (Right).

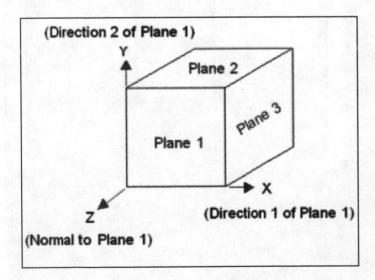

Where X- is Direction 1 of Plane 1, Y- is Direction 2 of Plane 1, and Z- is the Normal to Plane 1.

Local (Reference) coordinate systems are coordinate systems other than the Global coordinate system. You can specify restraints and loads in any desired direction. Example: Defining a force on a cylindrical face, you can apply it in the radial, circumferential, or axial directions. Similarly, if you choose a spherical face, you can choose the radial, longitude, or latitude directions. In addition, you can use reference planes and axes.

2 Dimensional Cartesian Coordinate System

A Cartesian coordinate system in two dimensions is commonly defined by two axes, at right angles to each other, forming a plane (an x,-y plane). The horizontal axis is normally labeled x, and the vertical axis is normally labeled y.

The axes are commonly defined as mutually orthogonal to each other (each at a right angle to the other). Early systems allowed "oblique" axes, that is, axes that did not meet at right angles, and such systems are occasionally used today, although mostly as theoretical exercises. All the points in a Cartesian coordinate system taken together form a so-called Cartesian plane. Equations that use the Cartesian coordinate system are called Cartesian equations.

The point of intersection, where the axes meet, is called the origin. The x and y axes define a plane that is referred to as the xy plane. Given each axis, choose a unit length, and mark off each unit along the axis, forming a grid. To specify a particular point on a two dimensional coordinate system, indicate the x unit first (abscissa), followed by the y unit (ordinate) in the form (x,-y), an ordered pair.

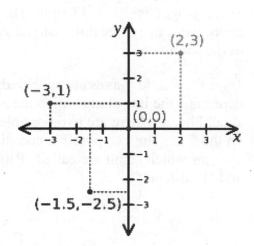

Example 1:

Example 1 displays an illustration of a Cartesian coordinate plane. Four points are marked and labeled with their coordinates: (2,3), (-3,1), (-1.5,-2.5) and the origin (0,0).

The intersection of the two axes creates four regions, called quadrants, indicated by the Roman numerals I, II, III and IV. Conventionally, the quadrants are labeled counter-clockwise starting from the upper right ("northeast") quadrant.

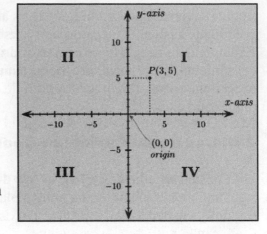

Example 2:

Example 2 displays an illustration of a Cartesian coordinate plane. Two points are marked and labeled with their coordinates: (3,5) and the origin (0,0) with four quadrants.

In the first quadrant, both coordinates are positive, in the second quadrant x-coordinates are negative and y-coordinates positive, in the third quadrant both coordinates are negative and in the fourth quadrant, x-coordinates are positive and y-coordinates negative.

3 Dimensional Cartesian Coordinate System

The three dimensional coordinate system provides the three physical dimensions of space: height, width and length. The coordinates in a three dimensional system are of the form (x,y,z).

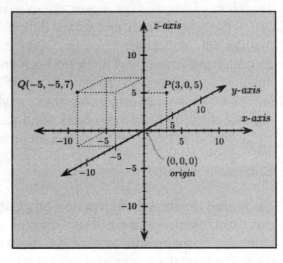

Once the x- and y-axes are specified, they determine the line along which the z-axis should lie, but there are two possible directions on this line. The two possible coordinate systems which result are called "Right-hand" and "Left-hand."

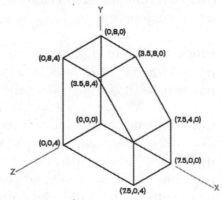

Cartesian coordinates are the foundation of analytic geometry, and provide enlightening geometric interpretations for many other branches of mathematics, such as linear algebra, complex analysis, differential geometry, multivariate calculus, group theory and more.

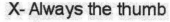

Most CAD systems use the Right-hand rule for a coordinate system. To use the Right-hand rule - point the thumb of your right hand in the positive direction for the x axis and your index finger in the positive direction for the y axis; your remaining fingers curl in the positive direction for the z axis as illustrated.

X- Always the thumb

  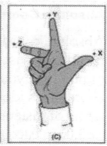

When the x,-y plane is aligned with the screen in a CAD system, the z axis is oriented horizontally (pointing towards you). In machining and many other applications, the z-axis is considered to be the vertical axis. In all cases, the coordinate axes are mutually perpendicular and oriented according to the Right-hand or Left-hand rule.

The Right-hand rule is also used to determine the direction of rotation. For rotation using the right-hand rule, point your thumb in the positive direction along the axis of rotation. Your fingers will curl in the positive direction for the rotation.

Some CAD systems use a Left-hand rule. In this case, the curl of the fingers on your left hand provides the positive direction for the z axis. In this case, when the face of your computer monitor is the x,-y plane, the positive direction for the z axis would extend into the computer monitor, not towards you.

Models and drawings created in SOLIDWORKS or a CAD system are defined and stored using sets of points in what is sometimes called World Space.

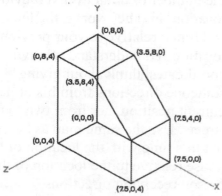

Each reference line is called a coordinate axis or just axis of the system, and the point where they meet is its origin. The coordinates can also be defined as the positions of the perpendicular projections of the point onto the two axes, expressed as a signed distance from the origin.

The origin ⚓ in SOLIDWORKS is displayed in blue in the center of the Graphics window. The origin represents the intersection of the three default reference planes: *Front Plane*, *Top Plane*, and *Right Plane* illustrated in the FeatureManager. The positive x-axis is horizontal and points to the right of the origin in the Front view. The positive y-axis is vertical and points upward in the Front view. The FeatureManager contains a list of features, reference geometry, and settings utilized in the part.

Absolute Coordinates

Absolute coordinates are the coordinates used to store the location of points in your CAD system. These coordinates identify the location in terms of distance from the origin (0,0,0) in each of the three axis (x, y, z) directions of the Cartesian coordinate system.

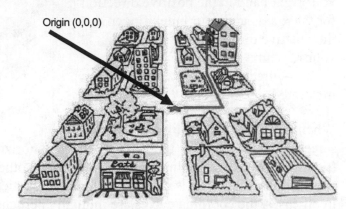

Origin (0,0,0)

As an example, someone provides directions to your house (or to a house in an area where the streets are laid out in nice rectangular blocks). A way to describe how to get to your house would be to inform the person how many blocks over and how many blocks up it is from two main streets (and how many floors up in the building, for 3D). The two main streets are like the x and y axes of the Cartesian coordinate system, with the intersection as the origin (0,0,0).

Relative Coordinates

Instead of having to specify each location from the origin (0,0,0), using relative coordinates allows you to specify a 3D location by providing the number of units from a previous location. In other words, the location is defined relative to your previous location. To understand relative coordinates, think about giving someone directions from his or her current position, not from two main streets. Use the same map as before but this time with the location of the house relative to the location of the person receiving directions.

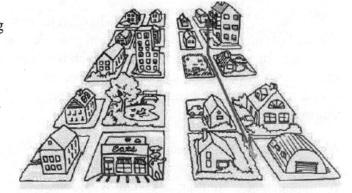

Polar Coordinates

Polar coordinates are used to locate
an object by providing an angle
(from the x axis) and a distance.
Polar coordinates can either be
absolute, providing the angle and
distance from the origin (0,0,0), or
they can be relative, providing the
angle and distance from the current
location.

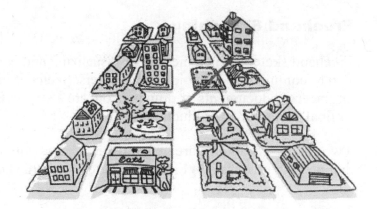

Picture the same situation of having to provide directions. You could inform the person to
walk at a specified angle from the crossing of the two main streets, and how far to walk.
In the illustration, it shows the angle and direction for the shortcut across the empty lot
using absolute polar coordinates. Polar coordinates can also be used to provide an angle
and distance relative to a starting point.

Cylindrical and Spherical Coordinates

Cylindrical and spherical coordinates are similar to polar
coordinates except that you specify a 3D location instead
of one on a single flat plane (such as a map). Cylindrical
coordinates specify a 3D location based on a radius, angle,
and distance (usually in the z axis direction). It may be
helpful to think about this as giving a location as though it
were on the edge of a cylinder. The radius tells how far the
point is from the center (or origin); the angle is the angle
from the x axis along which the point is located; and the
distance gives you the height where the point is located on
the cylinder. Cylindrical coordinates are similar to polar
coordinates, but they add distance in the z direction.

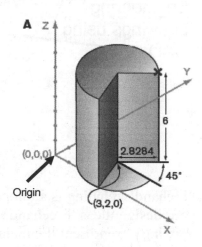

Spherical coordinates specify a 3D location by the radius,
an angle from the X axis, and the angle from the x,y plane.
It is helpful to think of locating a point on a sphere, where
the origin of the coordinate system is at the center of the
sphere. The radius gives the size of the sphere, and the
first angle gives a location on the equator. The second
angle gives the location from the plane of the equator to
the point on the sphere in line with the location specified
on the equator.

Freehand Sketching

Freehand sketching is a method of visualizing and conceptualizing your idea that allows you to communicate that idea with others. Sketches are not intended to be final engineering documents or drawings but are a step in the process from an idea or thought to final design or to production.

Two types of drawings are generally associated with the four key stages of the engineering process: (1) Freehand sketches and (2) Detailed Engineering Drawings.

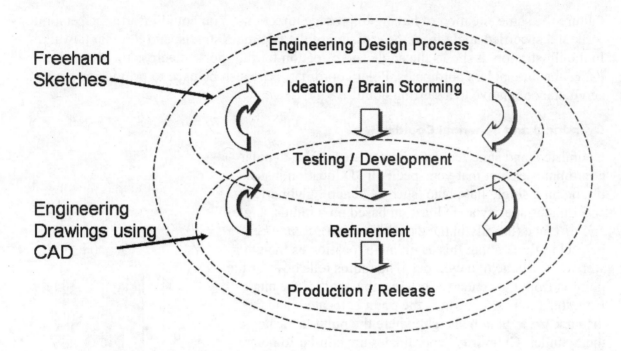

Freehand sketching is an important method to quickly document and to communicate your design ideas. Freehand sketching is a process of creating a rough, preliminary drawing to represent the main features of a design, whether it is a manufactured product, a chemical process or a structure.

Sketches take many forms and vary in level of detail. The designer or engineer determines the level of detail based on clarity and purpose of the sketch, as well as the intended audience. Sketches are important to record the fleeting thoughts associated with idea generation and brain storming in a group.

Freehand sketching is considered one of the most powerful methods to help develop visualization skills. The ability to sketch is helpful, not only to communicate with others, but also to work out details in ideas and to identify any potential problems.

Freehand sketching requires simple tools, a pencil, piece of paper, straight edge and can be accomplished almost anywhere. Creating freehand sketches does not require artistic ability, as some may assume.

General Sketching Techniques

Understand that it takes practice to perfect your skills in any endeavor, including freehand sketching. When sketching, you need to coordinate your eyes, hands (wrist and arm), and your brain. Chances are you have had little opportunity in recent years to use these together, so your first experience with freehand sketching will be taxing. Some tips to ease the process include:

- Orient the paper in a comfortable position.

- Determine the most comfortable drawing direction, such as left to right, or drawing either toward or away from your body.

- Relax your hand, arm and body.

- Use the edge of the paper as a guide for straight lines.

- When using pencil, work from the top left to the lower right corner (if you are right-handed). This helps avoid smudging your work (your hand is resting on blank paper, rather than on your work).

- Remember that sketches are generally drawn without dimensions, since you are trying to represent the main features of your design concept.

- Use a wooden pencil with soft HB lead or a mechanical pencil in 5mm or 7mm.

Today, you may not have a T-square available, but you can still sketch in your notebook and use good sketching techniques. You should also be prepared to sketch anywhere, even on the back of a napkin.

Geometric Entities

Points

Points are geometrical constructs. Points are considered to have no width, height, or depth. Points are used to indicate locations in space. When you represent a point in a freehand sketch, the convention is to make a small cross or a bar if it is along a line, to indicate the location of the point.

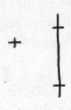

In CAD drawings, a point is located by its coordinates and usually shown with some sort of marker like a cross, circle, or other representation. Many CAD systems allow you to choose the style and size of the mark that is used to represent points. Most CAD systems offer three ways to specify a point:

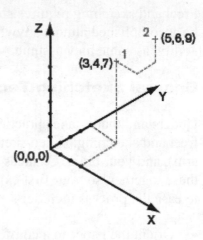

- End the coordinates for the point.

- Select a point in the Graphics window.

- Enter a point's location by its relationship to existing geometry. (Example: a centerpoint, an endpoint of a line, or an intersection of two lines).

Picking a point from the screen is a quick way to enter points when the exact location is not important, but the accuracy of the CAD database makes it impossible to enter a location accurately in this way.

Lines

A straight line is defined as the shortest distance between two points. Geometrically, a line has length, but no other dimension such as width or thickness. Lines are used in drawings to represent the edge view of a surface, the limiting element of a contoured surface, or the edge formed where two surfaces on an object join.

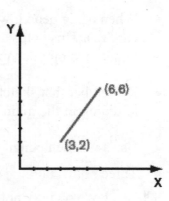

In CAD, 2D lines are typically stored by the coordinates (x,y) of their endpoints.

Planes

Planes are defined by:

- Two parallel lines.

- Three points not lying in a straight line.

- A point and a line.

- Two intersecting lines.

The last three ways to define a plane are all special cases of the more general case - three points not in a straight line. Knowing what can determine a plane can help you understand the geometry of solid objects - and use the geometry to work in CAD.

For example, a face on an object is a plane that extends between the vertices and edges of the surface. Most CAD programs allow you to align new entities with an existing plane. You can use any face on the object - whether it is normal, inclined, or oblique - to define a plane for aligning a new entity. The plane can be specified using existing geometry.

💡 Defining planes on the object or in 3D space is an important tool for working in 3D CAD. You will learn more about specifying planes to orient a user coordinate system to make it easy to create CAD geometry later in this text.

Circles

A circle is a set of points that are equidistant from a center point. The distance from the center to one of the points is the radius. The distance across the center to any two points on opposite sides is the diameter. The circumference of a circle contains 360° of arc. In a CAD file, a circle is often stored as a center point and radius. Most CAD systems allow you to define circles by specifying:

- Center and a radius.

- Center and a diameter.

- Two points on the diameter.

- Three points on the circle.

- Radius and two entities to which the circle is tangent.

- Three entities to which the circle is tangent.

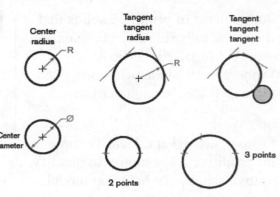

As with any points, the points defining a circle can be entered with absolute, relative, polar, cylindrical, or spherical coordinates; by picking points from the screen; or by specifying existing geometry.

Arcs

An arc is a portion of a circle. An arc can be defined by specifying:

- Center, radius, and angle measure (sometimes called the included angle or delta angle).

- Center, radius and arc length.

- Center, radius and chord length.

- Endpoints and arc length.

- Endpoints and a radius.

- Endpoints and one other point on the arc (3 points).

- Endpoints and a chord length.

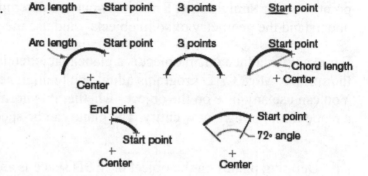

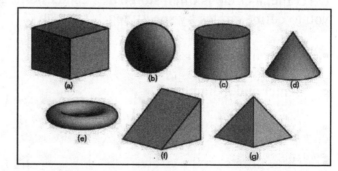

Solid Primitives

Many 3D objects can be visualized, sketched, and modeled in a CAD system by combining simple 3D shapes or primitives. Solid primitives are the building blocks for many solid objects. You should become familiar with these common primitive shapes and their geometry. The same primitives that helped you understand how to sketch objects can also help you create 3D models of them using your computer.

A common set of primitive solids that you can use to build more complex objects is illustrated: (a) box, (b) sphere, (c) cylinder, (d) cone, (e) torus, (f) wedge and (g) pyramid.

Look around and identify some solid primitives. The ability to identify primitive shapes can help you model features of the object.

Alphabet of Lines

The lines used in drafting (technical drawings) are referred to as the alphabet of lines.

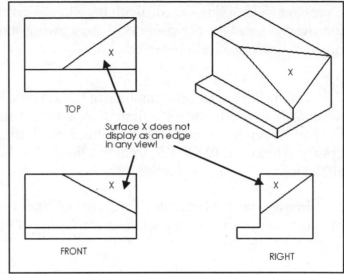

Line types and conventions for technical drawings are covered in ASME Y14.2M-1992 standard. There are four distinct thicknesses of lines: *Very Thick*, *Thick*, *Medium* and *Thin*. Thick lines are drawn using soft lead, such as F or HB. Thin lines are drawn using a harder lead, such as an H or 2H.

Every line on your drawing has a meaning. In other words, lines are symbols that mean a specific thing. The line type determines if the line is part of the object or conveys information about the object.

Below is a list of the most common line types and widths used in orthographic projection.

Visible lines: Visible lines (object or feature lines) are continuous lines used to represent the visible edges and contours (features) of an object. Since visible lines are the most important lines, they must stand out from all other secondary lines on the drawing. The line type is continuous and the line weight is thick (0.5 - 0.6mm).

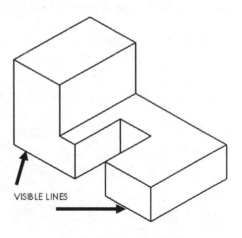

Hidden lines: Hidden lines are short-narrow dashed lines. They represent the hidden features of an object. Hidden lines should always begin and end with a dash, except when a dash would form a continuation of a visible line.

Dashes always meet at corners, and a hidden arc should start with dashes at the tangent points. When the arc is small, the length of the dash may be modified to maintain a uniform and neat appearance.

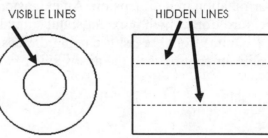

Excessive hidden lines are difficult to follow. Therefore, only lines or features that add to the clearness and the conciseness of the drawing should be displayed. Line weight is medium thick (0.35 - 0.45mm).

💡 Confusing and conflicting hidden lines should be eliminated. If hidden lines do not adequately define a part's configuration, a section should be taken that imaginarily cuts the part. Whenever possible, hidden lines are eliminated from the sectioned portion of a drawing. In SOLIDWORKS, to hide a line, right click the line in a drawing view, and click Hide.

Dimension lines: Dimension lines are thin lines used to show the extent and the direction of dimensions. Space for a single line of numerals is provided by a break in the dimension line.

If possible, dimension lines are aligned and grouped for uniform appearance and ease of reading. For example, parallel dimension lines should be spaced not less than (6mm) apart, and no dimension line should be closer than (10mm) to the outline of an object feature [(12mm) is the preferred distance].

All dimension lines terminate with an arrowhead on mechanical engineering drawings, a slash or a dot in architecture drawings. The preferred ending is the arrowhead to an edge or a dot to a face. Line weight is thin (0.3mm).

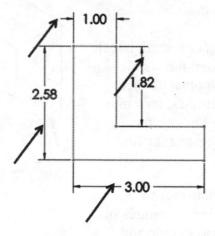

Extension lines: Extension lines are used to indicate the termination of a dimension. An extension line must not touch the feature from which it extends, but should start approximately (2 - 3mm) from the feature being dimensioned and extended the same amount beyond the arrow side of the last dimension line.

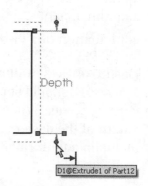

💡 In SOLIDWORKS, use the control points to create the needed extension line gap of ~1.5 - 2.5mm.

💡 In SOLIDWORKS, inserted dimensions in the drawing are displayed in gray. Imported dimensions from the part are displayed in black.

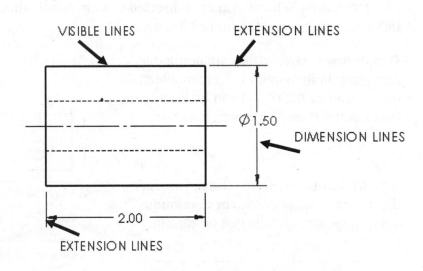

When extension lines cross other extension lines, dimension lines, leader lines, or object lines, they are usually not broken. When extension lines cross dimension lines close to an arrowhead, breaking the extension line is recommended for clarity. Line weight is thin (0.3mm).

Leader lines: A leader line is a continuous straight line that extends at an angle from a note, a dimension, or other reference to a feature. An arrowhead touches the feature at that end of the leader. At the note end, a horizontal bar (6mm) long terminates the leader approximately (3mm) away from mid-height of the note's lettering, either at the beginning or end of the first line.

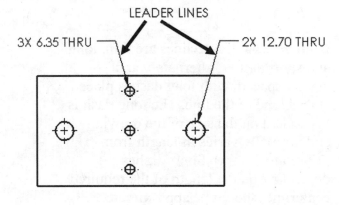

Leaders should not be bent to underline the note or dimension. Unless unavoidable, leaders should not be bent in any way except to form the horizontal terminating bar at the note end of the leader.

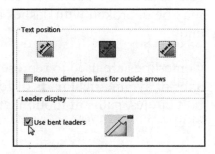

In SOLIDWORKS, use the dimension option to control Leader display.

Leaders usually do not cross. Leaders or extension lines may cross an outline of a part or extension lines if necessary, but they usually remain continuous and unbroken at the point of intersection. When a leader is directed to a circle or a circular arc, its direction should be radial. Line weight is thin (0.3mm).

Break lines: Break lines are applied to represent an imaginary cut in an object, so the interior of the object can be viewed or fitted to the sheet. Line weight is thick (0.5 - 0.6mm).

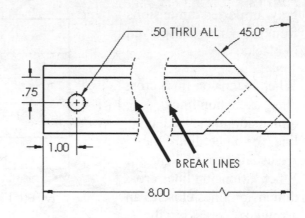

 In SOLIDWORKS, Break lines are displayed as short dashes or continuous solid lines, straight, curved or zig zag.

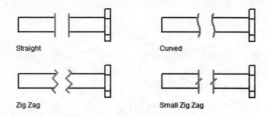

Centerlines: Centerlines are thin, long and short dashes, alternately and evenly spaced, with long dashes placed at each end of the line. The long dash is dependent on the size of the drawing and normally varies in length from (20mm to 50mm). Short dashes, depending on the length of the required centerline, should be approximately (1.5 to 3.0mm). Very short centerlines may be unbroken with dashes at both ends.

Centerlines are used to represent the axes of symmetrical parts of features, bolt circles, paths of motion, and pitch circles.

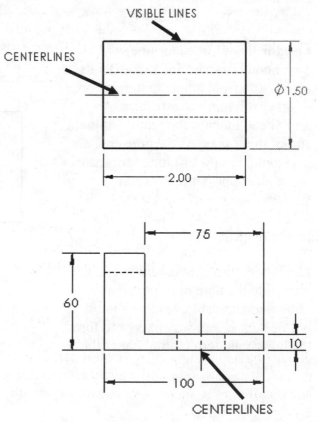

They should extend about (3mm) beyond the outline of symmetry, unless they are used as extension lines for dimensioning. Every circle, and some arcs, should have two centerlines that intersect at the center of the short dashes. Line weight is thin (0.3mm).

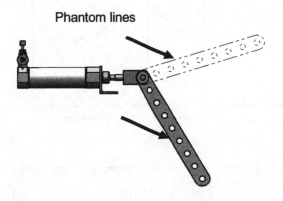

Phantom lines: Phantom lines consist of medium - thin, long and short dashes.

Phantom lines

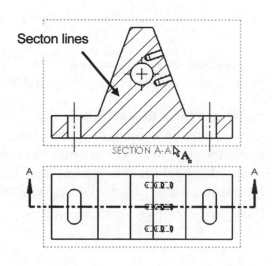

They are used to represent alternate positions of moving parts, adjacent positions of related parts, and repeated details. They are also used to show the cast, or the rough shape, of a part before machining. The line starts and ends with the long dash of (15mm) with about (1.5mm) space between the long and short dashes. Line weight is usually (0.45mm).

Section lines: Section lines are thin, uniformly spaced lines that indicate the exposed cut surfaces of an object in a sectional view.

Spacing should be approximately (3mm) and at an angle of 45°. The section pattern is determined by the material being "cut" or sectioned. Section lines are commonly referred to as "cross-hatching." Line weight is thin (0.3mm). Multiple parts in an assembly use different section angles for clarity.

In this text, you will concentrate on creating 3D models using SOLIDWORKS. Three-dimensional modeling is an integral part of the design, manufacturing and construction industry and contributes to increased productivity in all aspects of a project.

Section lines can serve the purpose of identifying the kind of material the part is made from.

Below are a few common section line types for various materials:

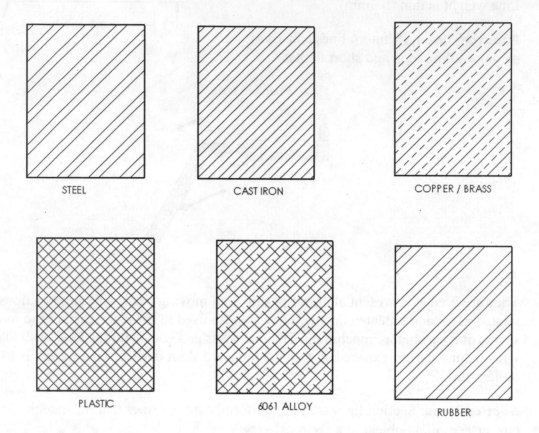

STEEL   CAST IRON   COPPER / BRASS

PLASTIC   6061 ALLOY   RUBBER

A Section lined area is always completely bounded by a visible outline.

Cutting Plane lines: Cutting Plane lines show where an imaginary cut has been made through an object in order to view and understand the interior features. Line type is phantom. Line weight is very thick (0.6 - 0.8mm).

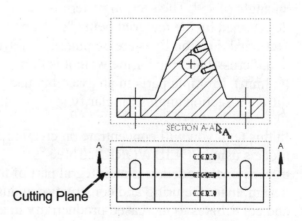

SECTION A-A

Cutting Plane

Arrows are located at the ends of the cutting plane line and the direction indicates the line of sight into the object.

Line weight (also called hierarchy) refers to thickness.

In hand drafting, the contrast in lines should be in the line weight and not in the density. All lines are of equal density except for Construction lines - Light Thin so they can be erased. Construction lines are drawn using 4H or 6H lead.

Precedence of Line Types

When creating Orthographic views, it is common for one line type to overlap another line type. When this occurs, drawing conventions have established an order of precedence. For example - perhaps a visible line type belongs in the same location as a hidden line type; since the visible features of a part (object lines) are represented by thick solid lines, they take precedence over all other lines.

If a centerline and cutting plane coincides, the more important one should take precedence. Normally the cutting plane line, drawn with a thicker weight, will take precedence.

The following list gives the preferred precedence of lines on your drawing:

1. **Visible (Object/Feature) Lines**.

2. **Hidden Lines**.

3. **Cutting Plane Lines**.

4. **Centerlines**.

5. **Phantom lines**.

6. **Break Lines**.

7. **Dimension Lines**.

8. **Extension Lines/Lead Lines**.

9. **Section Lines/Crosshatch Lines**.

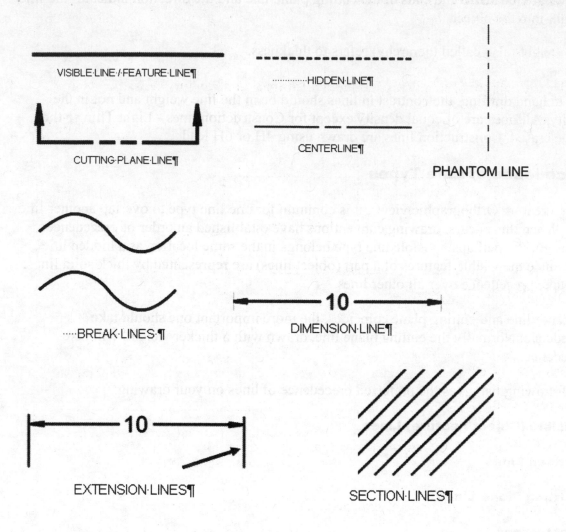

VISIBLE·LINE·/·FEATURE·LINE¶

HIDDEN·LINE¶

CUTTING·PLANE·LINE¶

CENTERLINE¶

PHANTOM LINE

BREAK·LINES·¶

DIMENSION·LINE¶

10

EXTENSION·LINES¶

SECTION·LINES¶

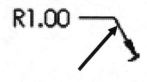

R1.00

LEADER LINE - BENT

View the presentations from the SOLIDWORKS-MODELS 2019\PPT Presentation folder for additional information.

Name
- Bracket
- Chapter 1 Homework
- Chapter 2 Homework
- Chapter 3 Homework
- Chapter 5 Homework
- Chapter 6 Homework
- Chapter 7 Homework
- Chapter 8 Homework
- Chapter 8 Models
- Chapter 9 Homework
- Chapter 10 CSWA Models
- Graph paper
- LOGO
- Multi-media
- MY-SHEETFORMATS
- MY-TEMPLATES
- PPT Presentation

Alphabet of Lines Exercises:

Identify the correct line types:

Exercise 1:

Identify the number of line types and the type of lines in the below view.

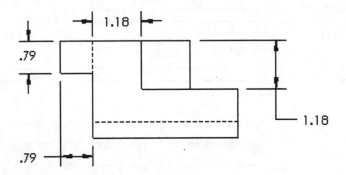

Number of Line Types:_____

Types of Lines: _____

_____

Exercise 2:

Identify the number of line types and the type of lines in the below view.

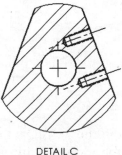

DETAIL C
SCALE 3 : 2

Number of Line Types:_____

Types of Lines: _____

_____

Exercise 3:

Identify the number of line types and the type of lines in the below view.

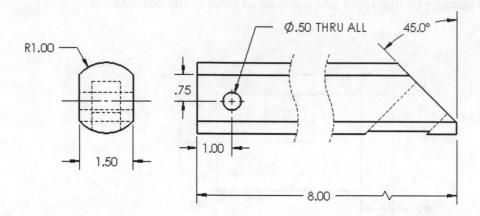

Number of Line Types:_____

Types of Lines: _____

_____

View the presentations from the SOLIDWORKS-MODELS 2019\PPT Presentation folder for additional information.

Name
Bracket
Chapter 1 Homework
Chapter 2 Homework
Chapter 3 Homework
Chapter 5 Homework
Chapter 6 Homework
Chapter 7 Homework
Chapter 8 Homework
Chapter 8 Models
Chapter 9 Homework
Chapter 10 CSWA Models
Graph paper
LOGO
Multi-media
MY-SHEETFORMATS
MY-TEMPLATES
PPT Presentation

Projections in General

To better understand the theory of projection, one must first become familiar with the general terminology and divisions of Projections as illustrated in the below figure.

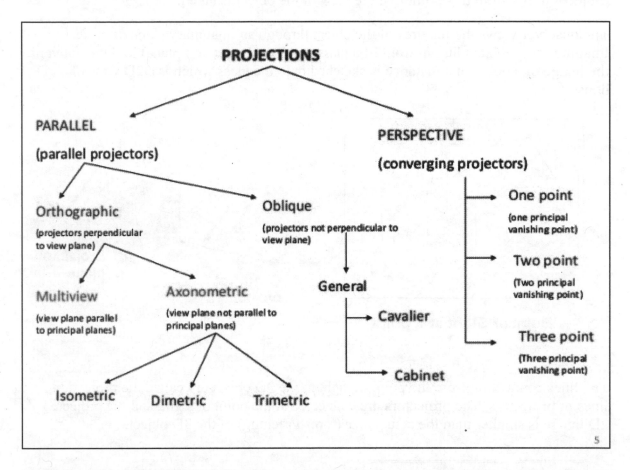

Projection from 3D to 2D is defined by straight Projection rays (projectors) emanating from the center of projection, passing through each point of the object and intersecting the projection plane to form a projection.

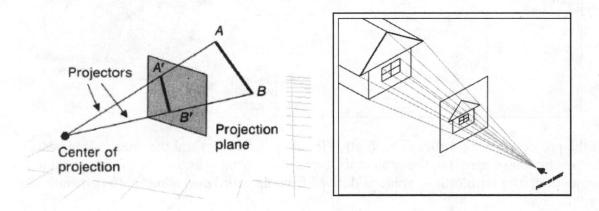

To better understand the theory of projection, you need to become familiar with the elements that are common to the principles of projection.

The point of sight is the position of the observer in relation to the object and the plane of projection. It's from this point that the view of the object is taken.

The observer views the features of the object through an imaginary plane of projection. Imagine yourself standing in front of a glass window (Projection plane) looking outward; the image of a house at a distance is sketched onto the glass which is a 2D view of a 3D house.

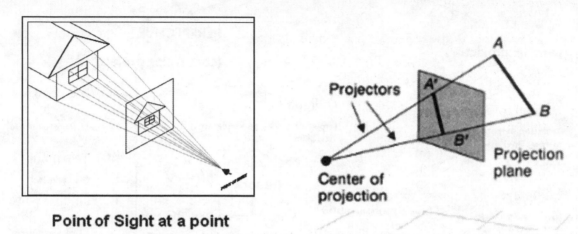

Point of Sight at a point

The lines connecting from the point of sight to the 3D object are called the projection lines or projectors. The projectors are connected at the point of sight, and the projected 2D image is smaller than the actual size (foreshortening) of the 3D object.

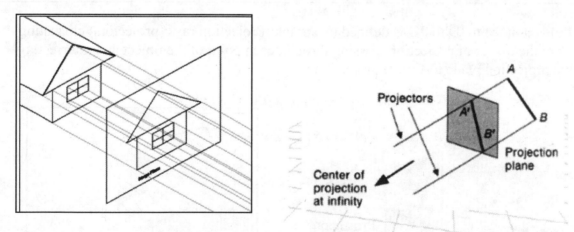

If the projectors are parallel to each other (no foreshortening) and the image plane is also perpendicular (normal) to the projectors, the result is what is known as an orthographic projection. The term orthographic is derived from the word *orthos* meaning perpendicular or 90°.

Projection Types

Parallel Projection

- It preserves relative proportion of object.

- Less realistic view because of not foreshortening.

- However, parallel lines remain parallel.

Perspective Projection

- Visual effect is similar to human visual system.

- Has perspective foreshortening.

- Projectors are rays (non-parallel).

- Vanishing points.

 See Chapter 2 for additional information.

There are two types of Parallel projections: Orthographic and Oblique.

Orthographic Projection

- When the projection is perpendicular to the view.

- Direction of projection = normal to the project plane.

- Projection is perpendicular to the view plane.

Oblique Projection

- When the projection is not perpendicular to the view plane.

- Direction of projection is not normal to the Projection plane.

- Not perpendicular.

Multi-view Projections

Multi-view drawings are based on parallel projection techniques and are used when there is a need to represent the features of an object more accurately than is possible with a single view. A multi-view drawing is a collection of flat 2D drawings that work together to provide an accurate representation of the overall model.

With a pictorial drawing, all three dimensions (**height**, **width**, and **depth**) of the object are represented in a single view. The disadvantage of this approach is that not all the features in all three dimensions can be illustrated with optimal clarity.

In a multi-view projection, however, each view concentrates on only two dimensions of the object, so particular features can be shown with a minimum of distortion. Enough views should be created to capture all the important features of the model.

A multi-view drawing should have the minimum number of views necessary to describe an object completely. Normally, three views are all that are needed; however, the three views chosen must be the most descriptive ones. The most descriptive views are those that reveal the most information about the features, with the fewest features hidden from view.

Orient and Select the Front View

When creating a multi-view drawing of a design, the selection and orientation of the front view is an important first step. The front view is chosen as the *most descriptive* of the object or model. It should be *positioned in a natural orientation* based on its function. For example, for an automobile, the normal or operation position is on its wheels rather than on its roof or bumper.

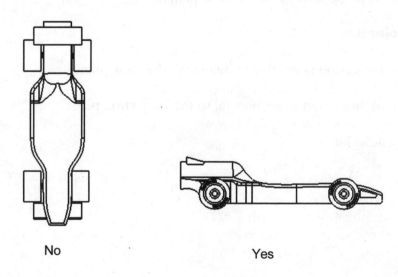

No   Yes

Orthographic Projection (Third Angle)

Before an object is drawn or created, it is examined to determine which views will best furnish the information required to manufacture the object. The surface, which is to be displayed as the observer looks at the object, is called the Front view.

To obtain the front view of an object, turn the object (either physically or mentally) so that the front of the object is all you see. The top and right-side views can be obtained in a similar fashion.

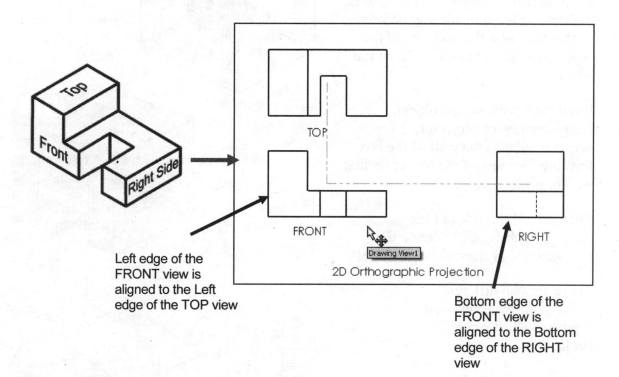

Left edge of the FRONT view is aligned to the Left edge of the TOP view

Bottom edge of the FRONT view is aligned to the Bottom edge of the RIGHT view

Orthographic projection is a common method of representing three-dimensional objects, usually by three two-dimensional drawings, in which each of the objects is viewed along parallel lines that are perpendicular to the plane of the drawing as illustrated. These lines remain parallel to the projection plane and are not convergent.

Orthographic projection provides the ability to represent the shape of an object using two or more views. These views together with dimensions and annotations are sufficient to manufacture the part.

The six principal views of an orthographic projection are illustrated. Each view is created by looking directly at the object in the indicated direction.

Glass Box and Six Principal Orthographic Views

The Glass box method is a traditional method of placing an object in an imaginary glass box to view the six principal views.

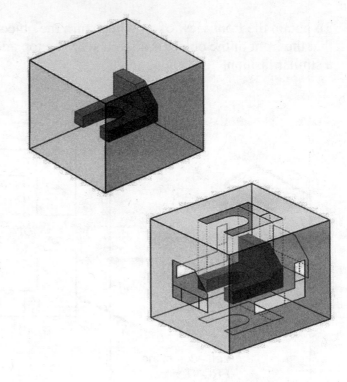

Imagine that the object you are going to draw is placed inside a glass box, so that the large flat surfaces of the object are parallel to the walls of the box.

From each point on the object, imagine a ray, or projector, perpendicular to the wall of the box forming the view of the object on that wall or projection plane.

Then unfold the sides of the imaginary glass box to create the orthographic projection of the object.

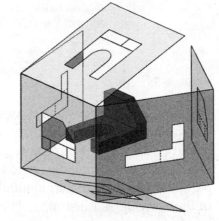

There are two different types of Angle Projection: First and Third Angle Projection.

- First Angle Projection is used in Europe and Asia.

- Third Angle Projection is used in the United States.

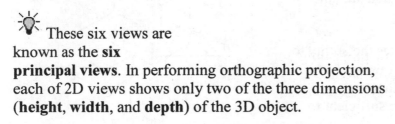

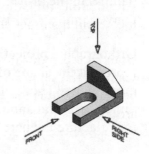

These six views are known as the **six principal views**. In performing orthographic projection, each of 2D views shows only two of the three dimensions (**height**, **width**, and **depth**) of the 3D object.

Third Angle Projection is used in the book. Imagine that the walls of the box are hinged and unfold the views outward around the front view. This will provide you with the standard arrangement of views.

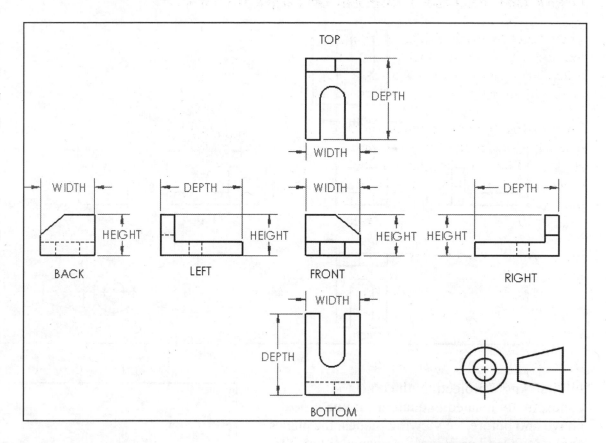

💡 SOLIDWORKS uses BACK view vs. REAR view.

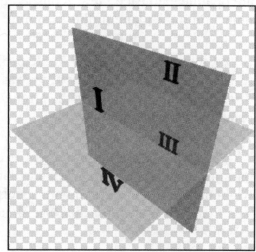

Modern orthographic projection is derived from Gaspard Monge's descriptive geometry. Monge defined a reference system of two viewing planes, horizontal H ("ground") and vertical V ("backdrop"). These two planes intersect to partition 3D space into four quadrants. In Third-Angle projection, the object is conceptually located in quadrant III.

Both First Angle and Third Angle projections result in the same six views; the difference between them is the arrangement of these views around the box.

Below is an illustration of First Angle Projection.

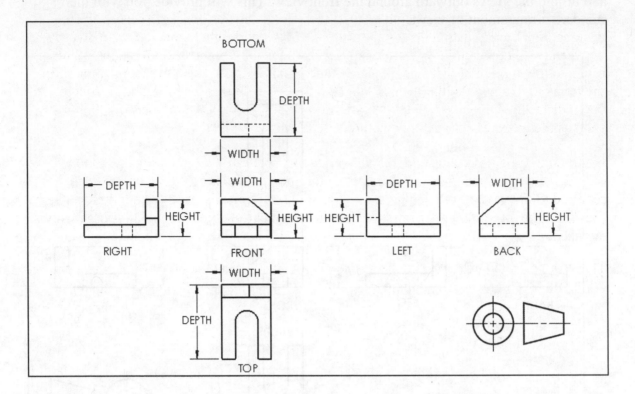

In First Angle projection, the object is conceptually located in quadrant I; i.e., it floats above and before the viewing planes, the planes are opaque, and each view is pushed through the object onto the plane furthest from it.

Both First Angle and Third Angle projections result in the same six principal views; the difference between them is the arrangement of the principal views around the box.

🔅 In SOLIDWORKS, when you create a new part or assembly, the three default Planes (Front, Right and Top) are aligned with specific views. The Plane you select for the Base sketch determines the orientation of the part or assembly.

🔅 https://www.youtube.com/watch?v=yGjVnXgUpQM. Great link for First vs Third angle project.

Height, Width, and Depth Dimensions

The terms height, width and depth refer to specific dimensions or part sizes. ANSI designations for these dimensions are illustrated above. Height is the vertical distance between two or more lines or surfaces (features) which are in horizontal planes. Width refers to the horizontal distance between surfaces in profile planes. In the machine shop, the terms length and width are used interchangeably. Depth is the horizontal (front to back) distance between two features in frontal planes. Depth is often identified in the shop as the thickness of a part or feature.

No orthographic view can show height, width and depth in the same view. Each view only depicts two dimensions. Therefore, a minimum of two projections or views are required to display all three dimensions of an object. Typically, most orthographic drawings use three standard views to accurately depict the object unless additional views are needed for clarity.

The Top and Front views are aligned vertically and share the same width dimension. The Front and Right side views are aligned horizontally and share the same height dimension.

When drawing orthographic projections, spacing is usually equal between each of the views. The Front, Top and Right views are most frequently used to depict orthographic projection.

The Front view should show the **most features** or characteristics of the object. It usually contains the least number of hidden lines. All other views are based (projected) on the orientation chosen for the front view.

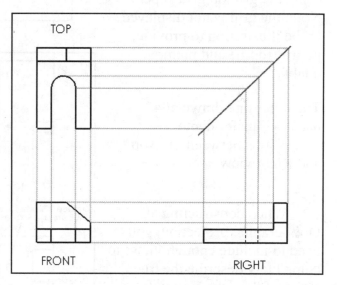

Transferring Dimensions

In SOLIDWORKS, you can view the projection lines from the Front view placement.

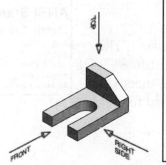
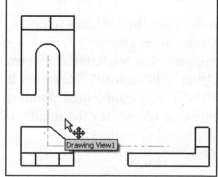

When transferring measurements between views, the width dimension can be projected from the Front view upward to the Top view or vice versa and the height dimension can be projected directly across from the Front view to the Right view.

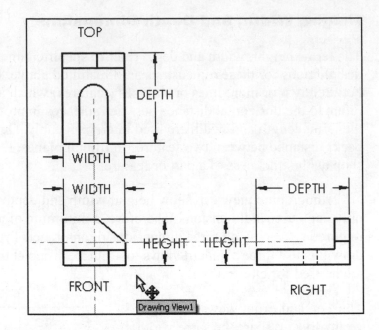

Depth dimensions are transferred from the Top view to the Right view or vice versa.

Height dimensions can be easily projected between two views using the grid on grid paper. Note: the grid is not displayed in the illustration to provide improved line and picture quality.

The miter line drawn at 45° is used to transfer depth dimensions between the Top and Right view.

When constructing an Orthographic projection, you need to include enough views to completely describe the true shape of the part. You will address this later in the book.

Sheet Media

Media are the surfaces upon which an engineer communicates graphical information. The American National Standards Institute (ANSI) has established standard sheet sizes and title blocks for the media used for technical drawings.

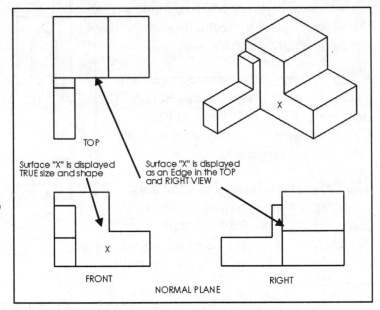

ANSI Standard Sheet Sizes:		
Metric (mm)	**US Standard**	**Architectural**
A4 210 x 297	A-Size 8.5" x 11"	9" x 12"
A3 297 x 420	B-Size 11" x 17"	12" x 18"
A2 420 x 594	C-Size 17" x 22"	18" x 24"
A1 594 x 841	D-Size 22" x 34"	24" x 36"
A0 841 x 1189	E-Size 34" x 44"	36" x 48"

Orthographic Projection Exercises:
Exercise 1:

Label the four remaining Principal views with the appropriate view name. Identify the Angle of Projection type.

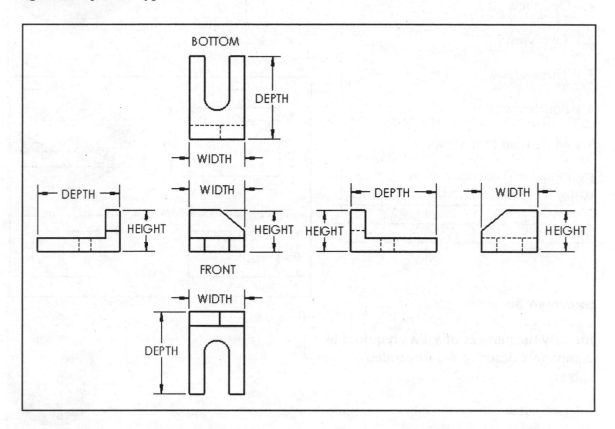

Angle of Projection type:_____

Describe the difference between the BOTTOM view and the TOP view._____

Describe the difference between the RIGHT view and the LEFT view._____

Which view should have the least amount of Hidden Lines?_____

💡 https://www.youtube.com/watch?v=Zptb2epQoEc. Great link for going from Isometric projecting to a three view drawing.

Exercise 2:

Identify the number of views required to completely describe the illustrated box.

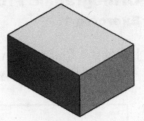

1.) One view

2.) Two views

3.) Three views

4.) Four views

5.) More than four views

Explain
Why._____

_____

_____

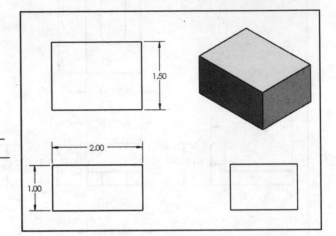

Exercise 3:

Identify the number of views required to completely describe the illustrated sphere.

1.) One view

2.) Two views

3.) Three views

4.) Four views

5.) More than four views

Explain
Why._____

_____

_____

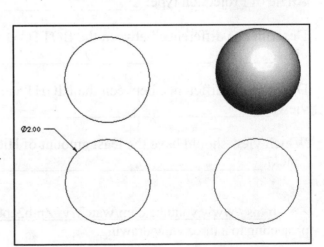

Exercise 4:

In Third Angle Projection, identify the view that **should** display the most information about the illustrated model.

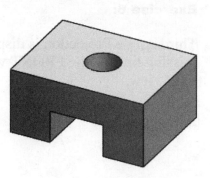

1.) FRONT view

2.) TOP view

3.) BOTTOM view

4.) RIGHT view

Explain
Why._____

_____

Exercise 5:

Third Angle Projection is displayed. Draw the Visible Feature Lines of the TOP view for the model. Fill in the missing lines in the FRONT view, RIGHT view and TOP view.

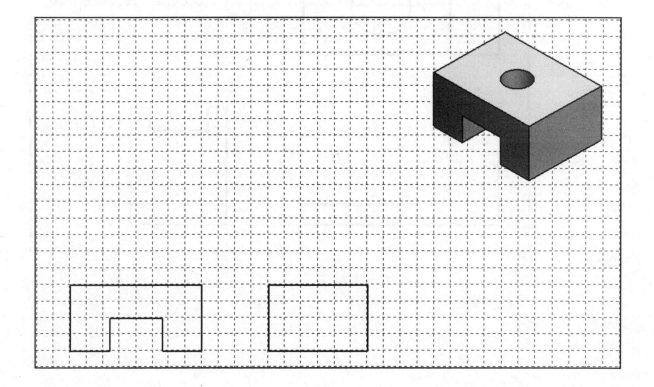

Exercise 6:

Third Angle Projection is displayed. Fill in the
missing lines in the FRONT view, RIGHT view, and
TOP view.

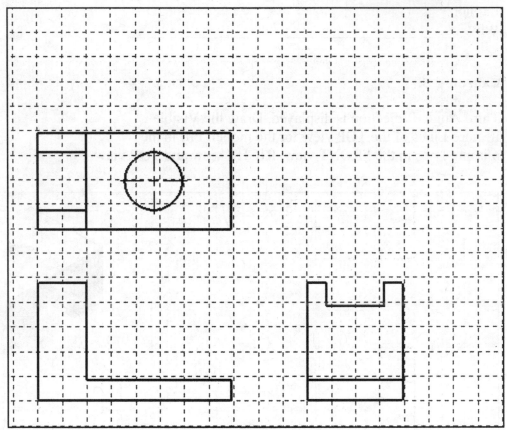

Exercise 7:

Third Angle Projection is displayed. Fill in the missing lines in the FRONT view, RIGHT view, and TOP view.

Tangent Edges are displayed for educational purposes.

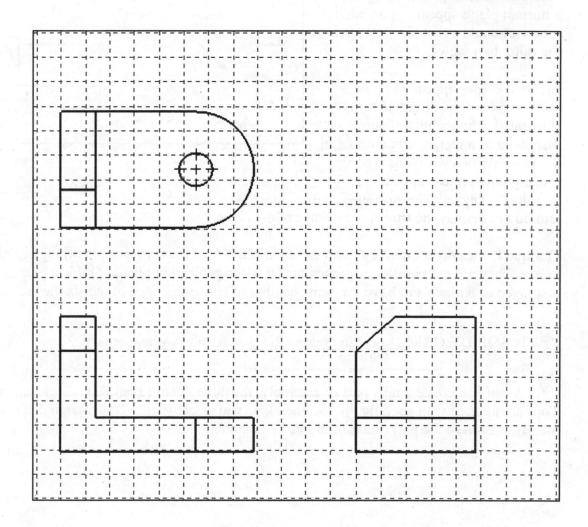

Planes (Normal, Inclined and Oblique)

Each type of plane (Normal, Inclined and Oblique) has unique characteristics when viewed in orthographic projection. To understand the three basic planes, each is illustrated.

Normal planes appear as an edge in two views and true sized in the remaining view when using three views such as the Front, Top, and Right side views.

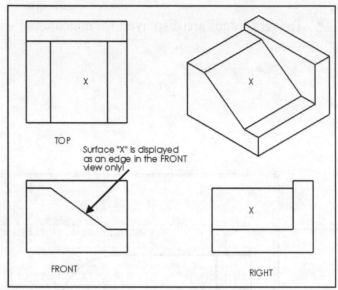

Surface "X" is displayed as an edge in the FRONT view only!

When viewing the six possible views in an orthographic projection, a normal plane appears as an edge in four views and a true sized plane in the other two views.

Inclined Planes appear as an edge view in one of the three views as illustrated. The inclined plane is displayed as a rectangular surface in the other two views. Note: The two rectangular surfaces appear "normal"; they are foreshortened and do not display the true size or shape of the object.

Oblique Planes do not display as an edge view in any of the six principal orthographic views. They are not parallel or perpendicular to the projection planes. Oblique planes are displayed as a plane and have the same number of corners in each of the six views.

💡 In SOLIDWORKS, you can create a 2D sketch on any plane or face.

💡 When you create a new part or assembly, the three default Planes (Front, Right and Top) are aligned with specific views. The Plane you select for the Base sketch determines the orientation of the part or assembly in the document.

Plane Exercises:
Exercise 1:

Identify the surfaces with the appropriate letter that will appear in the FRONT view, TOP view and RIGHT view.

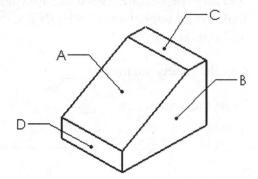

FRONT view surfaces:_____

TOP view surfaces:_____

RIGHT view surfaces:_____

Exercise 2:

Estimate the size; draw the FRONT view, TOP view, and RIGHT view of the illustrated part in Exercise 1 on graph paper.

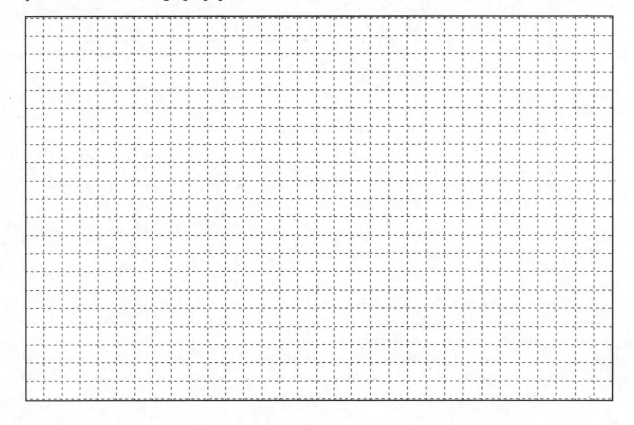

Exercise 3:

Identify the surfaces with the appropriate letter
that would appear in the FRONT view, TOP view,
and RIGHT view.

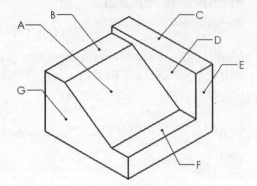

FRONT view surfaces:_____

TOP view surfaces:_____

RIGHT view surfaces:_____

Exercise 4:

Estimate the size; draw the FRONT view, TOP view, and RIGHT view of the illustrated
part in Exercise 3 on graph paper.

Exercise 5:

Identify the surfaces with the appropriate letter that would appear in the FRONT view, TOP view, and RIGHT view.

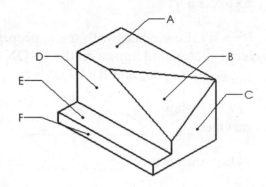

FRONT view surfaces:_____

TOP view surfaces:_____

RIGHT view surfaces:_____

Exercise 6:

Estimate the size; draw the FRONT view, TOP view, and RIGHT view of the illustrated part in Exercise 5 on graph paper.

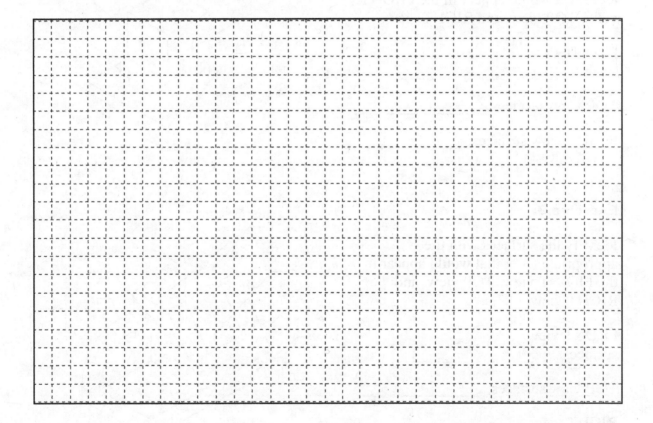

Exercise 7:

Identify the surfaces with the appropriate letter that would appear in the FRONT view, TOP view, and RIGHT view.

FRONT view
surfaces:_____

TOP view surfaces:_____

RIGHT view surfaces:_____

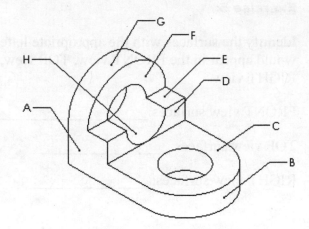

Exercise 8:

Identify the surfaces with the appropriate letter that would appear in the FRONT view, TOP view, and RIGHT view.

FRONT view
surfaces:_____

TOP view surfaces:_____

RIGHT view surfaces:_____

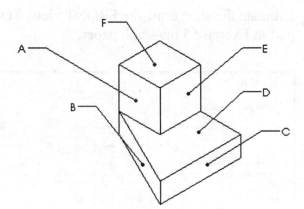

Exercise 9:

Identify the surfaces with the appropriate letter that would appear in the FRONT view, TOP view, and RIGHT view.

FRONT view
surfaces:_____

TOP view surfaces:_____

RIGHT view
surfaces:_____

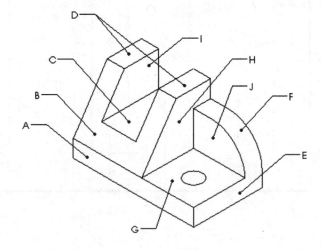

Exercise 10:

Fill in the following table for the below object.

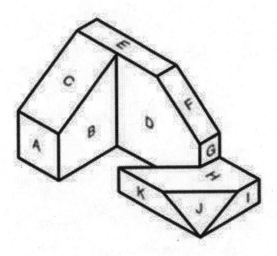

SURFACE	TOP	FRONT	RIGHT
A			
B			
C			
D			
E			
F			
G			
H			
I			
J			
K			

Exercise 11:

Fill in the following table for the below object.

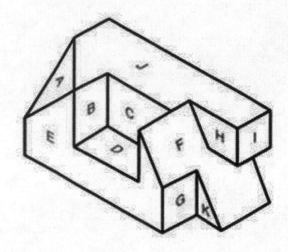

SURFACE	TOP	FRONT	RIGHT
A			
B			
C			
D			
E			
F			
G			
H			
I			
J			
K			

Chapter Summary

Chapter 1 provided a short discussion on the history of Engineering Graphics and the evolution of manual drawing/drafting. You were introduced to general sketching techniques and the 2D and 3D Cartesian Coordinate system. In engineering graphics there is a specific alphabet of lines that represent different types of geometry.

Freehand sketching is an important method to quickly document and to communicate your design ideas. Freehand sketching is a process of creating a rough, preliminary drawing to represent the main features of a design, whether it is a manufactured product, a chemical process, or a structure.

All orthographic views must be looked at together to comprehend the shape of the three-dimensional object. The arrangement and relationship between the views are therefore very important in multi-view drawings.

In creating multi-view orthographic projection, different systems of projection can be used to create the necessary views to fully describe the 3D object. In the figure below, two perpendicular planes are established to form the image planes for a multi-view orthographic projection.

The angles formed between the horizontal and the vertical planes are called the **first, second, third**, and **fourth angles** as indicated in the figure.

For engineering drawings, both **First angle projection** and **Third angle projection** are commonly used.

In first-angle projection, the object is placed in **front** of the image planes and the views are formed by projecting to the image plane located at the back.

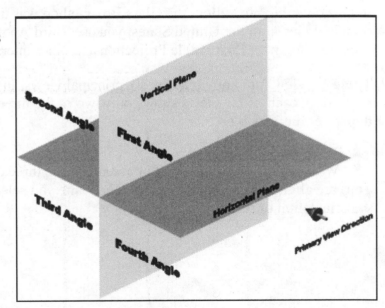

 Great link for going from Isometric projecting to a three view drawing:
https://www.youtube.com/watch?v=Zptb2epQoEc

 Great link for Third vs. First angle project.
https://www.youtube.com/watch?v=yGjVnXgUpQM

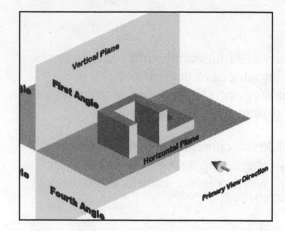 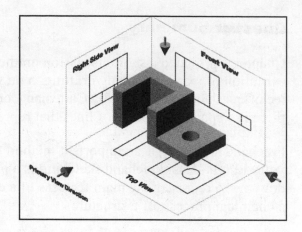

In order to draw all three views of the object on the same plane, the horizontal (Top view) and profile (Right Side view) are rotated into the same plane as the primary image plane (Front view).

In third-angle projection, the image planes are placed in between the object and the observer. And the views are formed by projecting to the image plane located in front of the object.

In Orthographic projection, the Glass Box method was used to distinguish the six principal views. In the United States, you use Third Angle Projection. However, it is important to know First Angle Projection and other international standards.

These six views are known as the six principal views. In performing orthographic projection, each of 2D views shows only two of the three dimensions (**height**, **width**, and **depth**) of the 3D object.

When you create a new part or assembly, the three default Planes (Front, Right, and Top) are aligned with specific views. The Plane you select for the Base sketch determines the orientation of the part or assembly.

Questions

1. Describe the Cartesian coordinate system.

2. Name the point of intersection, where the axes meet _____.

3. Explain the Right-hand rule in drafting.

4. Why is freehand sketching important to understand?

5. Describe the difference between First and Third Angle Projection type.

6. True or False. First Angle Projection type is used in the United States.

7. Explain the Precedent of Line types. Provide a few examples.

8. True or False. A Hidden Line has precedence over a Visible/Feature line.

9. True or False. The intersection of the two axes creates four regions, called quadrants, indicated by the Roman numerals I, II, III, and IV.

10. Explain the Glass Box method in Standard Orthographic Projection.

11. True or False. Both First Angle and Third Angle Projection type result in the same six views; the difference between them is the arrangement of these views.

12. True or False. Section lines can serve the purpose of identifying the kind of material the part is made from.

13. True or False. All dimension lines terminate with an arrowhead on mechanical engineering drawings.

14. True or False. Break lines are applied to represent an imaginary cut in an object, so the interior of the object can be viewed or fitted to the sheet. Provide an example.

15. True or False. The Front view should show the most features or characteristics of the object. It usually contains the least number of hidden lines. All other views are based (projected) on the orientation chosen for the front view. Explain your answer.

Exercises

Exercise 1.1: Third Angle Projection type is displayed. Name the six illustrated views in the below document.

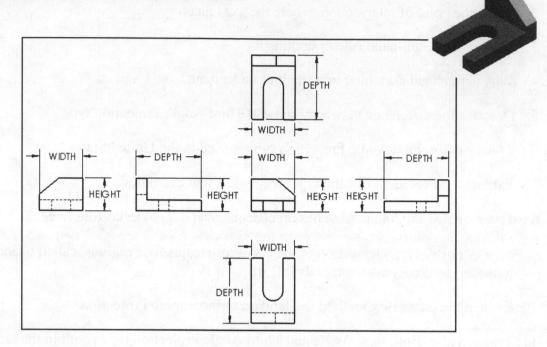

Exercise 1.2: First Angle Projection type is displayed. Name the six illustrated views in the below document.

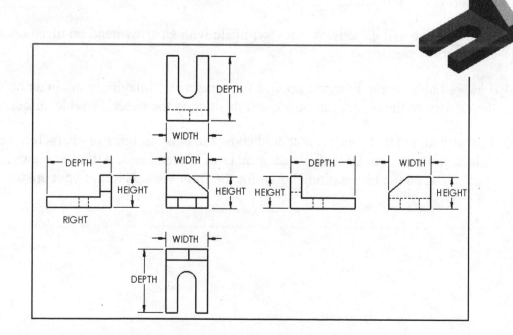

Exercise 1.3: Identify the various Line types in the below model.

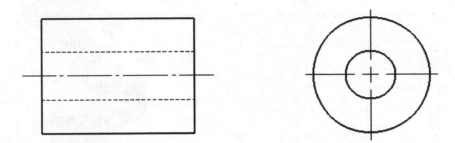

Exercise 1.4: Third Angle Projection type is displayed. Estimate the size using graph paper. Draw the Visible Feature Lines of the Top view for the model. Draw any Hidden lines or Centerlines if needed. Identify the view that should display the most information about the illustrated model.

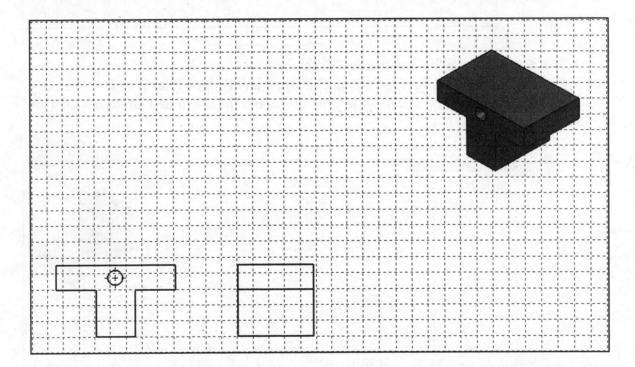

Exercise 1.5: Third Angle Projection type is displayed. Estimate the dimensions. Draw the Visible Feature Lines of the Right view for the model. Draw any Hidden lines if needed. Identify the view that should display the most information about the illustrated model.

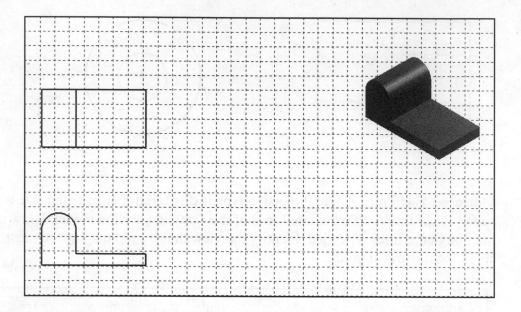

Exercise 1.6: Estimate the dimensions. Draw the Front view, Top view and Right view. Draw the Visible Feature Lines for the model. Draw any Hidden lines if needed. Identify the view that should display the most information about the illustrated model. Note: Third Angle Projection.

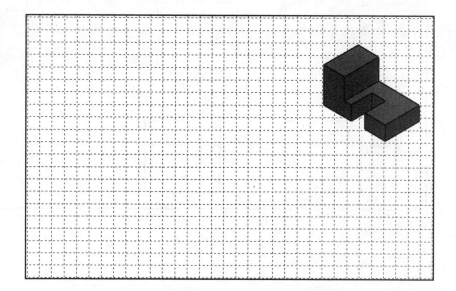

Exercise 1.7: Estimate the dimensions. Draw the Front view, Top view and Right view. Draw the Visible Feature Lines for the model. Draw any Hidden lines or Centerlines if needed. Which view displays the most information about the illustrated model? Note: Third Angle Projection type.

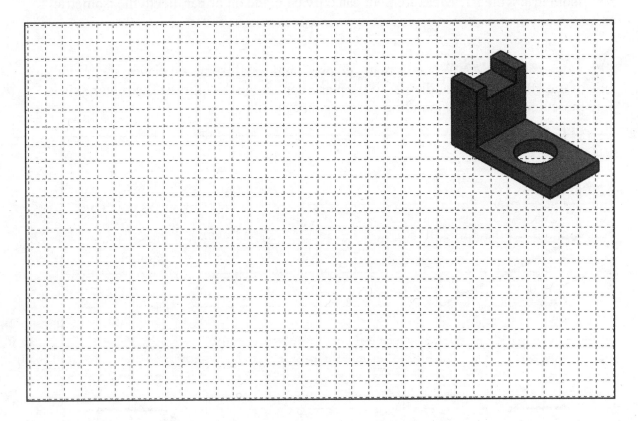

Exercise 1.8: Draw the Isometric view from the provided views (Front, Top, Right).
Note: Third Angle Projection.

Isometric Rule #1: Measurement can only be made on or parallel to the isometric axis.

Isometric Rule #2: When drawing ellipses on normal isometric planes, the minor axis of the ellipse is perpendicular to the plane containing the ellipse. The minor axis is perpendicular to the corresponding normal isometric plane.

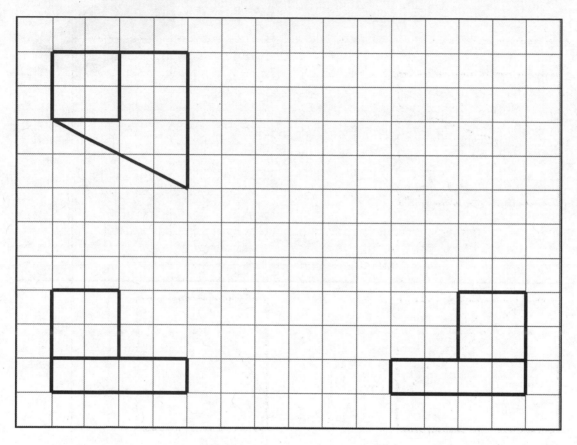

Exercise 1.9: Draw the Isometric view from the provided views (Front, Top, Right).
Note: Third Angle Projection.

Isometric Rule #1: Measurement can only be made on or parallel to the isometric axis.

Isometric Rule #2: When drawing ellipses on normal isometric planes, the minor axis of the ellipse is perpendicular to the plane containing the ellipse. The minor axis is perpendicular to the corresponding normal isometric plane.

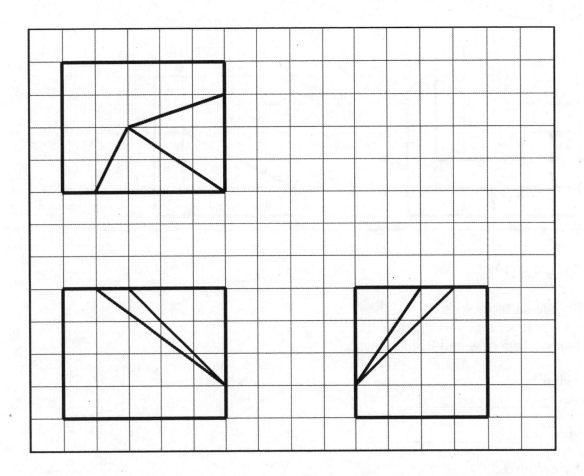

Exercise 1.10: Draw the Isometric view from the provided views (Front, Top, Right).
Note: Third Angle Projection.

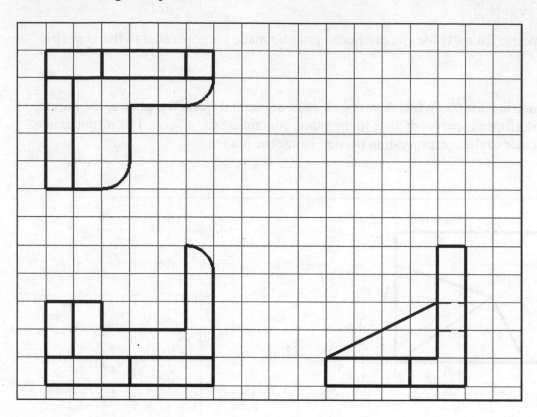

Exercise 1.11:

Identify the surfaces with the appropriate letter that will appear in the Front view, Top view and Right view.

Front view
surfaces:_____

Top view surfaces:_____

Right view surfaces:_____

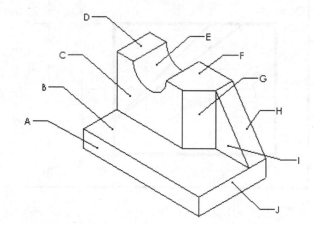

Chapter 2

Isometric Projection and Multi View Drawings

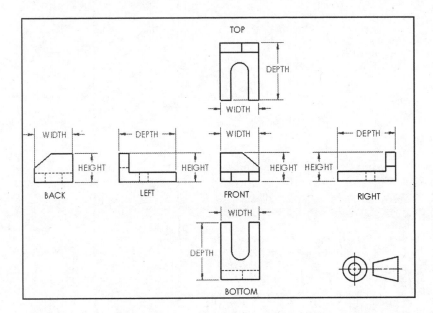

Below are the desired outcomes and usage competencies based on the completion of Chapter 2.

Desired Outcomes:	Usage Competencies:
• Understand Isometric Projection and 2D sketching.	• Identify the three main projection divisions in freehand engineering sketches and drawings: o Axonometric, Oblique, and Perspective
• Knowledge of additional Projection views and arrangement of drawing views.	• Create one and two view drawings.
• Comprehend the history and evolution of CAD and the development of SOLIDWORKS. • Recognize Boolean operations and feature based modeling.	• Identify the development of historic CAD systems and SOLIDWORKS features, parameters and design intent of a sketch, part, assembly, and drawing. • Apply the Boolean operation: Union, Difference, and Intersection.

Notes:

Chapter 2 - Isometric Projection and Multi View Drawings

Chapter Overview

Chapter 2 provides a general introduction into Isometric Projection and Sketching along with Additional Projections and arrangement of views. It also covers advanced drawing views and an introduction from manual drafting to CAD.

On the completion of this chapter, you will be able to:

- Understand and explain Isometric Projection.

- Create an Isometric sketch.

- Identify the three main projection divisions in freehand engineering sketches and drawings:

  o   Axonometric.

  o   Oblique.

  o   Perspective.

- Comprehend the history and evolution of CAD.

- Recognize the following Boolean operations: Union, Difference, and Intersection.

- Understand the development of SOLIDWORKS features, parameters, and design intent of a sketch, part, assembly, and drawing.

Isometric Projections

There are three main projection divisions commonly used in freehand engineering sketches and detailed engineering drawings: 1.) Axonometric, with its divisions in Isometric, Dimetric and Trimetric, 2.) Oblique and 3.) Perspective. Let's review the three main divisions.

Axonometric is a type of parallel projection, more specifically a type of Orthographic projection, used to create a pictorial drawing of an object, where the object is rotated along one or more of its axes relative to the plane of projection.

There are three main types of axonometric projection: *Isometric*, *Dimetric*, and *Trimetric* projection depending on the exact angle at which the view deviates from the Orthogonal.

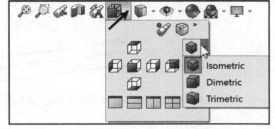

To display Isometric, Dimetric, or Trimetric of a 3D SOLIDWORKS model, select the drop-down arrow from the View Orientation icon in the Heads-up view toolbar.

Axonometric drawings often appear distorted because they ignore the foreshortening effects of perspective (foreshortening means the way things appear to get smaller in both height and depth as they recede into the distance). Typically, Axonometric drawings use vertical lines for those lines representing height and sloping parallel edges for all other sides.

- *Isometric Projection*. Isometric projection is a method of visually representing three-dimensional objects in two dimensions, in which the three coordinate axes appear equally foreshortened and the angles between them are 120°.

The term "Isometric" comes from the Greek for "equal measure" reflecting that the scale along each axis of the projection is the same (this is not true of some other forms of graphical projection).

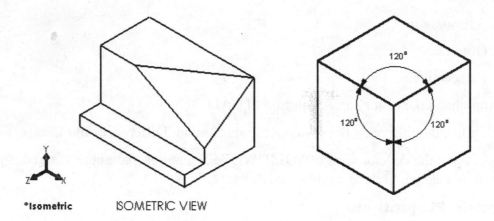

*Isometric   ISOMETRIC VIEW

- *Dimetric Projection*. A Dimetric projection is created using 3 axes, but only two of the three axes have equal angles. The smaller these angles are, the less we see of the top surface. The angle is usually around 105°.

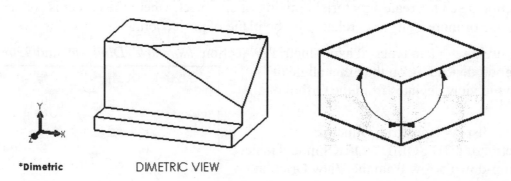

*Dimetric   DIMETRIC VIEW

- *Trimetric Projection.* A Trimetric projection is created using 3 axes where each of the angles between them is different (there are no equal angles). The scale along each of the three axes and the angles among them are determined separately as dictated by the angle of viewing. Approximations in trimetric drawings are common.

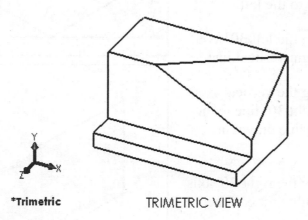

*Trimetric   TRIMETRIC VIEW

Isometric Sketching

Isometric sketches provide a 3D dimensional pictorial representation of an object. Isometric sketches help in the visualization of an object.

The surface features or the axes of the object are drawn around three axes from a horizontal line: a vertical axis, a 30° axis to the right, and a 30° axis to the left. All three axes intersect at a single point on the horizontal line.

All horizontal lines in an Isometric sketch are always drawn at 30° and parallel to each other and are either to the left or to the right of the vertical.

For this reason, all shapes in an Isometric sketch are not true shapes; they are distorted shapes.

All vertical lines in an Isometric sketch are always drawn vertically, and they are always parallel to each other as illustrated in the following example.

Example 1:

Exercise: Draw an Isometric sketch of a cube.

1. Draw a light horizontal axis (construction line) as illustrated on graph paper. Draw a light vertical axis. Draw a light 30° axis to the right. Draw a light 30° axis to the left.

2. Measure the length along the left 30° axis, make a mark and draw a light vertical line.

3. Measure the height along the vertical axis, make a mark and draw a light 30° line to the left to intersect the vertical line drawn in step 2.

4. Measure the length along the right 30° axis, make a mark and draw a light vertical line.

5. From the height along the vertical axis, make a mark and draw a light 30° line to the right to intersect the vertical line drawn in step 4.

6. Draw a light 30° line to the right and a light 30° line to the left to complete the cube. Once the sketch is complete, darken the shape.

🔆 In an Isometric drawing, the object is viewed at an angle, which makes circles appear as ellipses.

🔆 Isometric Rule #1: Measurement can only be made on or parallel to the isometric axis.

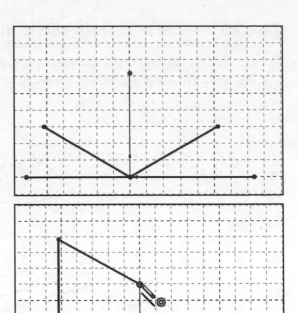

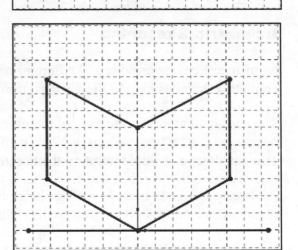

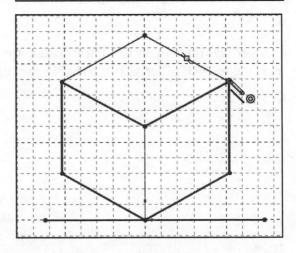

Circles drawn in Axonometric view

A circle drawn on a sloping surface in Axonometric projection will be drawn as an ellipse. An ellipse is a circle turned through an angle. All the examples shown above were box shapes without any curved surfaces. In order to draw curved surfaces we need to know how to draw an ellipse.

If you draw a circle and rotate it slowly, it will become an ellipse. As it is turned through 90° - it will eventually become a straight line. Rotate it 90° again, and it will eventually be back to a circle.

Example 1:

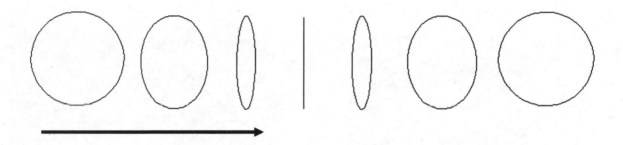

An ellipse has a major axis and a minor axis. The major axis is the axis about which the ellipse is being turned. The minor axis becomes smaller as the angle through which the ellipse is turned approaches 90°.

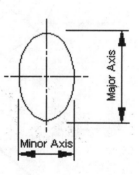

You can draw a cylinder using the technique shown below. The ellipses can either be sketched freehand or drawn using an ellipse template.

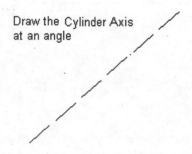

Draw the Cylinder Axis
at an angle

Draw the Major Axis
of the first ellipse at
right angles to the
cylinder axis

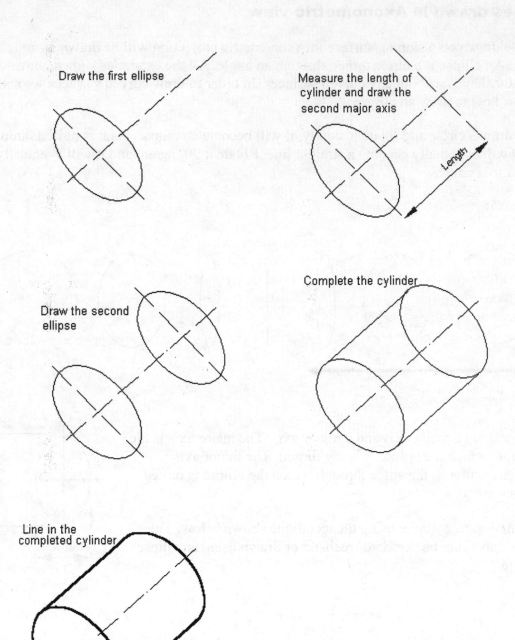

Draw the first ellipse

Measure the length of cylinder and draw the second major axis

Length

Draw the second ellipse

Complete the cylinder

Line in the completed cylinder

💡 Isometric Rule #2: When drawing ellipses on normal isometric planes, the minor axis of the ellipse is perpendicular to the plane containing the ellipse. The minor axis is perpendicular to the corresponding normal isometric plane.

Additional Projections

Oblique Projection: In Oblique projections, the front view is drawn true size, and the receding surfaces are drawn on an angle to give it a pictorial appearance. This form of projection has the advantage of showing one face (the front face) of the object without distortion. Generally, the face with the greatest detail faces the front.

There are two types of Oblique projection used in engineering design.

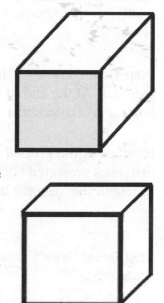

- *Cavalier*: In Cavalier Oblique drawings, all lines (including receding lines) are created to their true length or scale (1:1).

- *Cabinet*: In Cabinet Oblique drawings, the receding lines are shortened by one-half their true length or scale to compensate for distortion and to approximate more closely what the human eye would see. It is for this reason that Cabinet Oblique drawings are the most used form of Oblique drawings.

In Oblique drawings, the three axes of projection are vertical, horizontal, and receding. The front view (vertical & horizontal axis) is parallel to the frontal plane and the other two faces are oblique (receding). The direction of projection can be top-left, top-right, bottom-left, or bottom-right. The receding axis is typically drawn at 60º, 45º or 30º.

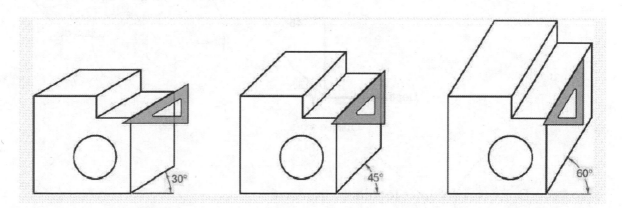

In the oblique pictorials coordinate system, only one axis is at an angle. The most commonly used angle is 45º.

Isometric Rule #1: A measurement can only be made on or parallel to the isometric axis. Therefore, you cannot measure an isometric inclined or oblique line in an isometric drawing because they are not parallel to an isometric axis.

Example: Drawing cylinders in Oblique projection is quite simple if the stages outlined below are followed. In comparison with other ways of drawing cylinders (for example, perspective and isometric) using Oblique projection is relatively easy.

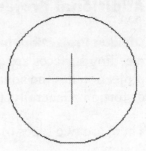

Step One: Draw vertical and horizontal centerlines to indicate the center of a circle, then use a compass to draw the circle itself.

Step Two: Draw a 45° line to match the length on the cylinder. At the end of this line, draw vertical and horizontal centerlines.

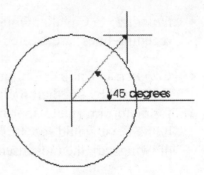

Remember the general rule for Oblique is to half all distances projected backwards. If the cylinder is 100mm in length the distance back must be drawn to 50mm.

Step Three: Draw the second circle with a compass as illustrated.

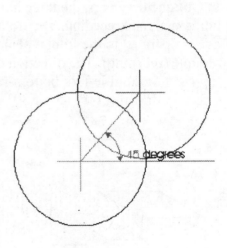

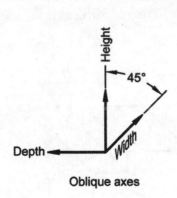

Oblique axes

Step Four: Draw two 45° lines to join the front and back circles.

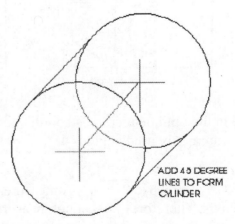

ADD 45 DEGREE LINES TO FORM CYLINDER

Step Five: Go over the outline of the cylinder with a fine pen or sharp pencil. Add shading if required.

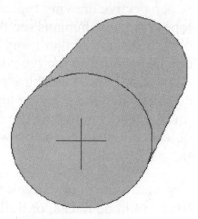

Perspective Projection: If you look along a straight road, the parallel sides of the road appear to meet at a point in the distance. This point is called the vanishing point and has been used to add realism. Suppose you want to draw a road that vanishes into the distance. The rays from the points a given distance from the eye along the lines of the road are projected to the eye. The angle formed by the rays decreases with increasing distance from the eye.

To display a Perspective view in SOLIDWORKS of a 3D model, click View, Display, Perspective from the Main toolbar.

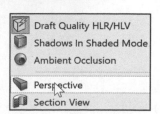

A perspective drawing typically aims to reproduce how humans see the world: objects that are farther away seem smaller, etc. Depending on the type of perspective (1-pt, 2-pt, 3-pt), vanishing points are established in the drawing towards which lines recede, mimicking the effect of objects diminishing in size with distance from the viewer.

One vanishing point is typically used for roads, railroad tracks, or buildings viewed so that the front is directly facing the viewer as illustrated above.

Any objects that are made up of lines either directly parallel with the viewer's line of sight or directly perpendicular (the railroad slats) can be represented with one-point perspective.

The selection of the locations of the vanishing points, which is the first step in creating a perspective sketch, will affect the looks of the resulting images.

Two-point perspective can be used to draw the same objects as one-point perspective, rotated: looking at the corner of a house, or looking at two forked roads shrink into the distance, for example. One point represents one set of parallel lines, the other point represents the other. Looking at a house from the corner, one wall would recede towards one vanishing point; the other wall would recede towards the opposite vanishing point as illustrated.

Two Point
Perspective

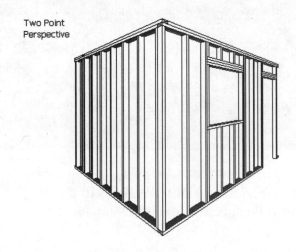

Three-point perspective is usually used for buildings seen from above (or below). In addition to the two vanishing points from before, one for each wall, there is now one for how those walls recede into the ground. This third vanishing point will be below the ground. Looking up at a tall building is another common example of the third vanishing point. This time the third vanishing point is high in space.

💡 One-point, two-point and three-point perspectives appear to embody different forms of calculated perspective. Despite conventional perspective drawing wisdom, perspective basically just means "position" or "viewpoint" of the viewer relative to the object.

Arrangement of Views

The main purpose of an engineering drawing is to provide the manufacturer with sufficient information needed to build, inspect or assemble the part or assembly according to the specifications of the designer. Since the selection and arrangement of views depends on the complexity of a part, only those views that are needed should be drawn.

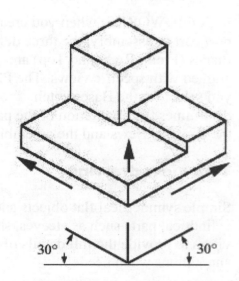

The average part drawing which includes the Front view, Top view and Right view are known as a three-view drawing. However, the designation of the views is not as important as the fact that the combination of views must give all the details of construction in a clear, correct and concise way.

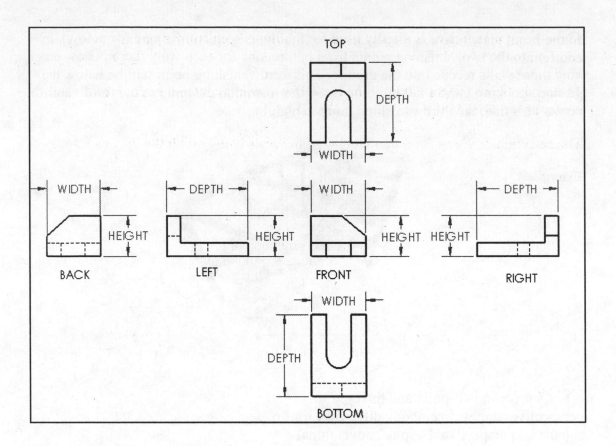

Third "3rd" Angle Projection is displayed and used in the book.

In SOLIDWORKS, when you create a new part or assembly, the three default Planes (Front, Right and Top) are aligned with specific views. The Plane you select for the Base sketch determines the orientation of the part, the drawing views and the assembly.

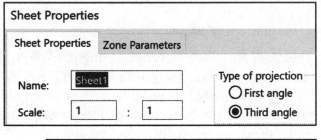

Two-view drawing

Simple symmetrical flat objects and

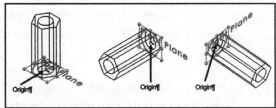

cylindrical parts such as sleeves, shafts, rods and studs most of the time only require two views to provide the full details of construction and or inspection. Always use annotations when needed.

In the Front view below, a centerline runs through the axis of the part as a horizontal centerline.

The second view (Right view) of the two-view drawing contains a center mark at the center of the cylinders.

The selection of views for a two-view drawing rests largely with the designer/engineer.

Example 1:

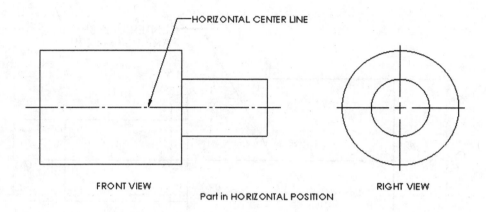

FRONT VIEW   RIGHT VIEW

Part in HORIZONTAL POSITION

Example 2:

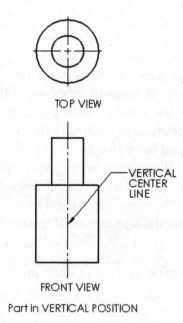

TOP VIEW

VERTICAL CENTER LINE

FRONT VIEW

Part in VERTICAL POSITION

One-view drawing

Parts that are uniform in shape often require only one view to describe them adequately. This is particularly true of cylindrical objects where a one-view drawing saves time and simplifies the drawing.

When a one-view drawing of a cylindrical part is used, the dimension for the diameter (according to ANSI standards) must be preceded by the symbol Ø, as illustrated.

Example 1:

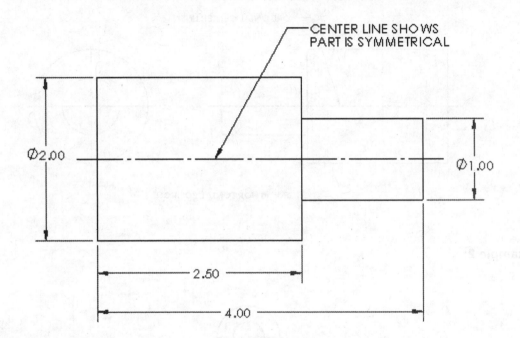

The one-view drawing is also used extensively for flat (Sheet metal) parts. With the addition of notes to supplement the dimensions on the view, the one view furnishes all the necessary information for accurately describing the part. In the first illustration below, you have two views: Front view and Top view. In the section illustration below, you replace the Top view with a Note: MATERIAL THICKNESS .125 INCH. Utilize Smart Notes when using a 3D software package. The note is linked to the dimension of the model.

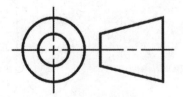

 Third Angle Projection type symbol is illustrated.

Example 1: No Note Annotation

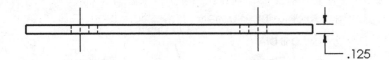

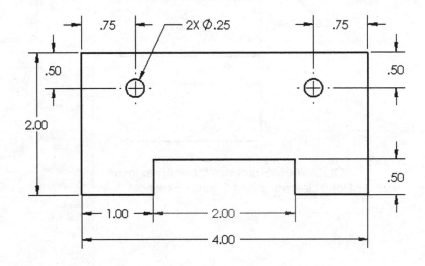

Example 2: Note Annotation to replace the Top view

MATERIAL THICKNESS .125 INCH

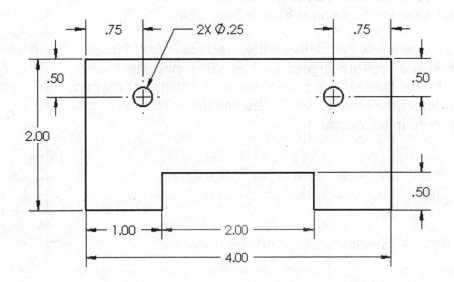

Example 3: Note Fastener Annotation

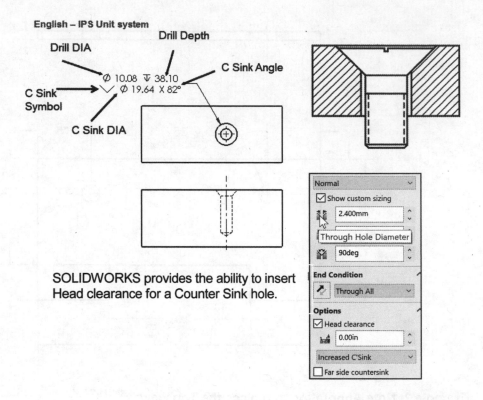

SOLIDWORKS provides the ability to insert Head clearance for a Counter Sink hole.

A multi-view drawing should have the minimum number of views necessary to describe an object completely. The most descriptive views are those that reveal the most information about the features, with the fewest features hidden from view.

When you create a new part or assembly, the three default Planes (Front, Right and Top) are aligned with specific views. The Plane you select for the Base sketch determines the orientation of the part, the orientation of the Front drawing view and the orientation of the first component in the assembly.

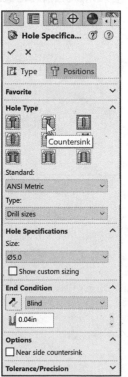

Exercises:

Exercise 1:

Draw freehand the Isometric view of the illustrated model on graph paper. Approximate the size of the model.

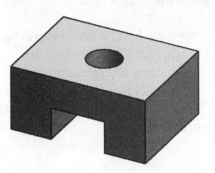

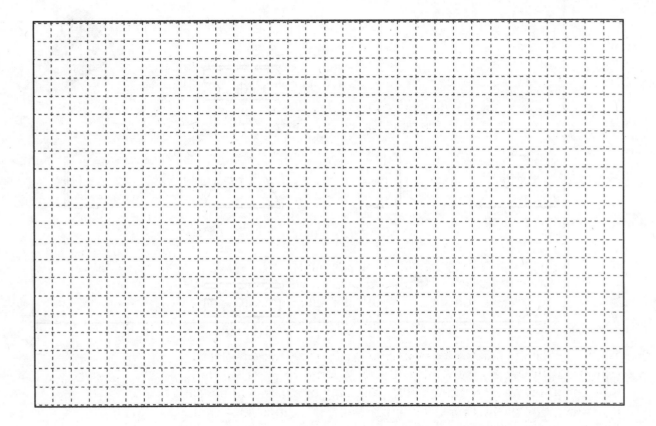

Exercise 2:

Name each view and insert the Width, Height and Depth name. No dimensions are required in this exercise. Note: Centerlines are not displayed. Third Angle Projection is used.

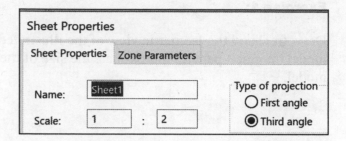

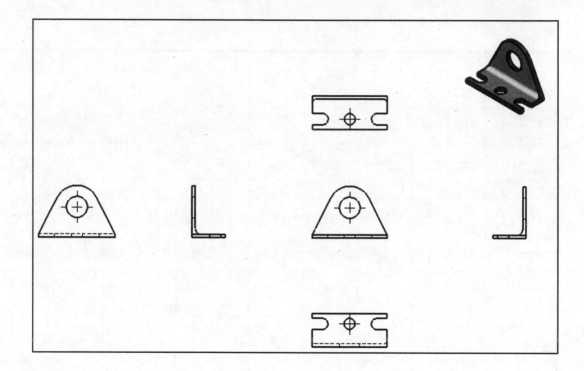

Drawing Views - Advanced

The standard views used in an orthographic projection are Front view, Top view, Right view and Isometric view. Non-standard orthographic drawing views are used when the six principal views do not fully describe the part for manufacturing or inspection. Below are a few non-standard orthographic drawing views.

Section view

Section views are used to clarify the interior of a part that can't clearly be seen by hidden lines in a view.

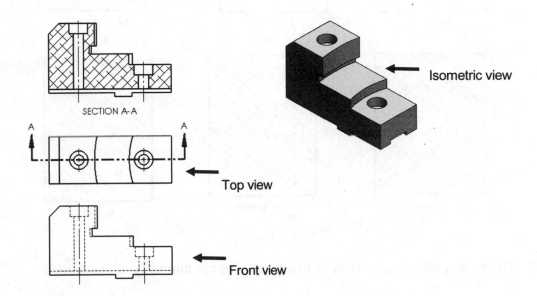

SECTION A-A

Isometric view

Top view

Front view

Think of an **imaginary** cutting (Plane) through the object and removing a portion. Imaginary is the key word.

A Section view is a child of the parent view. The Cutting Plane arrows used to create a Section view indicate the direction of sight. Section lines in the Section view are bounded by visible lines.

Section lines in the Section view can serve the purpose of identifying the kind of material the part is made from. Below are a few examples:

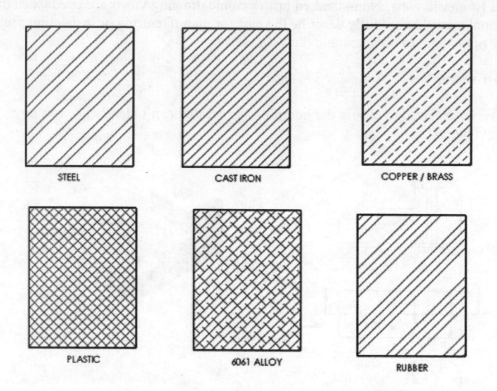

To avoid a false impression of thickness, ribs are normally not sectioned.

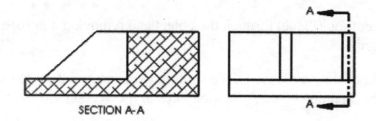

Detail View

The Detail view provides the ability to add a portion of a view, usually at an enlarged scale. A Detail view is a child of the parent view. Create a detail view in a drawing to display or highlight a portion of a view.

A Detail view may be of an Orthographic view, a non-planar (isometric) view, a Section view, a Crop view, an Exploded assembly view or another detail view.

Example 1:

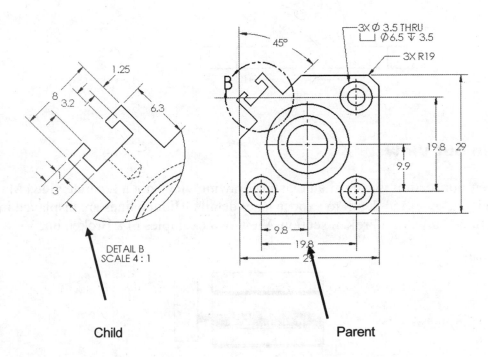

Child   Parent

If the Detail view has a different scale than the sheet, the scale needs to be supplied as an annotation as illustrated.

Example 2:

Below is a Detail view of a Section view. The Detail view is a child view of the parent view (Detail view). The Section view cannot exist without the Detail view.

Parent   Child

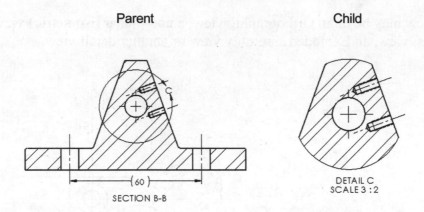

SECTION B-B

DETAIL C
SCALE 3 : 2

Broken out View

A Broken-out section is part of an existing drawing view, not a separate view. Material is removed to a specified depth to expose inner details. Hidden lines are displayed in the non-sectioned area of a broken section. View two examples of a Broken out View below.

Example 1:

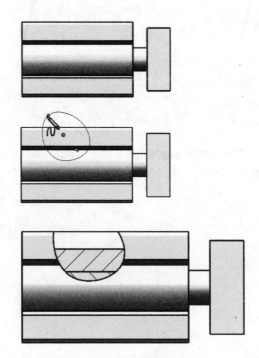

Example 2:

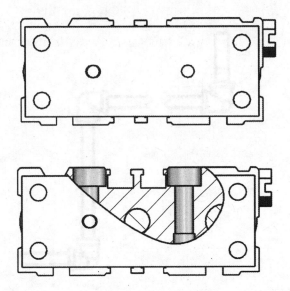

Break or Broken View

A Break view is part of an existing drawing view, not a separate view. A Break view provides the ability to add a break line to a selected view. Create a Broken view to display the drawing view in a larger scale on a smaller drawing sheet size. Reference dimensions and model dimensions associated with the broken area reflect the actual model values.

Example 1:

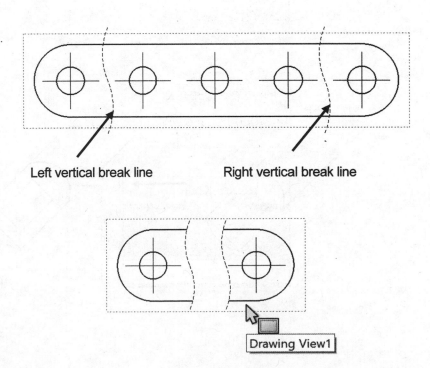

Example 2:

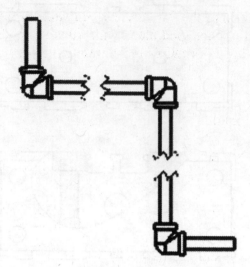

Crop View

A Crop view is a Child of the Parent view. A Crop view provides the ability to crop an existing drawing view. You cannot create a Crop view on a Detail view, a view from which a Detail view has been created or an Exploded view.

Create a Crop view to save time. Example: instead of creating a Section view and then a Detail view, then hiding the unnecessary Section view, create a Crop view to crop the Section view directly.

Example 1:

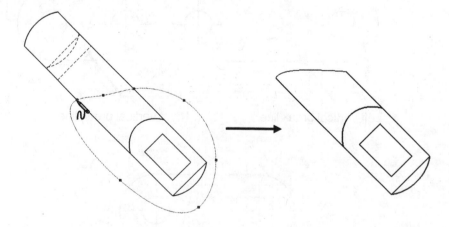

Auxiliary View

An Auxiliary view is a Child of the Parent view. An Auxiliary view provides the ability to display a plane parallel to an angled plane with true dimensions. A primary Auxiliary view is hinged to one of the six principle orthographic views.

Example 1:

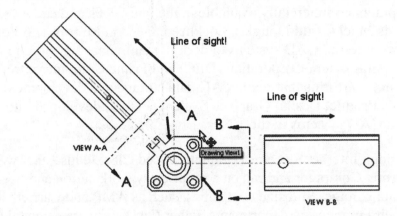

Exercises:

Exercise 1:

Label all of the name views below. Note: Third Angle projection.

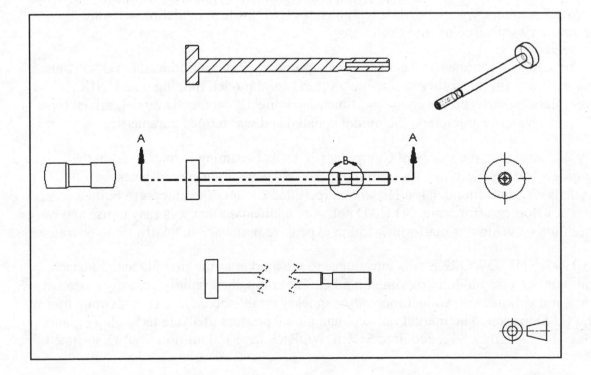

History of Computer Aided Design (CAD)

In 1963, Ivan Sutherland of MIT developed "Sketchpad," a graphical communication system, where with a light pen, Sutherland was able to select and modify geometry on a Cathode Ray System (CRT) and input values through a key pad. Geometric relationships were made between lines and arc and geometry could be moved and copied.

With aerospace and automotive technologies becoming more complex and IBM mainframe computers commercially available in the late 1960s and early 1970s, companies such as MacDonald-Douglas, Lockheed, General Motors, and Ford were utilizing their own internal CAD systems to design, manipulate and store models and drawings. Digital Equipment Corporation (DEC) and Prime Computer introduced computer hardware platforms that made CAD data storage and development more affordable. Ford's Product Design Graphics System (PDGS) developed into one of the largest integrated CAD systems in the 1980s.

By 1980, Cambridge Interact Systems (UK) introduced CIS Medusa, that was bought and distributed by Prime Computer and ran on a proprietary workstation and used Prime mini computers for data storage. Mid size companies, such as AMP and Carrier, were now using CAD in their engineering departments. Other CAD software companies also introduced new technology. Computervision utilized both proprietary hardware and SUN workstations and became a leader in 2D drafting technology.

But in the early 80s, 3D CAD used Boolean algorithms for solid geometry that were a challenge for engineers to manipulate. Other major CAD players were Integraph, GE Calma, SDRC, and IBM (Dassault Systèmes). Dassault Systèmes, with its roots in the aerospace industry, expanded development in CAD surface modeling software technology with Boeing and Ford.

In the late 80s, Parametric Technology Corporation (PTC) introduced CAD software to the market with the ability to manipulate a 3D solid model, running on a UNIX workstation platform. By changing dimensions directly on the 3D model, driven by dimensions and parameters, the model updated and was termed parametric.

By the early 90s, the Personal Computer (PC) was becoming incorporated in the engineer's daily activities for writing reports and generating spreadsheets. In 1993, SOLIDWORKS founder Jon Hirschtick recruited a team of engineers to build a company and develop an affordable, 3D CAD software application that was easy to use and ran on an intuitive Windows platform, without expensive hardware and software to operate.

In 1995, SOLIDWORKS was introduced to the market as the first 3D feature based, parametric solid modeler running on a PC. The company's rapidly growing customer base and continuous product innovation quickly established it as a strong competitor in the CAD market. The market noticed, and global product lifecycle technology giant Dassault Systèmes S.A. acquired SOLIDWORKS for $310 million in stock in June of 1997.

SOLIDWORKS went on to run as an independent company, incorporating finite element analysis (FEA) which has advanced dynamics, nonlinear, fatigue, thermal, steady state and turbulent fluid flow (CFD) and electromagnetic analysis capabilities, as well as design optimization. SOLIDWORKS open software architecture has resulted in over 1500 partner applications such as Computer Aided Manufacturing (CAM), robot simulation software, and process management. Today, SOLIDWORKS software has the most worldwide users in production - more than 5,500,000 users at over 600,000 locations in more than 300 countries.

Note: There are many university researchers and commercial companies that have contributed to the history of computer aided design. We developed this section on the history of CAD based on the institutions and companies that we worked for and worked with over our careers and as it relates to the founders of SOLIDWORKS.

Boolean operations

To understand the difference between parametric solid modeling and Boolean based solid modeling you will first review Boolean operations. In the 1980s, one of the key advancements in CAD was the development of the Constructive Solid Geometry (CSG) method. Constructive Solid Geometry describes the solid model as combinations of basic three-dimensional shapes or better known as primitives. Primitives are typically simple shapes: cuboids, cylinders, prisms, pyramids, spheres and cones.

Two primitive solid objects can be combined into one using a procedure known as the Boolean operations. There are three basic Boolean operations:

- Boolean Union

- Boolean Difference

- Boolean Intersection

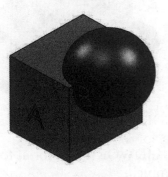

Boolean Operation:	Result:
Boolean Union - The merger of two separate objects into one. A + B	
Boolean Difference - The subtraction of one object from another. A - B	
Boolean Intersection - The portion common to both objects. A ∩ B	

Even today, Boolean operations assist the SOLIDWORKS designer in creating a model with more complex geometry by combining two bodies together with a Boolean intersection.

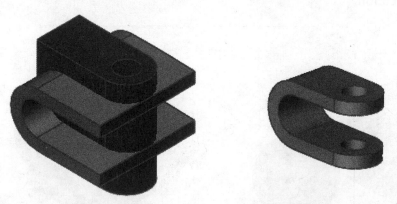

Use the SOLIDWORKS Tutorial to view Multi-body Parts, Boolean model examples.

What is SOLIDWORKS?

SOLIDWORKS® is a mechanical design automation software package used to build parts, assemblies and drawings that takes advantage of the familiar Microsoft® Windows graphical user interface.

SOLIDWORKS is an easy to learn design and analysis tool (SOLIDWORKS Simulations, SOLIDWORKS Motion, SOLIDWORKS Flow Simulation etc.), which makes it possible for designers to quickly sketch 2D and 3D concepts, create 3D parts and assemblies and detail 2D drawings.

In SOLIDWORKS, you create 2D and 3D sketches, 3D parts, 3D assemblies and 2D drawings. The part, assembly and drawing documents are related. Additional information on SOLIDWORKS and its family of products can be obtained at their URL, www.SOLIDWORKS.com.

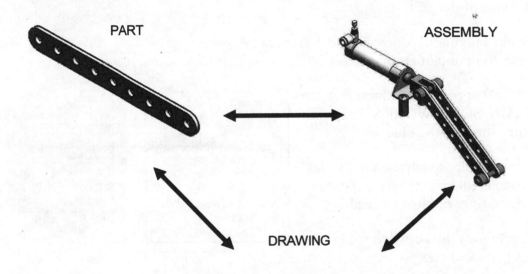

PART   ASSEMBLY

DRAWING

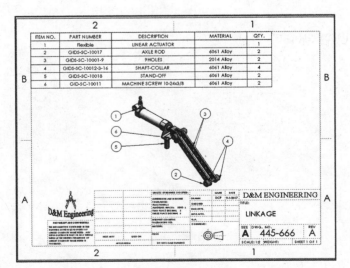

ITEM NO.	PART NUMBER	DESCRIPTION	MATERIAL	QTY.
1	Flexible	LINEAR ACTUATOR		1
2	GIDS-SC-10017	AXLE ROD	6061 Alloy	2
3	GIDS-SC-10001-9	9 HOLES	2014 Alloy	2
4	GIDS-SC-10012-3-16	SHAFT-COLLAR	6061 Alloy	4
5	GIDS-SC-10018	STAND-OFF	6061 Alloy	2
6	GID-SC-10011	MACHINE SCREW 10-24x3/8	6061 Alloy	2

D&M ENGINEERING

TITLE:

LINKAGE

SIZE A   DWG. NO. 445-666   REV A

Features are the building blocks of parts. Use features to create parts, such as Extruded Boss/Base and Extruded Cut. Extruded features begin with a 2D sketch created on a Sketch plane.

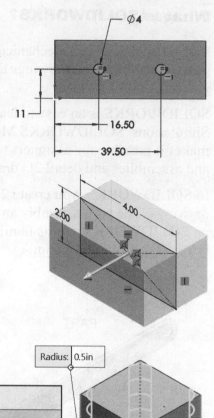

The 2D sketch is a profile or cross section. Sketch tools such as lines, arcs and circles are used to create the 2D sketch. Sketch the general shape of the profile. Add Geometric relationships and dimensions to control the exact size of the geometry.

Create features by selecting edges or faces of existing features, such as a Fillet. The Fillet feature rounds sharp corners.

Dimensions drive features. Change a dimension, and you change the size of the part.

Apply Geometric relationships: Vertical, Horizontal, Parallel, etc. to maintain Design intent.

Create a hole that penetrates through a part (Through All). SOLIDWORKS maintains relationships through the change.

The step-by-step approach used in this text allows you to create parts, assemblies and drawings by doing, not just by reading.

The book provides the knowledge to modify all parts and components in a document.

Change is an integral part of design.

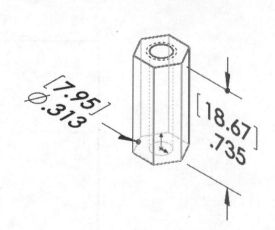

Design Intent

What is design intent? All designs are created for a purpose. Design intent is the intellectual arrangements of features and dimensions of a design. Design intent governs the relationship between sketches in a feature, features in a part and parts in an assembly.

The SOLIDWORKS definition of design intent is the process in which the model is developed to accept future modifications. Models behave differently when design changes occur.

Design for change. Utilize geometry for symmetry, reuse common features, and reuse common parts. Build change into the following areas that you create:

- Sketch

- Feature

- Part

- Assembly

- Drawing

When editing or repairing geometric relations, it is considered best practice to edit the relation vs. deleting it.

Design Intent in a sketch

Build design intent in a sketch as the profile is created. A profile is determined from the Sketch Entities. Example: Rectangle, Circle, Arc, Point, Slot, etc. Apply symmetry into a profile through a sketch centerline, mirror entity and position about the reference planes and Origin.

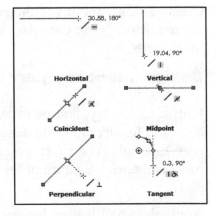

Build design intent as you sketch with automatic Geometric relations. Document the decisions made during the up-front design process. This is very valuable when you modify the design later.

A rectangle (Center Rectangle Sketch tool) contains Horizontal, Vertical and Perpendicular automatic Geometric relations.

Apply design intent using added Geometric relations if needed. Example: Horizontal, Vertical, Collinear, Perpendicular, Parallel, Equal, etc.

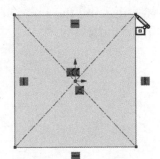

Example A: Apply design intent to create a square profile. Sketch a rectangle. Apply the Center Rectangle Sketch tool. Note: No construction reference centerline or Midpoint relation is required with the Center Rectangle tool. Insert dimensions to fully define the sketch.

Example B: Develop a rectangular profile. Apply the Corner Rectangle Sketch tool. The bottom horizontal midpoint of the rectangular profile is located at the Origin. Add a Midpoint relation between the horizontal edge of the rectangle and the Origin. Insert two dimensions to fully define the rectangle as illustrated.

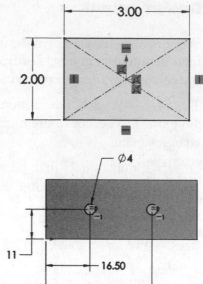

Design intent in a feature

Build design intent into a feature by addressing symmetry, feature selection, and the order of feature creation.

Example A: The Extruded Base feature remains symmetric about the Front Plane. Utilize the Mid Plane End Condition option in Direction 1. Modify the depth, and the feature remains symmetric about the Front Plane.

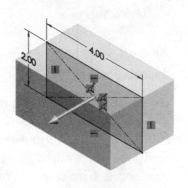

Example B: Create 34 teeth in the model. Do you create each tooth separately using the Extruded Cut feature? No.

Create a single tooth and then apply the Circular Pattern feature. Modify the Circular Pattern from 32 to 24 teeth.

Design intent in a part

Utilize symmetry, feature order and reusing common features to build design intent into a part. Example A: Feature order. Is the entire part symmetric? Feature order affects the part.

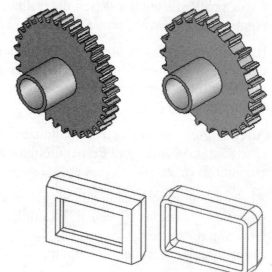

Apply the Shell feature before the Fillet feature and the inside corners remain perpendicular.

Design intent in an assembly

Utilizing symmetry, reusing common parts and using the Mate relation between parts builds the design intent into an assembly.

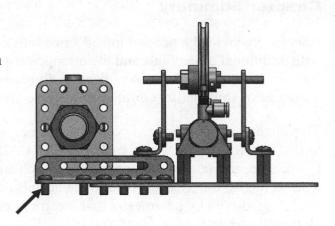

Example A: Reuse geometry in an assembly. The assembly contains a linear pattern of holes. Insert one screw into the first hole. Utilize the Component Pattern feature to copy the machine screw to the other holes.

Design intent in a drawing

Utilize dimensions, tolerance and notes in parts and assemblies to build the design intent into a drawing.

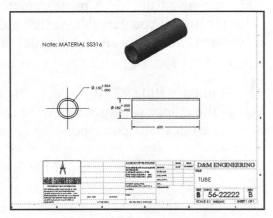

Example A: Tolerance and material in the drawing. Insert an outside diameter tolerance +.000/-.002 into the TUBE part. The tolerance propagates to the drawing.

Define the Custom Property Material in the Part. The Material Custom Property propagates to your drawing.

Create a sketch on any of the default planes: Front, Top, Right or a created plane.

Additional information on design process and design intent is available in SOLIDWORKS Help.

The book is designed to expose the new user to many tools, techniques and procedures. It does not always use the most direct tool or process.

Chapter Summary

Chapter 2 provided a general introduction into isometric projection and sketching along with additional projections and the arrangement of standard views and advanced views. You explored the three main projection divisions in freehand engineering sketches and drawings: Axonometric, Oblique and Perspective.

This chapter also introduced you to the history of CAD and the development of DS SOLIDWORKS Corp. From early Boolean CAD software, you explored Union, Difference, and Intersection operations which are modeling techniques still used today. You were also introduced to the fundamentals of SOLIDWORKS, its feature based modeling, driven by parameters that incorporates your design intent into a sketch, part, assembly and drawing.

Isometric Rule #1: A measurement can only be made on or parallel to the isometric axis. Therefore, you cannot measure an isometric inclined or oblique line in an isometric drawing because they are not parallel to an isometric axis.

Isometric Rule #2: When drawing ellipses on normal isometric planes, the minor axis of the ellipse is perpendicular to the plane containing the ellipse. The minor axis is perpendicular to the corresponding normal isometric plane.

Questions

1. Name the three main projection divisions commonly used in freehand engineering sketches and detailed engineering drawings: _____ ,, _____ and _____

2. Name the projection divisions within Axonometric projection: _____, _____, and _____ .

3. True or False: In oblique projections the front view is drawn true size, and the receding surfaces are drawn on an angle to give it a pictorial appearance.

4. Name the two types of Oblique projection used in engineering design: _____, _____ .

5. Describe Perspective Projection. Provide an example.

6. True or False: Parts that are uniform in shape often require only one view to describe them adequately.

7. True or False: The designer usually selects as a Front view of the object that view which best describes the general shape of the part. This Front view may have no relationship to the actual front position of the part as it fits into an assembly.

8. True or False: When a one-view drawing of a cylindrical part is used, the dimension for the diameter (according to ANSI standards) must be preceded by the symbol Ø.

9. Draw a Third Angle Projection Symbol.

10. Draw a First Angle Projection Symbol.

11. Describe the difference between First and Third Angle Projection.

12. True or False. First Angle Projection is used in the United States.

13. True or False. Section lines can serve the purpose of identifying the kind of material the part is made from.

14. True or False. All dimension lines terminate with an arrowhead on mechanical engineering drawings.

15. True or False. Break lines are applied to represent an imaginary cut in an object, so the interior of the object can be viewed or fitted to the sheet. Provide an example.

Exercises

Exercise 2.1: Hand draw the Isometric view from the illustrated model below.

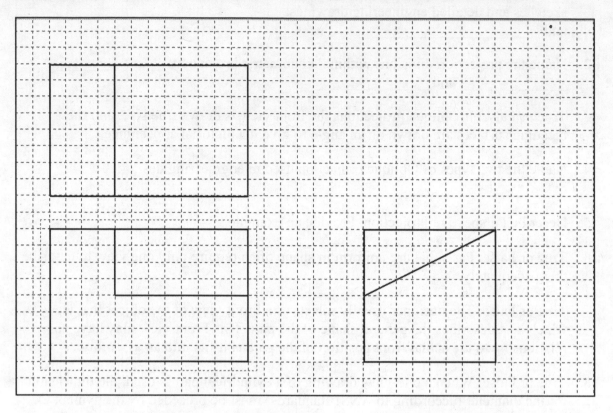

Exercise 2.2: Hand draw the Isometric view for the following models. Approximate the size of the model.

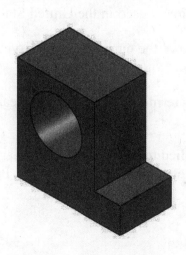

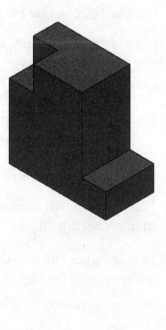

Exercise 2.3: Hand draw the Isometric view for the following models. Approximate the size of the model.

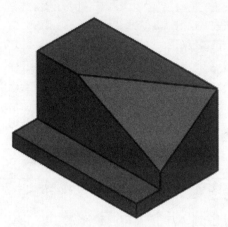

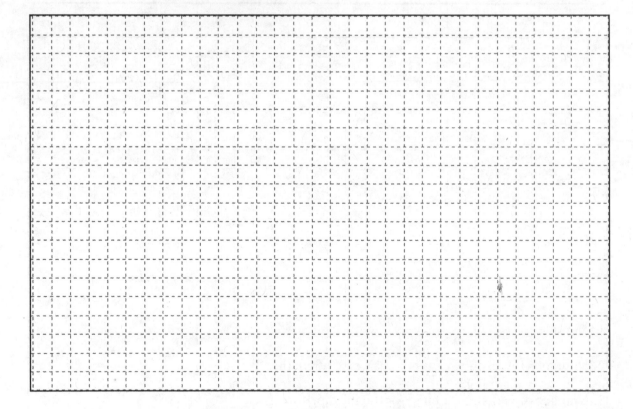

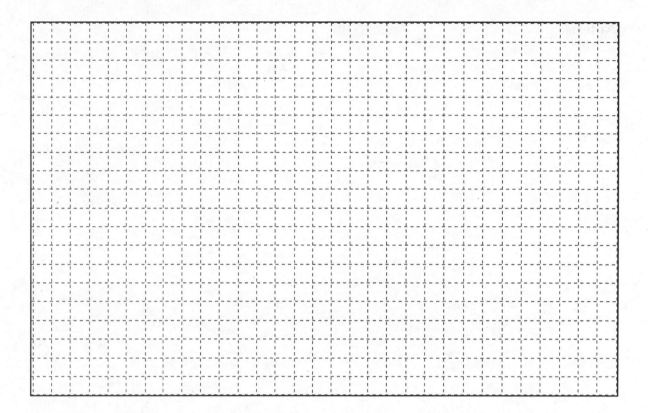

Exercise 2.4: Identify the number of vanishing points for each picture.

1. Number of vanishing points for the first picture._____

2. Number of vanishing points for the second picture._____

3. Number of vanishing points for the third picture. _____

Notes:

Chapter 3

Dimensioning Practices, Scales, Tolerancing and Fasteners

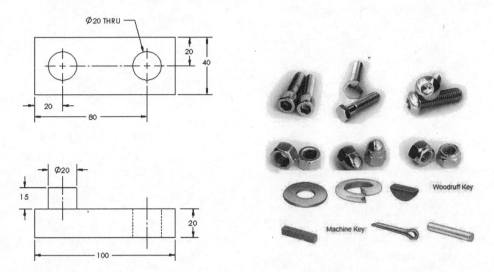

Below are the desired outcomes and usage competencies based on the completion of Chapter 3.

Desired Outcomes:	Usage Competencies:
• Knowledge of dimensioning and the ASME Y14.5-2009 standard.	• Ability to correctly dimension the following features, objects and shapes: rectangle, cone, sphere, hole, cylinder, angle, point or center, arc, chamfer and more.

• Awareness of measurement units.
• Understand Scale type.
	• Apply the following measurement system:
	  o Metric system (MMGS).
	  o English system (IPS).
	• Engineer's scale, Architect's scale, Linear scale, Vernier scale, and Linear encoder.

• Understand Tolerancing for a drawing. • Comprehend Fasteners and hole dimensioning. • Recognize Fit type.	• Apply dimension and drawing Tolerances. • Read and understand general Fastener and hole annotation. • Apply Fit type.

Notes:

Chapter 3 - Dimensioning Practices, Tolerancing and Fasteners

Chapter Overview

Chapter 3 provides a general introduction into dimensioning systems and the ANSI Y14.5 2009 standards along with fasteners, fits, and general tolerancing practices.

On the completion of this chapter, you will be able to:

- Understand and apply the ASME Y14.5-2009 dimensioning standard.

- Awareness of measurement units:

  o Metric system (MMGS).

  o English system (IPS).

- Familiarity of dual dimensioning:

  o (Primary vs. Secondary).

- Understand Scale type:

  o Engineer's scale, Architect's scale, Linear scale, Vernier scale and Linear encoder.

- Ability to correctly dimension the following features, objects and shapes: rectangle, Cone, Sphere, hole, cylinder, angle, point or center, arc, chamfer and more.

- Understand and apply part and drawing Tolerance.

- Read and understand Fastener notation.

- Recognize single, double and triple thread.

- Distinguish between Right-handed and Left-handed thread.

- Recognize annotations for a simple hole, Counterbore and Countersink in a drawing.

- Identify Fit type.

American National Standards Institute (ANSI)

To ensure some measure of uniformity in industrial drawings, the American National Standards Institute (ANSI) has established drafting standards; these standards are called the language of drafting and are in general use throughout the United States.

ANSI was originally formed in 1918, when five engineering societies and three government agencies founded the American Engineering Standards Committee (AESC). In 1928, the AESC became the American Standards Association (ASA). In 1966, the ASA was reorganized and became the United States of America Standards Institute (USASI). The present name was adopted in 1969 and the standards are published by the American Society of Mechanical Engineers (ASME).

While these drafting standards or practices may vary in some respects between industries, the principles are basically the same. The practices recommended by ANSI for dimensioning and for marking notes are followed in this book.

Dimensioning

Dimensioning is the process of defining the size, form and location of geometric features and components on an engineering drawing. The principle for dimensioning a drawing is clarity and accuracy. To promote clarity and accuracy, ANSI developed standard practices.

There are two key types of dimensions:

- **Location dimension** - locates a horizontal position, vertical position, center of a hole, slot, chamfer or other model features.

- **Size dimension** - provides a horizontal position, vertical position, angle, diameter, radius etc.

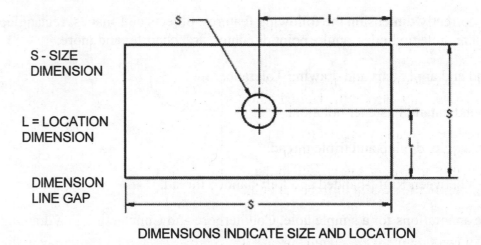

DIMENSIONS INDICATE SIZE AND LOCATION

Dimensions should not be duplicated or the same information given in two different ways. If a reference dimension is used, the size value is placed within parentheses (X).

Measurement

Measurement is the process or the result of determining the ratio of a physical quantity, such as a length, time, temperature etc., to a unit of measurement, such as the meter, second or degree Celsius.

The metric system is a decimal system of measurement based on its units for length, the meter and for mass, the kilogram.

Other names for the Metric system include:

- International System of Units SI.

- International Organization for Standardization ISO units.

The Metric system is the most commonly used system of measurement in the world.

The Metric system is based on the meter as the standard unit of reference. A meter (approximately 39.37 inches in length) is subdivided into 10 equal parts called decimeters. Each decimeter is divided into 10 parts called centimeters and each centimeter is divided into 10 parts called millimeters.

The Metric system is a very coherent system because it is exclusively a decimal system and therefore has a common multiplier and divisor of 10. Regular fractions are not used in the metric system. Instead the metric system uses only decimal fractions.

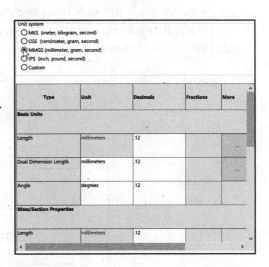 The default SOLIDWORKS (SI) metric unit system is millimeter, gram, second (MMGS).

The United States Customary units are a system of measurements commonly used in the United States.

Many U.S. units are virtually identical to their imperial counterparts, but the U.S. customary system developed from English units used in the British Empire before the system of imperial units was standardized in 1824.

The U.S system is a system of measurement based on its units for length, the inch and for mass, the pound.

The Inch system is based on the foot as the standard unit of reference. A foot is divided into 12 equal parts called inches. Each inch is subdivided into a variety of fractions and decimals.

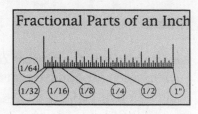

Parts of a foot cannot be easily expressed as decimal inches. For example, in the metric system 7 millimeters is 0.7 centimeters which is 0.07 decimeters which is 0.007 meters. But 7 inches is 0.583333 feet which is 0.19444 yards and so on. This is a clear advantage for the metric system.

🔅 The default SOLIDWORKS English system is inch, pound, second (IPS).

🔅 Architectural drawings using Metric units are based on the meter. Architectural drawings using English units are based on feet.

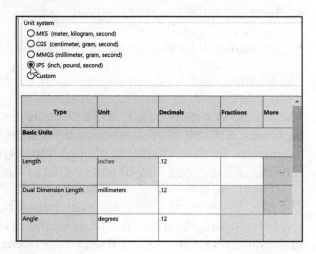

Dual dimensioning

Working drawings are usually drawn with all U.S. or all metric dimensions. Sometimes the object manufactured requires using both the U.S. and metric measuring system. In the illustration, the primary units are in inches and the secondary units (mm) are displayed in parentheses.

Example 1:

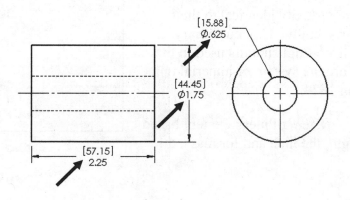

Scale

An engineer's scale is a tool for measuring distances and transferring measurements at a fixed ratio of length. A few of the common scale types are:

- **Architect's scale**: A ruler-like device which facilitates the production of technical drawings.

- **Engineer's scale**: A ruler-like device similar to the Architect's scale, helpful when drawing rooms.

- **Linear encoder**: A kind of linear scale used in precision manufacturing for positioning.

- **Linear scale**: A means of showing the scale of a map, chart, or drawing.

- **Vernier scale**: A scale that allows for higher precision than a uniformly-divided straight or circular measurement scale.

Never measure a drawing directly. Its dimensions can be printed at any scale or the drawing may be drawn one way and dimensioned another.

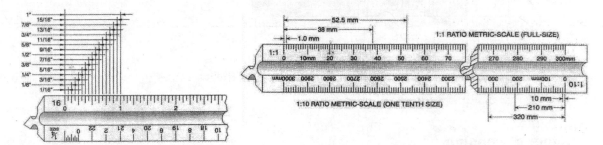

Engineering Scale - English
Full inches are measured to the right while fractions of an inch are measured to the left.

Engineering Scale - Metric
Note the Scale - Full Size: (1:1)

Visit the following sites for additional information on scales:
http://www.usfa.dhs.gov/downloads/pdf/nfa/engineer-architect-scales.pdf

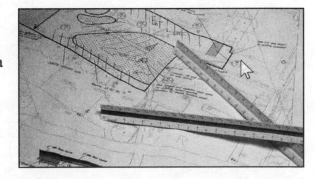

Standards for Dimensioning

All drawings should be dimensioned completely so that a minimum of computation is necessary, and the part can be built without scaling the drawing. However, there should not be a duplication of dimensions unless such dimensions make the drawing clearer and easier to read. These dimensions are called reference dimensions and are enclosed in parentheses.

Linear dimension

A Linear dimension is a dimension that is either horizontal or vertical to the dimensioning plane. Read a drawing from left to right.

Example 1:

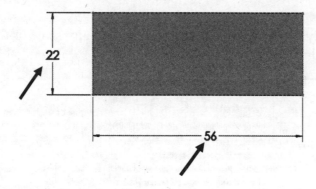

Stagger dimension

The general practice is to stagger the dimension text on <u>parallel dimensions</u> (small to large) as illustrated.

Example 1:

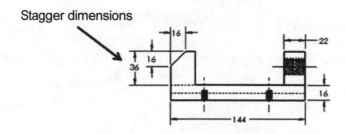

Aligned dimension

Aligned dimension is a style of dimensioning in which text is placed parallel to the dimension line. The aligned method of dimensioning is **not** approved by the current ANSI standards but may be seen on older drawings.

Example 1:

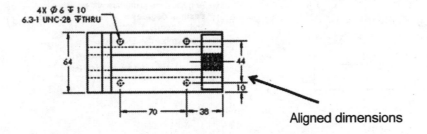

Aligned dimensions

Angular dimension

The design of a part may require some lines to be drawn at an angle. The amount of the divergence (the amount the lines move away from each other) is indicated by an angle measured in degrees or fractional parts of a degree. The degree is indicated by a symbol ° placed after the numerical value of the angle.

Example 1:

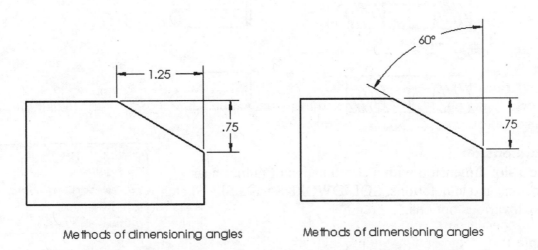

Methods of dimensioning angles   Methods of dimensioning angles

The dimension line for an angle should be an arc whose ends terminate in arrowheads.

The numeral indication, the number of degrees in the angle, is read in a horizontal position, except where the angle is large enough to permit the numerals to be placed along the arc.

Example 2:

- Angular dimensions are shown either in **decimal degrees** or in **degrees**, **minutes**, and **seconds**.
- The symbol used for degrees is °
- The symbol used for minutes is '
- The symbol used for seconds is "

Chamfer dimension

A chamfer connects two objects to meet in a flattened or beveled corner.
In SOLIDWORKS, the chamfer tool creates a beveled feature on selected edges, faces or a vertex.

Example 1:

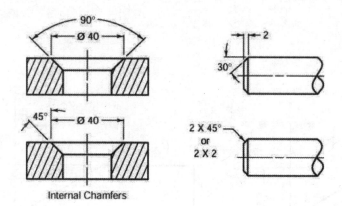

Internal Chamfers

Slot dimension

Create a slot dimension with a combination of radii, linear dimensions and annotations. SOLIDWORKS has a Slot Sketch tool with various options.

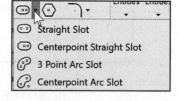

Straight Slot
Centerpoint Straight Slot
3 Point Arc Slot
Centerpoint Arc Slot

Example 1:

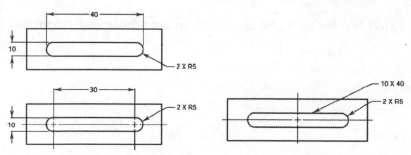

Radius dimension (Leader line)

The Radius leader line is a continuous straight line that extends at an angle from a note, a dimension, or other reference to a feature. A leader line for a diameter or radius should be radial. A radial line is one that passes through the center of the circle or arc if extended.

If an arc is less than half a circle, the radius is specified, preceded by an R. The leader dimension line for a radius shall have a single arrowhead touching the arc as illustrated.

Example 1:

Simple Hole dimension (Leader line)

The simple hole dimension leader line is a continuous straight line that extends at an angle from a note, a dimension, or other reference to a feature.

Example 1:

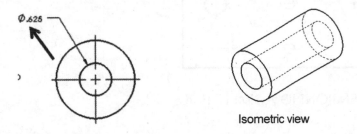

Isometric view

ANSI standard requires that full cylinders (holes and bosses) must always be measured by their diameter. The diameter symbol must precede the numerical value to indicate that the dimension shows the diameter of a circle or cylinder. The symbol used is the Greek letter phi (Ø).

A feature should be dimensioned only once. Each feature should be dimensioned or identified with a note. Dimension features or surfaces should be done to a logical reference point.

Enough space should be provided to avoid crowding and misinterpretation. Extension lines and object lines should not overlap.

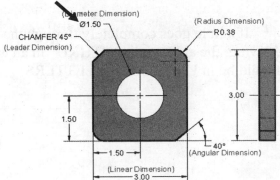

The standard size ratio for all arrowheads on mechanical drawings is ~2.5:1 (Length to width).

Simple Hole dimension

A simple hole is created by drilling, reaming or punching. A simple hole must be measured by its diameter. The diameter symbol must precede the numerical value followed by a note (annotation) indicating the operation to be performed and the number of holes to be produced, as illustrated.

Example 1:

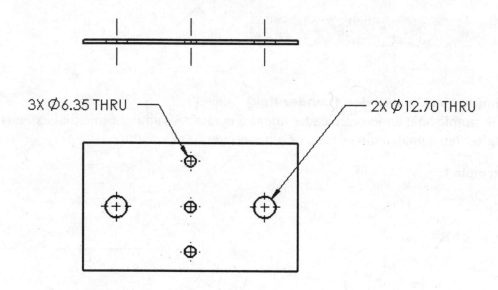

3X ⌀6.35 THRU   2X ⌀12.70 THRU

DIMENSIONING A SIMPLE HOLE

Repetitive features or dimensions can be specified by using the symbol "X" along with the number of times the feature is repeated as illustrated above. There is no space between the number of times the feature is repeated and the "X" symbol; however, there is a space between the symbol "X" and the dimension.

If a hole goes completely through the feature and it is not clearly shown on the drawing, the abbreviation "THRU" in all upper case follows the dimension. All notes should be in UPPER CASE LETTERS.

Fastener Hole dimension (Annotations)

Denote drilled hole information by a bent leader line as illustrated.

Example 1:

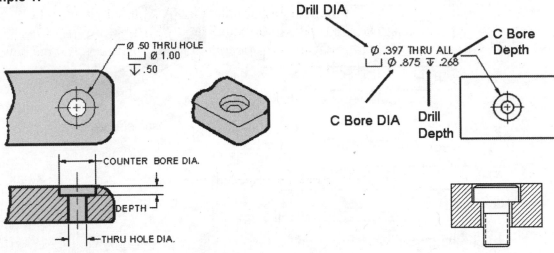

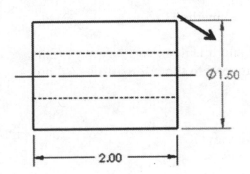

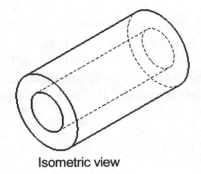

If a hole goes completely through the feature and it is not clearly shown on the drawing, the abbreviation **"THRU"** or **"THRU ALL"** in all upper case follows the dimension. Additional information is presented later in the chapter on Fasteners.

Cylindrical dimension

Full cylinders (holes and bosses) must always be measured by their diameter. The diameter symbol must precede the numerical value to indicate that the dimension shows the diameter of a circle or cylinder. The symbol used is the Greek letter phi (ø).

The length and diameter of the cylinder are usually placed in the view which shows the cylinder as a rectangle as illustrated below.

Example 1:

Example 2:

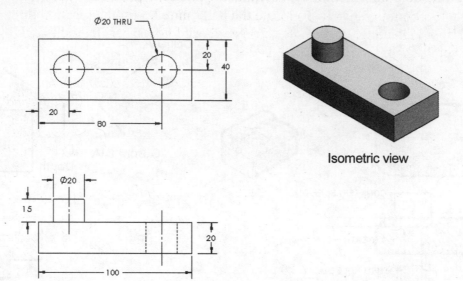

Isometric view

Example 3:

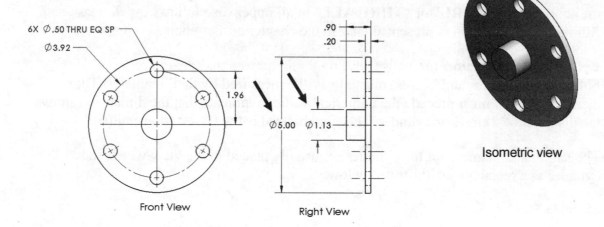

Front View   Right View   Isometric view

Your choice of dimensions will directly influence the method used to manufacture the part.

If there is room, position the diameter dimension up and to the left, off the model in a circular view.

🔆 The Front view should be the most descriptive view if appropriate.

🔆 Dimension lines should not cross, if avoidable.

Equally spaced hole dimension

Dimension equally spaced holes on a cylinder. The exact location of the first hole is given by a location dimension. To locate the remaining holes, the location dimension is followed by 1.) Diameter of the holes, 2.) Number of holes, 3.) Notation EQUALLY SPACED or "EQ SP" as illustrated.

Example 1:

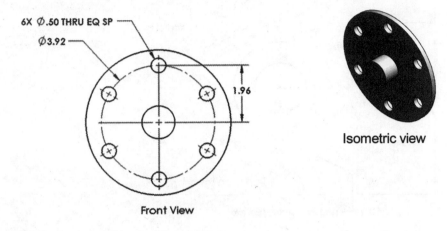

Front View

Isometric view

Hole dimension location

Holes are often dimensioned in relation to one another and to a finished surface. Dimensions are usually given, in such cases, in the view in which the shape of the hole is, that is, square, round or elongated. The preferred method of placing these dimensions is illustrated below.

Example 1:

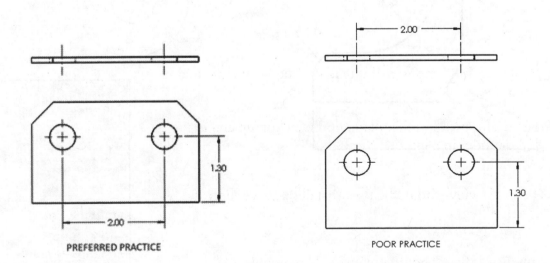

PREFERRED PRACTICE   POOR PRACTICE

Point/center of a circle dimension

Center of an arc or circle can be found by creating vertical and horizontal center lines from the machined surfaces.

Example 1:

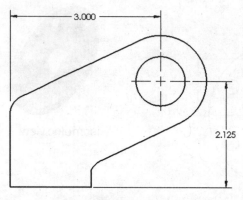

Dimensioning the center of a circle

Arc dimension

Arc dimensions measure the distance along an arc or polyline arc segment. Apply Foreshortened leader lines for large arcs as illustrated. Typical uses of arc length dimensions include measuring the travel distance around a cam or indicating the length of a cable.

Example 1:

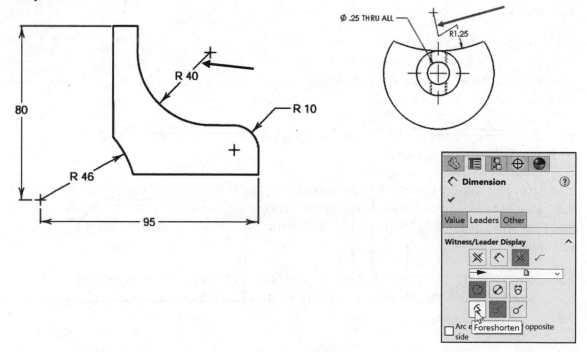

Order of Preference - Linear Dimension line

There is an order and style of preference for a linear dimension line using arrowheads. The first order of preference is that the dimension and arrowheads are drawn between the extension line if space is available. If space is limited, see the below order of preference.

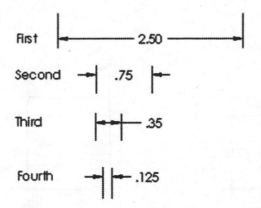

1. Arrows in / dimension in
2. Arrows out / dimension in
3. Arrows in / dimension out
4. Arrows out / dimension out

Precision

In the fields of science and engineering the accuracy of a measurement system is the degree of closeness of measurements of a quantity to that quantity's actual (true) value.

The precision of a measurement system, also called reproducibility or repeatability, is the degree to which repeated measurements under unchanged conditions show the same results. Although the two words precision and accuracy can be synonymous in colloquial use, they are deliberately contrasted in the context of the scientific method.

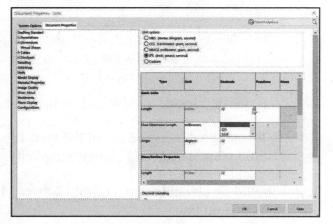

Precision of a dimension on an engineering drawing is the number of digits located after the decimal point. Dimensions may appear on drawings as a one-place decimal, two-place decimal or more. SOLIDWORKS provides the ability to set both primary and secondary precision.

Three and four place decimal dimensions continue to be used for more precise dimensions requiring machining accuracies in thousandths or ten-thousandths of an inch.

Size Dimension

Every solid or part has three size dimensions: width or length, height and depth. In the case of the Glass box method, two of the dimensions are usually placed on the Principal view and the third dimension is located on one of the other (Top, Right) views.

Example 1:

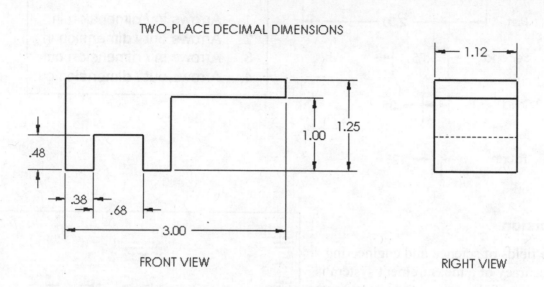

TWO-PLACE DECIMAL DIMENSIONS

FRONT VIEW   RIGHT VIEW

Do not dimension inside a view and do not over dimension the model in a drawing.

Always locate the dimension off of the view if possible. Only place dimensions on the inside of the view if they add clarity, simplicity, and ease of reading.

There should be a visible gap ~1.5mm between the object (feature) line and the beginning of each extension line.

When you create a new part or assembly, the three default Planes (Front, Right and Top) are aligned with specific views. The Plane you select for the Base sketch determines the orientation of the part, the drawing views and the assembly.

Continuous Dimensions

Sets of dimension lines and dimensions should be located on drawings close enough so they may be read easily without any possibility of confusing one dimension with another. If a series of dimensions is required, the dimensions should be placed in a line as continuous dimensions (chain dimensioning or point-to-point dimensioning) as illustrated below. This method is preferred over the staggering of dimensions, because of ease in reading, appearance, and simplified dimensioning. Note: Tolerance stack-up can be an issue with this method.

Example 1:

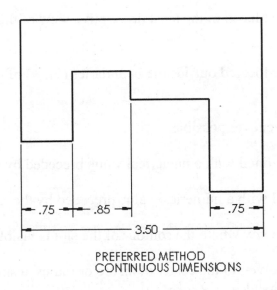

PREFERRED METHOD
CONTINUOUS DIMENSIONS

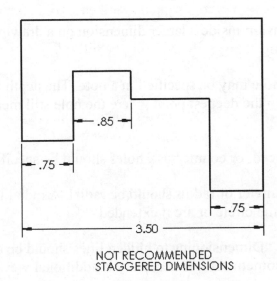

NOT RECOMMENDED
STAGGERED DIMENSIONS

Spacing between dimension lines should be uniform throughout the drawing.

Principles of good dimensioning

The overriding principle of dimension is clarity. Below are a few key rules that you should know when applying dimensions:

- A dimension is a numerical value shown on a drawing to define the size of an object or a part of an object.

- Each feature of an object is dimensioned once and only once.

- Dimensions should be selected to suit the function of the model or feature.

- Dimensions should be placed in the most descriptive view of the feature being dimensioned.

- Dimensions should be located outside the boundaries (view) of the object whenever possible.

- Group dimensions whenever possible.

- Diameters are dimensioned with a numerical value preceded by the diameter symbol.

- Radii are dimensioned with a numerical value preceded by the radius symbol.

- Dimension a slot in a view where the contour of the slot is visible.

- When a dimension is given to the center of an arc or radius, a small cross (center mark) should be displayed.

- Place a smaller dimension inside a larger dimension on a drawing view to avoid dimension line crossing.

- The depth of a blind hole may be specified in a note. The depth is measured from the surface of the object to the deepest point where the hole still measures a full diameter in width.

- Counter bored, spotfaced, or countersunk holes should be specified in a note.

- A leader line for a diameter or radius should be radial. A radial line is one that passes through the center of the circle or arc if extended.

- ANSI standard states, "Dimensioning to hidden lines should be avoided wherever possible." However, sometimes it is necessary if additional views are needed to fully define the model.

Example 1:

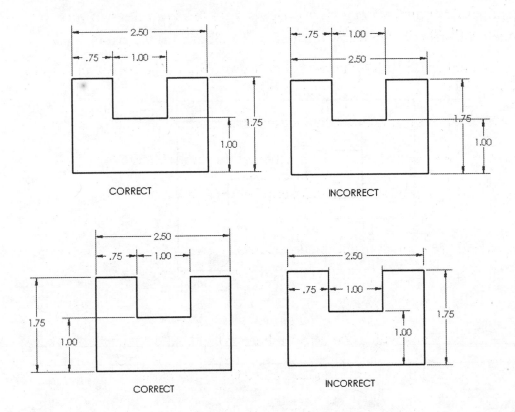

There is an order and style of preference for a linear dimension line using arrowheads. The first order of preference is that the dimension and arrowheads are drawn between the extension line if space is available. If space is limited, see the below order of preference.

Example 1:

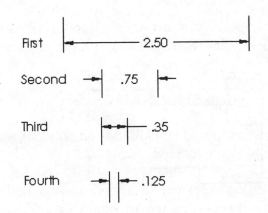

1.	Arrows in / dimension in
2.	Arrows out / dimension in
3.	Arrows in / dimension out
4.	Arrows out / dimension out

Dimension Exercises:
Exercise 1:

Identify the dimension errors in the below illustration. Circle and list the errors.

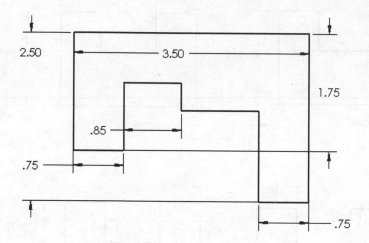

Errors:_____

_____

Exercise 2:

Identify the dimension errors in the below illustration. Circle and list the errors.

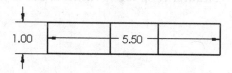

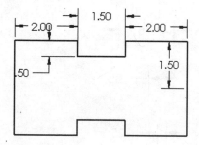

Errors:_____

_____

Exercise 3:

Identify the duplicate dimensions and cross out the ones that you feel should be omitted. Explain why. Are there any dimensioning mistakes in this drawing? Explain.

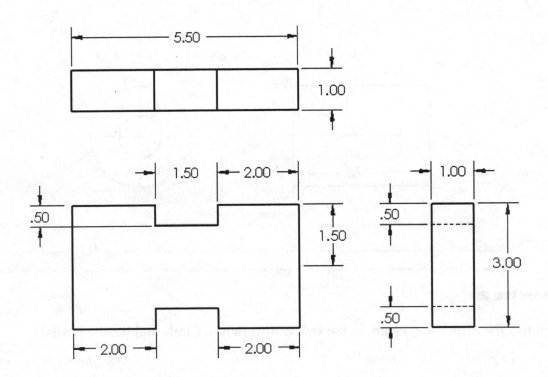

Explain:

_____

_____

_____

Dimension Exercises:

Exercise 1:

Identify the dimension errors in the below illustration. Circle and list the errors.

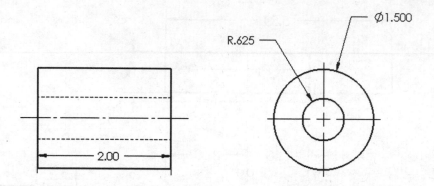

Errors:_____

_____

Exercise 2:

Identify the dimension errors in the below illustration. Circle and list the errors.

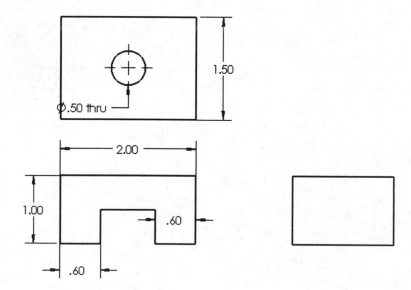

Errors:_____

_____

Exercise 3:

Identify the dimension errors in the below illustration. Circle and list the errors.

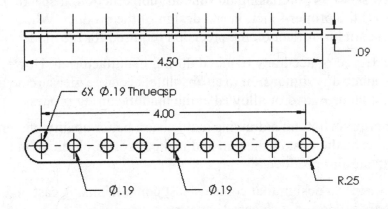

Explain:

_____

_____

Exercise 4:

Identify the dimension errors in the below illustration. Circle and list the errors.

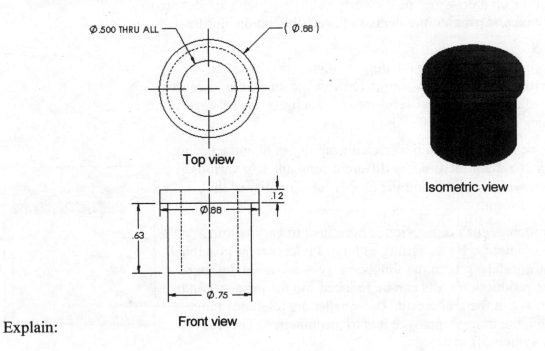

Explain:

_____

_____

Precision and Tolerance

In a manufacturing environment, quality and cost are two of the main considerations for an engineer or designer. Engineering drawings with local and general notes and dimensions often serve as purchasing documents, construction, inspection, and legal contracts to ensure the proper function and design of the product. When dimensioning a drawing, it is essential to reflect on the precision required for the model.

Precision is the degree of accuracy required during manufacturing. However, it is unfeasible to produce any dimension to an absolute, accurate measurement. Some discrepancy must be provided or allowed in the manufacturing process.

Specifying higher precision on a drawing may ensure better quality of a product, but doing so can increase the cost of the part and make it cost prohibitive in being competitive with similar products.

For example, consider a design that contains cast components. A cast part usually has two types of surfaces: 1.) mating surfaces, and 2.) non-mating surfaces.

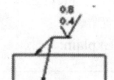

Mating surfaces work together with other surfaces, typically machined to a specified finish. Mating surfaces typically require higher precision on all corresponding dimensions.

Non-mating surfaces are usually left in the original rough-cast form. They have no significant connection with other surfaces. The dimensions on a drawing must clearly indicate which surfaces are to be finished and provide the degree of precision needed for the finishing.

The method of specifying the degree of precision is called Tolerancing. Tolerance in simple terms is the amount of size variation permitted and provides a useful means to achieve the precision necessary in a design.

Tolerancing makes certain interchangeability in manufacturing. Parts can be manufactured by different companies in various locations while maintaining the proper functionality of the intended design.

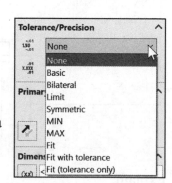

In tolerancing, each dimension is permitted to vary within a specified amount. By assigning as large a tolerance as possible, without interfering with the functionality or intended design of a part, the production costs can be reduced and the product can be competitive in the real world. The smaller the tolerance range specified, the more expensive it is to manufacture. There is always a trade off in design.

Tolerance for a drawing

The two most common Tolerance Standard agencies are American National Standards Institute (ANSI)/(ASME) and the International Standards Organization (ISO). This book covers the ANSI (US) standards.

In this section we will discuss Dimensional Tolerances vs. Geometric Tolerances.

General Tolerance - Title Block

General tolerances are typically provided in the Title Block. General tolerances are applied to the dimensions in which tolerances are not given in the drawing.

As a part is designed, the engineer should consider: 1.) function either as a separate unit or as a component relation to other components in an assembly, 2.) manufacturing operations, 3.) material, 4.) quantity (run size), 5.) sustainability and 6.) cost.

The dimensions displayed on a drawing (obtained from the part) indicate the accuracy limits for manufacturing. The limits are called tolerances and are normally displayed in decimal notation. Tolerances can be specified in various unit systems. ANSI specifications are normally specified either in English (IPS) or Metric (MMGS).

Tolerances on decimal dimensions are expressed in terms of one, two, three, or more decimal places. This information can be documented on a drawing in several ways. One of the common methods of specifying a tolerance that applies to all dimensions is to use a general note in the Title block as illustrated.

Example 1 & 2:

UNLESS OTHERWISE SPECIFIED:	UNLESS OTHERWISE SPECIFIED:
DIMENSIONS ARE IN INCHES TOLERANCES: ANGULAR: ± 1° ONE PLANE DECIMAL ± .1 TWO PLACE DECIMAL ± .01 THREE PLACE DECIMAL ± .005	DIMENSIONS ARE IN MILLIMETERS TOLERANCES: ANGULAR: MACH± 0°30' ONE PLACE DECIMAL ±0.5 TWO PLACE DECIMAL ±0.15

Local Tolerance - Dimension

A Local Tolerance note indicates a special situation which is not covered by the General Title box. A Local Tolerance is located on the drawing (Not in the Title box) with the dimension.

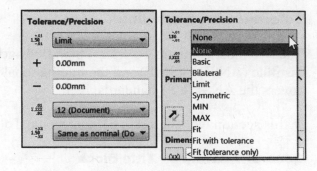

The three most common Tolerance types are *Limit, Bilateral* and *Unilateral*.

Limit Tolerance is when a dimension has a high (upper) and low (Lower) limits stated. In a limit tolerance, the higher value is placed on top, and the lower value is placed on the bottom as illustrated.

Limits are the maximum and minimum size that a part can obtain and still pass inspection and function in the intended

$$\emptyset^{1.001}_{.999} \quad \text{or} \quad \emptyset.999 - 1.001$$

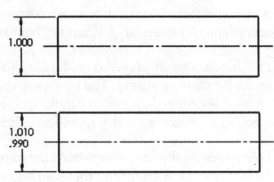

assembly. When both limits are placed on a single line, the lower limit precedes the higher limit. The tolerance for the dimension illustrated above is the total amount of variation permitted or .002.

In the angle example the dimension may vary between 60°and 59°45'.

Note: Each degree is one three hundred and sixtieth of a circle (1/360). The degree (°) may be divided into smaller units called minutes ('). There are 60 minutes in each degree. Each minute may be divided into smaller units called seconds ("). There are 60 seconds in each minute. To simplify the dimensioning of angles, symbols are used to indicate degrees, minutes and seconds as illustrated below.

Name	Symbol
Degrees	°
Minutes	'
Seconds	"

Unilateral Tolerance is the variation of size in a single direction - either (+) or (-). The examples of Unilateral tolerances shown below indicate that the first part meets standards of accuracy when the nominal or target dimension varies in one direction only and is between 3.000" and 3.025".

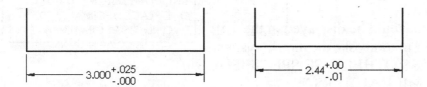

Bilateral Tolerance is the variation of size in both directions. The dimensions may vary from a larger size (+) to a smaller size (-) than the basic dimension (nominal size). The basic 2.44" dimension as illustrated with a bilateral tolerance of +-.01" is acceptable within a range of 2.45" and 2.43".

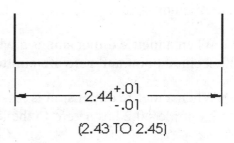

Specify a tolerance with the degree of accuracy that is required for the design to work properly and is cost effective.

You can also create a note on the drawing referring to a specific dimension or specifying general tolerances in the Title block.

Formatting Inch Tolerances

The basic dimension and the plus and minus values should have the same number of decimal places. Below are examples of *Unilateral* and *Bilateral* tolerances.

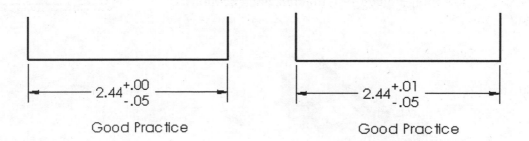

Metric Dimension Specifications

For Metric dimension specification, the book uses the Metric International System of Units (SI). The millimeter is the common unit of measurement used on engineering drawings made to the metric system.

UNLESS OTHERWISE SPECIFIED:
DIMENSIONS ARE IN MILLIMETERS TOLERANCES: ANGULAR: MACH ± 0°30' ONE PLACE DECIMAL ±0.5 TWO PLACE DECIMAL ±0.15

In industry, a general note would be displayed in the Title block section of the drawing to invoke the metric system. A general note is "UNLESS OTHERWISE SPECIFIED: DIMENSIONS ARE IN MILLIMETERS."

Three conventions are used when specifying dimensions in metric units:

1.) When a metric dimension is a whole number, the decimal point and zero are omitted.

2.) When a metric dimension is less than 1 millimeter, a zero precedes the decimal point. Example - 0.2 has a zero to the left of the decimal point.

3.) When a metric dimension is not a whole number, a decimal point with the portion of a millimeter (10ths or 100ths) is specified.

Tolerance Parts and Important Terms

The illustration below shows a system of two parts with tolerance dimensions. The two parts are an example of ASME Y14.5 2009 important terms.

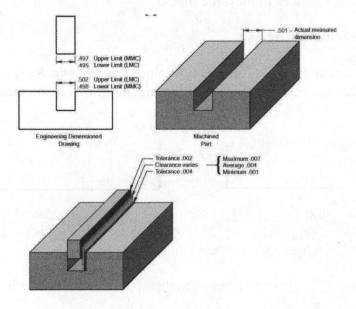

- **Nominal size** - a dimension used to describe the general size, usually expressed in common fractions. The slot in the above illustration has a nominal size of ½″.

- **Basic size** - the theoretical size used as a starting point for the application of tolerances. The basic size of the slot is .500″.

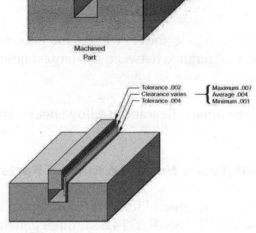

- **Actual size** - the measured size of the finished part after machining is .501″.

- **Limits** - the maximum and minimum sizes shown by the tolerance dimension. The slot has limits of .502″ and .498″, and the mating part has limits of .495″ and .497″. The larger value for each part is the upper limit, and the smaller value is the lower limit.

- **Allowance** - the minimum clearance or maximum interference between parts, or the tightest fit between two mating parts. In the illustration, the allowance is .001″, meaning that the tightest fit occurs when the slot is machined to its smallest allowable size of .498″ and the mating part is machined to its largest allowable size of .497″. The difference between .498″ and .497″ or 001″ is the allowance.

- **Tolerance** - the total allowable variance in a dimension; the difference between the upper and lower limits. The tolerance of the slot is .004″. (.502″ - .498″ = .004″) and the tolerance of the mating parts is .002″ (.497″ - .495″ = .002″).

- **Maximum material condition (MMC)** - the condition of a part when it contains the greatest amount of material. The MMC of an external feature, such as a shaft, is the upper limit. The MMC of an internal feature, such as a hole, is the lower limit.

- **Least material condition (LMC)** - the condition of a part when it contains the least amount of material possible. The LMC of an external feature is the lower limit. The LMC of an internal feature is the upper limit.

- **Piece tolerance** - the difference between the upper and lower limits of a single part.

- **System tolerance** - the sum of all the piece tolerances.

Fit - Hole Tolerance

In the figure below, what is the minimum clearance (Allowance)? Minimum clearance is the minimum amount of space which exists between the hole and the shaft.

Example 1:

Minimum Clearance (Allowance) = $(0.49d_{hole}) - (0.51D_{shaft}) = -0.02in.$

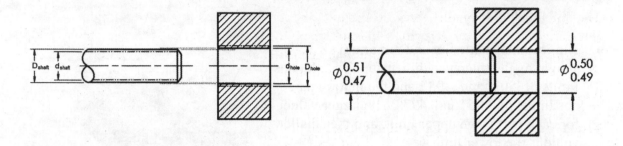

In the figure above, what is the maximum clearance (Allowance)? Maximum clearance is the difference between the largest hole diameter D_{hole} and the smallest shaft diameter d_{shaft}.

Maximum Clearance (Allowance) = $(0.50D_{hole}) - (0.47d_{shaft}) = 0.03in.$

Fit Types between Mating Parts

Fit is the general term used to signify the range of tightness in the design of mating parts. In ANSI/ASME Y 14.5M, three general types of fits are designated for mating parts:

1. **Clearance Fit**

2. **Interference Fit**

3. **Transition Fit**

A basic hole and shaft system will be used to apply English unit tolerances to parts in the following examples.

Clearance Fit: The difference between the hole and shaft sizes before assembly is positive. Clearance fits have limits of size prearranged such that a clearance always results when the mating parts are assembled. Clearance fits are intended for accurate assembly of parts and bearings. The parts can be assembled by hand because the hole is always larger than the shaft. Min. Clearance > 0. Two examples: Lock and Key, Door and Door frame.

Interference Fit: (also referred to as Force fit or Shrink fit) - interference fit has limits of size that always result in interference between mating parts. The hole is always smaller than the shaft. Interference fits are for permanent assemblies of parts which require rigidity and alignment, such as dowel pins and bearings in casting, hinge pin or a pin in a bicycle chain. Max. Clearance ≤ 0.

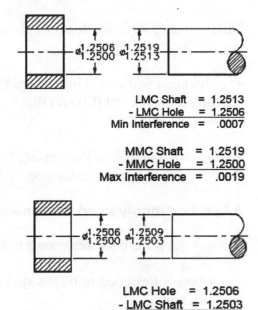

```
LMC Shaft        = 1.2513
- LMC Hole       = 1.2506
Min Interference =   .0007

MMC Shaft        = 1.2519
- MMC Hole       = 1.2500
Max Interference =   .0019
```

Transition Fit: May provide either clearance or interference, depending on the actual value of the tolerance of individual parts. Transition fits are a compromise between the clearance and Interference fits. They are used for applications where accurate location is important, but either a small amount of clearance or interference is permissible. Max. Clearance > 0, Min. Clearance < 0.

```
LMC Hole                        =  1.2506
- LMC Shaft                     =  1.2503
Positive Clearance              =    .0003
MMC Shaft                       =  1.2509
- MMC Hole                      =  1.2500
Negative Allowance (Interference) =  .0003
```

Why is this information important? By specifying the correct allowances and tolerances, mating components in an assembly can be completely interchangeable.

Sometimes the desired fit may require very small allowances and tolerances, and the production cost may become too high and cost prohibitive. In these cases, either manual or computer-controlled selective assembly is often used. The manufactured parts are then graded as small, medium and large based on the actual sizes. In this way, very satisfactory fits are achieved at a much lower cost than manufacturing all parts to very accurate dimensions.

Fasteners in General

Fasteners include Bolts and Nuts (threaded), Set screws (threaded), Washers, Keys, and Pins to name a few. Fasteners are not a permanent means of assembly such as welding or adhesives.

Fasteners and threaded features should be specified on your engineering drawing.

Woodruff Key

Machine Key

- Threaded features: Threads are specified in a thread note. In SOLIDWORKS, apply the Hole Wizard feature.

- General Fasteners: Purchasing information must be given to allow the fastener to be ordered correctly.

A few commonly used Fasteners

What is the difference between a bolt and a screw?

Fully Threaded

- A bolt is designed to be inserted through a hole and secured with a nut.

Partially Threaded

- A screw is designed to be used in a threaded hole - sometimes with a nut.

See American Society of Mechanical Engineers (ASME) standard B18.2.1 (1996) for additional information.

Hex Bolt:

A bolt is a fastener having a head on one end and a thread on the other end. A bolt is used to hold two parts together by means of passing through aligned clearance holes with a nut screwed on the threaded end.

Carriage Bolt:

A carriage bolt is mostly used in wood with a domed shape top and a square under the head, which is pulled into the wood as the nut is tightened.

Studs:

A stud is a rod with threaded ends.

Cap Screws:

A hexagon cap screw is similar to a bolt except it is used without a nut and generally has a longer thread. Cap screws are available in a variety of head styles and materials.

Machine Screws:

A machine screw is similar to the slot-head cap screw but smaller, available in many styles and materials. A machine screw is also commonly referred to as a stove bolt. From left to right: Slotted, Phillips and Square.

Wood Screws:

A tapered shank screw is for use exclusively in wood. Wood screws are available in a variety of head styles and materials. From left to right: Slotted, Phillips and Hex.

Sheet Metal Screws:

Highly versatile fasteners designed for thin materials. Sheet metal screws can be used in wood, fiberglass and metal, also called self-tapping screws, available in steel and stainless steel. From left to right: Slotted, Square and Phillips.

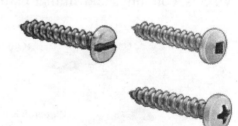

Socket Screws:

Socket screws, also known as Allen head, are fastened with a hexagonal Allen wrench, available in several head styles and materials.

Allen wrench

Set Screws:

Set screws are used to prevent relative motion between two parts. A set screw is screwed into one part so that its point is pushed firmly against the other part, available in a variety of point styles and materials.

Nuts:

Nuts are used to attach machine thread fasteners. From left to right there is a Hex nut, Locked nut, Slotted nut and a Wing nut.

Washers:

Washers provide a greater contact surface under the fastener. This helps prevent a nut, bolt or screw from breaking through the material. Shown is a Flat vs. Split Lock washer.

Keys:

Keys are used to prevent relative motion between shafts and wheels, couplings and similar parts attached to shafts.

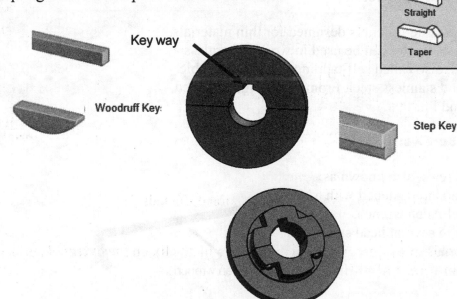

Key way

Woodruff Key

Step Key

Rivets:

Rivets are generally used to hold sheet metal
parts together. Rivets are generally
considered as permanent fasteners and are
available in a variety of head styles and
materials.

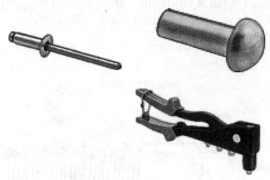

🔆 Threads are only symbolically
represented on drawings; therefore, thread
notes are needed to provide the required
information. A thread note must be included
on all threaded parts, with a leader line to the external or internal
in a circular view.

Representing External (Male) Threads

Screw threads are used widely (1) to fasten two or more parts together in position, (2) to
transmit power such as a feed screw on a machine, and (3) to move a scale on an
instrument used for precision measurements.

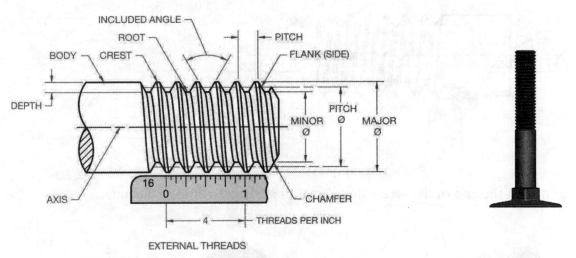

EXTERNAL THREADS

*Profile of UNIFIED and the American National Threads.

🔆 An edge of a uniform section in the form of a helix on the **external** surface of a
cylinder or cone. **"A"** suffix.

Cutting External (Male) Threads

Start with a shaft the same size as the major diameter. An external thread is cut using a die or a lathe as illustrated below.

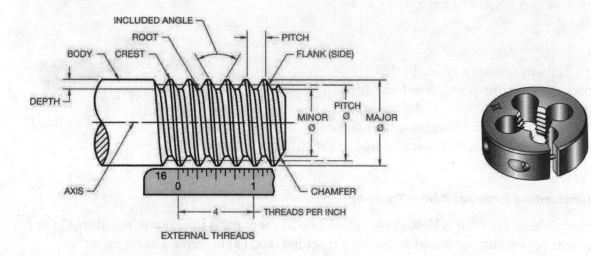

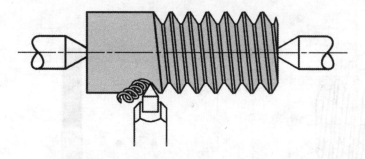

A chamfer on the end of the screw thread makes it easier to engage the nut.

Representing Internal (Female) Threads

An internal thread is a ridge of a uniform section in the form of a helix on the **internal** surface of a cylinder or cone. **"B"** suffix.

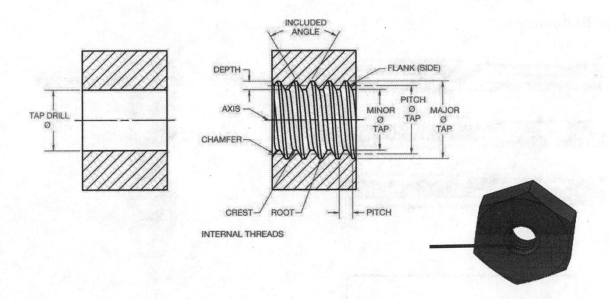

Cutting Internal (Female) Threads

In general, a tap drill hole is cut with a twist drill. The tap drill hole is a little larger than the minor diameter. Start with a shaft the same size as the major diameter as illustrated below.

💡 **Minor Diameter**: The smallest diameter (fractional diameter or number) of a screw thread.

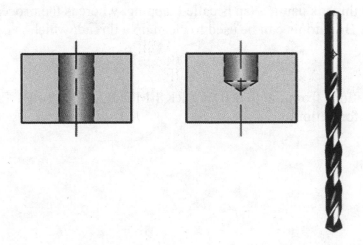

Then the threads are cut using a tap. Major tap types:

- **Taper** tap.

- **Plug** tap.

- **Bottoming** tap.

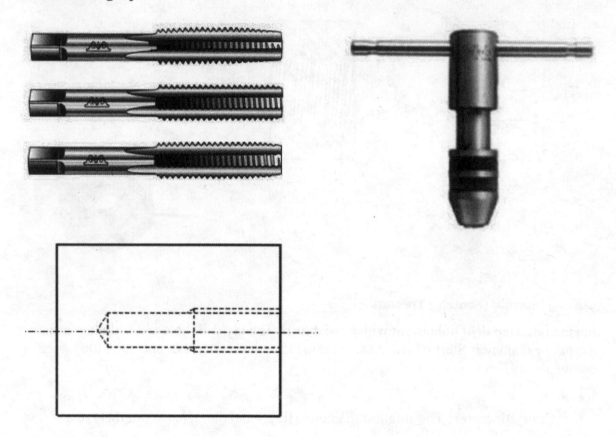

The process of cutting threads using a tap is called tapping, whereas the process using a die is called threading. Both tools can be used to clean up a thread, which is called chasing.

🔆 View the presentations from the SOLIDWORKS-MODELS 2019\PPT Presentation folder for additional information.

American National Standard and Unified Screw threads

The basic profile is the theoretical profile of the thread. An essential principle is that the actual profiles of both the nut and bolt threads must never cross or transgress the theoretical profile. So bolt threads will always be equal to, or smaller than, the dimensions of the basic profile. Nut threads will always be equal to, or greater than, the basic profile. To ensure this in practice, tolerances and allowances are applied to the basic profile.

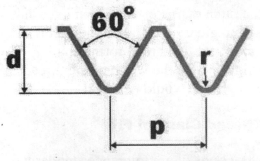

The most common screw thread form is the symmetrical V-Profile with an included angle of 60 degrees. This form is prevalent in the Unified National Screw Thread Series (UN, UNC, UNF, UNRC, UNRF) form as well as the ISO/Metric.

A thread may be either right-hand or left-hand. A right-hand thread on an external member advances into an internal thread when turned clockwise. Threads are always considered to be Right-handed unless otherwise specified.

A left-hand thread advances when turned counterclockwise (bike pedals, older propane tanks, etc.). All left-hand threads are labeled LH.

RIGHT HANDED   LEFT HANDED

Single vs. Double or Triple Threads

If a single helical groove is cut or formed on a cylinder, it is called a single-thread screw. Should the helix angle be increased sufficiently for a second thread to be cut between the grooves of the first thread, a double thread will be formed on the screw. Double, triple, and even quadruple threads are used whenever a rapid advance is desired, as on valves.

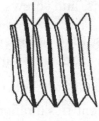
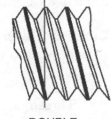

SINGLE   DOUBLE

 To designate a multiple thread the word "DOUBLE" (or "TRIPLE," and so on) is placed after the class of fit, like this: 3/8x1-16 UNC 2B DOUBLE.

Pitch and Major Diameter

Pitch and major diameter designates a thread. Lead is the distance advanced parallel to the axis when the screw is turned one revolution.

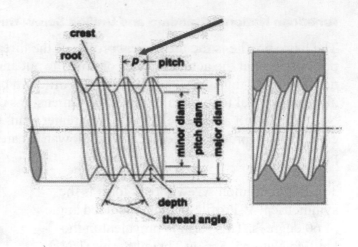

For a single thread, *lead is equal to the pitch*; for a double thread, lead is twice the pitch. For a straight thread, the pitch diameter is the diameter of an imaginary coaxial cylinder that would cut the thread forms at a height where the width of the thread and groove would be equal.

Thread Class of Fit

Classes of fit are tolerance standards; they set a plus or minus figure that is applied to the pitch diameter of bolts or nuts. The classes of fit used with almost all bolts sized in inches are specified by the ANSI/ASME Unified Screw Thread standards (which differ from the previous American National standards). There are three major Thread classes of fits:

Class 1: The loosest fit. Used on parts which require assembly with a minimum of binding. Only found on bolts ¼ inch in diameter and larger.

Class 2: By far the most common class of fit. General purpose threads for bolts, nuts, and screws used in mass production.

Class 3: The closest fit. Used in precision assemblies where a close fit is desired to withstand stress and vibration.

Thread class identifies a range of thread tightness or looseness.

Classes for **External (male)** threads have an "**A**" suffix, for example, "2A," and classes for **Internal threads** have a "**B**" suffix.

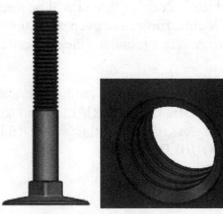

General Thread Notes

The Thread note is usually applied to a drawing with a leader in the view where the thread is displayed as a circle for internal threads as illustrated below. External threads can be dimensioned with a leader with the thread length given as a dimension or at the end of the note.

English – ISP Unit system

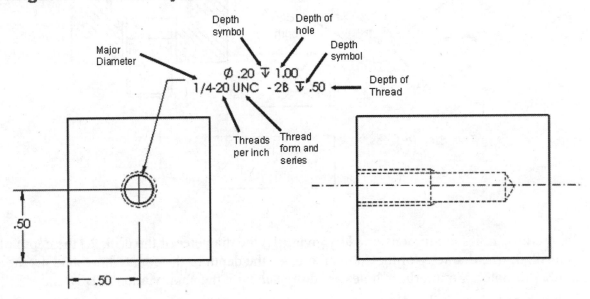

Per the illustration to the right: Major Diameter - .25in, 20 threads per inch or a Pitch - 1/20, threads 2 inches long, Unified National Series Course, Thread Class 2, A - male. Note: If not stated, the thread is always Right-handed, Single.

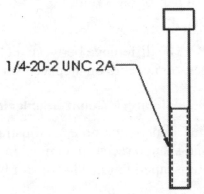

🔆 Threads are only symbolically represented on drawings; therefore, thread notes are needed to provide the required information. A thread note must be included on all threaded parts, with a leader line to the external or internal in a circular view.

Dimensioning a Counterbore Hole

A Counterbore hole is a cylindrical flat bottom hole that has been machined to a larger diameter for a specified depth, so that a bolt or pin will fit into the recessed hole. The Counterbore provides a flat surface for the bolt or pin to seat against. In SOLIDWORKS, use the Hole Wizard to insert the hole callout for a Counterbore.

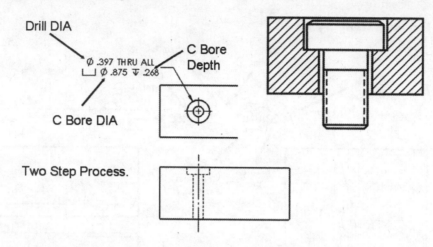

Counterbore holes are dimensioned by giving 1.) the diameter of the drill, 2.) the depth of the drill, 3.) the diameter of the counterbore, 4.) the depth of the counterbore, and 4.) the number of holes. Counterbore holes are displaced with the abbreviation C'BORE, C BORE or the symbol ⊔.

The difference between a C'BORE and a SPOTFACE is that the machining operation occurs on a curved surface.

Dimensioning a Countersunk Hole

The Countersunk hole, as illustrated below, is a cone-shaped recess machined in a part to receive a cone-shaped flat head screw or bolt.

A Countersunk hole is dimensioned by giving 1.) the diameter of the hole, 2.) the depth of the hole 3.) the diameter of the Countersunk, 4.) the angle at which the hole is to be Countersunk, 5.) the Counterbore diameter, 6.) the depth of the Counterbore, and 7.) the number of holes to be Countersunk.

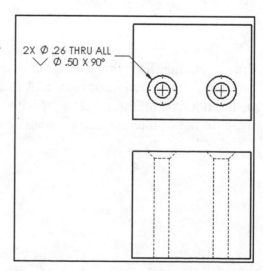

Adding a Counterbore for head clearance to a Countersink is optional. In SOLIDWORKS, the Head Clearance option is located in the Hole Wizard Property Manager. The symbol for a Countersunk hole on a drawing annotation is CSK or ⌵.

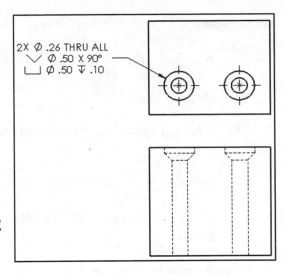

Chapter Summary

In Chapter 3 you reviewed basic dimensioning practices and were introduced to general tolerancing terminology according to the ASME Y14.5-2009 standard. You reviewed various dimensioning systems and fits and were presented with the right and wrong way to dimension simple shapes, lines, angles, circles and arcs.

Dimensioning a drawing is a means to communicate the requirements to manufacture a part. It requires special annotations for fasteners, threads, countersunk holes, counterbored holes and other types of holes.

Tolerances determine the maximum and minimum variation that a dimension on a part is manufactured to. By understanding the required tolerance, you can save both time and money to create a part from your drawing.

Although SOLIDWORKS automatically generates most annotations for a part, it is up to the designer to determine if all the required information is available to manufacture the part. The annotations must be presented according to a dimensioning standard. There is no partial credit in the machine shop.

Questions

1. True or False: Dimensions should not be duplicated or the same information given in two different ways. If a reference dimension is used, the size value is placed within parentheses (X).

2. The U.S. unit system is also known as the (IPS) unit system. What does IPS stand for?

3. The diameter of a hole is placed in the view in which the hole is shown as a _____.

4. The length and diameter of cylinder are usually placed in the view which shows the cylinder as a _____.

5. Dimension a hole by its _____.

6. True or False: Dimensioning to hidden lines should be avoided wherever possible.

7. If a hole goes completely through the feature and it is not clearly shown on the drawing, the abbreviation _____ follows the dimension.

8. True or False: A dimension is said to have a *Unilateral* (single) tolerance when the total tolerance is in one direction only, either (+) or (-).

9. The degree (°) may be divided into smaller units called _____. There are 60 _____ in each degree. Each minute may be divided into smaller units called _____.

10. Classes for an **External (male thread)** have a _____ suffix.

11. Classes for an **Internal (female thread)** have a _____ suffix.

12. There are three major Thread classes of fits; they are: _____, _____, _____. Explain the differences.

13. Identify the pitch of the following Thread note: 3/8-16 UNC 2B DOUBLE_____.

14. Identify the symbol of a Counterbore and <u>Countersunk</u>:_____, _____.

Exercises

Exercise 3.1: Estimate the dimensions in a whole number. Dimension the below illustration. Note: There is more than one way to dimension an angle.

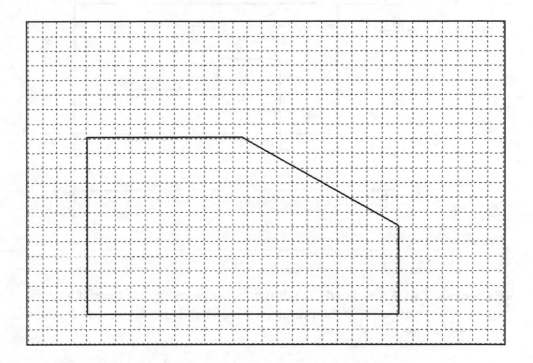

Exercise 3.2: Estimate the dimensions in a whole number. Dimension the illustration. Note: There is more than one way to dimension an angle.

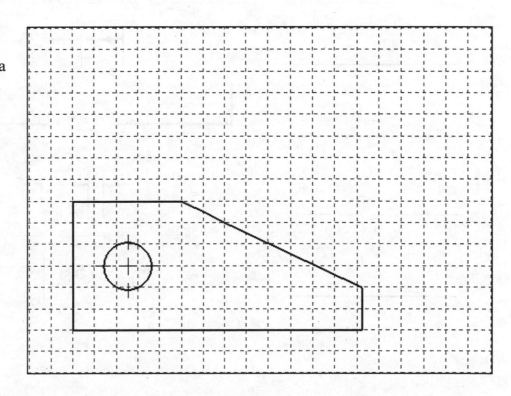

Exercise 3.3: Identify the dimension errors in the below illustration. Circle and list the errors.

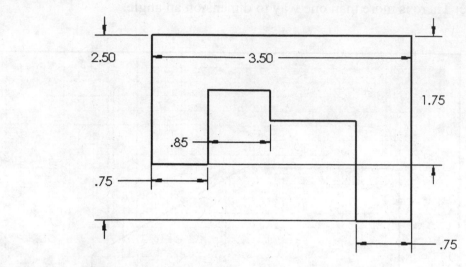

Errors:_____

_____

Exercise 3.4: Arrowheads are drawn between extension lines if space is available. If space is limited, what is the preferred Arrowhead and dimension location order? List the preferred order.

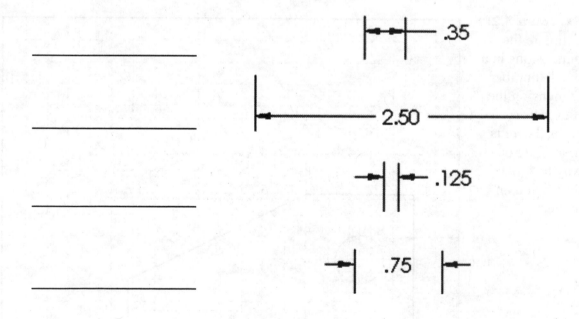

Exercise 3.5: Identify the dimension errors in the below illustration. Circle and list the errors.

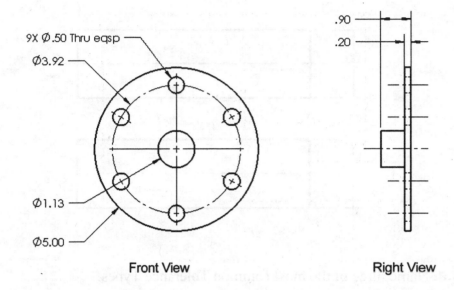

Front View   Right View

Errors:_____

_____

Exercise 3.6: Identify the dimension errors in the below illustration. Circle and list the errors.

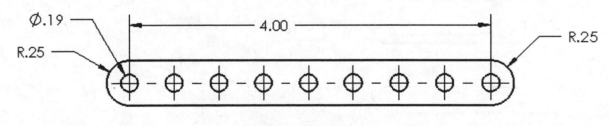

material thickness .10

Front View

Errors:_____

_____

Exercise 3.7: Place a *limit* tolerance of 002 on the below model.

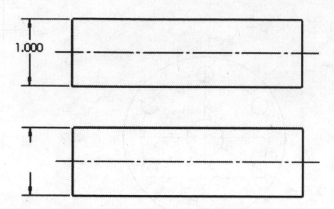

Exercise 3.8: Name three of the most common Tolerance Types.

1._____

2._____

3._____

Exercise 3.9: Identify the following symbols.

_____ $\vee$

_____ $\sqcup$

_____ $\varnothing$

_____ $\downarrow\!\!\!\!\top$

Exercise 3.10: Describe the following hole callouts (symbols and meanings) in detail.

⌀ .2500 THRU ALL
⌴ ⌀ .5000 ▼ .1250

⌀ .3970 THRU ALL
∨ ⌀ .7731 X 82°
⌴ ⌀ .7731 ▼ .0402

Exercise 3.11: True/False - The loosest fit is a Class 1 fit. A Class 1 fit is used on parts which require assembly with a minimum of binding.

Exercise 3.12: The two most common Tolerance Standard agencies are American National Standards Institute (ANSI)/(ASME) and the International Standards Organization (ISO). In the ANSI (US) standard: This is a two-part question.

True or False:

T   F   The higher limit is placed below the lower limit.

T   F   When both limits are placed on one line, the lower limit precedes the higher limit.

Exercise 3.13: There are basically two types of dimensioning systems used in creating parts and drawings - **U.S.** and **Metric**.

True or False: The U.S. system uses the decimal inch value. When the decimal inch system is used, a zero is not used to the left of the decimal point for values less than one inch and trailing zeros are not used.

True or False: The Metric system normally is expressed in millimeters. When the millimeter system is used, the number is rounded to the nearest whole number. Trailing zeros are used.

Exercise 3.14: Identify the illustrated Thread Note.
Remember units.

1/4-20-2 UNC 2A

1.) Pitch of the Thread:_____

2.) Major Thread Diameter:_____

3.) Internal or External Threads:_____

4.) Left Handed or Right Handed Threads:_____

5.) Number of Threads per inch:_____

6.) Identify the Thread class:_____

7.) Length of the Thread:_____

Exercise 3.15: In the figure below, the **tightest fit** between the two parts will be when the **largest shaft** is fit inside the **smallest hole**. Calculate the Allowance (MMC).

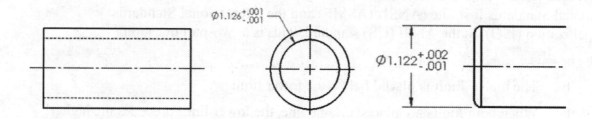

Exercise 3.16: In the figure below, the **loosest fit** between the two parts below will be when the **smallest shaft** is fit inside the **largest hole**. Calculate the maximum clearance between the two parts.

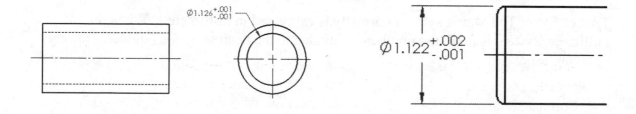

Chapter 4

Overview of SOLIDWORKS 2019 and the User Interface

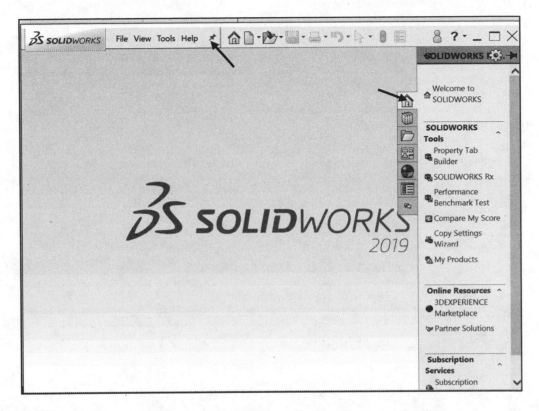

Below are the desired outcomes and usage competencies based on the completion of Chapter 4.

Desired Outcomes:	Usage Competencies:
• A comprehensive understanding of the SOLIDWORKS 2019 User Interface (UI) and CommandManager.	• Ability to establish a SOLIDWORKS session. • Aptitude to utilize the following items: Menu bar toolbar, Menu bar menu, Drop-down menus, Context toolbars, Consolidated drop-down toolbars, System feedback icons, Confirmation Corner, Heads-up View toolbar, Document Properties and more. • Knowledge to zoom, rotate and maneuver a three button mouse in the SOLIDWORKS Graphics window.

Notes:

Chapter 4 - Overview of SOLIDWORKS and the User Interface

Chapter Objective

Provide a comprehensive understanding of the SOLIDWORKS default User Interface and CommandManager: Menu bar toolbar, Menu bar menu, Drop-down menu, Right-click Pop-up menus, Context toolbars/menus, Fly-out tool button, System feedback icons, Confirmation Corner, Heads-up View toolbar and more.

On the completion of this chapter, you will be able to:

- Utilize the SOLIDWORKS Welcome dialog box.

- Establish a SOLIDWORKS 2019 session.

- Comprehend the SOLIDWORKS 2019 User Interface.

- Recognize the default Reference Planes in the FeatureManager.

- Utilize SOLIDWORKS Help and SOLIDWORKS Tutorials.

- Open a new and existing SOLIDWORKS part.

- Zoom, rotate and maneuver a three button mouse in the SOLIDWORKS Graphics window.

What is SOLIDWORKS®?

- SOLIDWORKS® is a mechanical design automation software package used to build parts, assemblies and drawings that takes advantage of the familiar Microsoft® Windows graphical user interface.

- SOLIDWORKS is an easy to learn design and analysis tool (SOLIDWORKS Simulation, SOLIDWORKS Motion, SOLIDWORKS Flow Simulation, Sustainability, etc.), which makes it possible for designers to quickly sketch 2D and 3D concepts, create 3D parts and assemblies and detailed 2D drawings.

- Model dimensions in SOLIDWORKS are associative between parts, assemblies and drawings. Reference dimensions are one-way associative from the part to the drawing or from the part to the assembly.

Start a SOLIDWORKS 2019 Session

Start a SOLIDWORKS session and familiarize yourself with the SOLIDWORKS User Interface. As you read and perform the tasks in this chapter, you will obtain a sense of how to use the book and the structure. Actual input commands or required actions in the chapter are displayed in bold.

The book does not cover starting a SOLIDWORKS session in detail for the first time. A default SOLIDWORKS installation presents you with several options. For additional information, visit http://www.SOLIDWORKS.com.

Activity: Start a SOLIDWORKS 2019 Session.

Start a SOLIDWORKS 2019 session.

1) Type **SOLIDWORKS** in the Search window.

2) Click **All Programs.**

3) Click the **SOLIDWORKS 2019** application (or if available, **double-click** the SOLIDWORKS icon on the desktop). The SOLIDWORKS Welcome dialog box is displayed by default.

The Welcome dialog box provides a convenient way to open recent documents (Parts, Assemblies and Drawings), view recent folders, access SOLIDWORKS resources, and stay updated on SOLIDWORKS news.

4) **View** your options. Do not open a document at this time.

5) If the Welcome dialog box is not displayed, click the **Welcome to SOLIDWORKS** 🏠 icon from the Standard toolbar or click **Help** > **Welcome to SOLIDWORKS** from the Main menu.

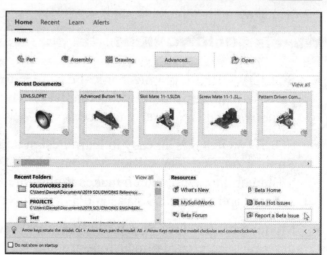

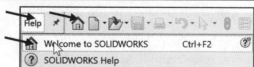

Home Tab

The Home tab lets you open new and existing documents, view recent documents and folders, and access SOLIDWORKS resources (*Part, Assembly, Drawing, Advanced, Open*).

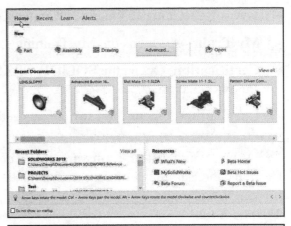

Recent Tab

The Recent tab lets you view a longer list of recent documents and folders. Sections in the Recent tab include *Documents* and *Folders*.

The Documents section includes thumbnails of documents that you have opened recently.

Click a thumbnail to open the document, or hover over a thumbnail to see the document location and access additional information about the document. When you hover over a thumbnail, the full path and last saved date of the document appears.

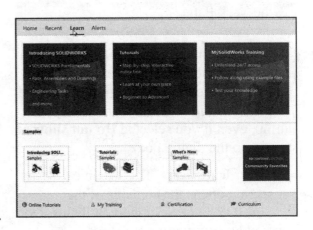

Learn Tab

The Learn tab lets you access instructional resources to help you learn more about the SOLIDWORKS software.

Sections in the Learn tab include:

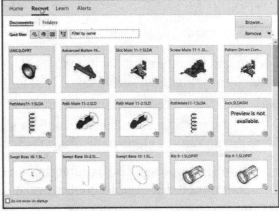

- **Introducing SOLIDWORKS**. Opens the Introducing SOLIDWORKS book.

- **Tutorials**. Opens the step-by-step tutorials in the SOLIDWORKS software.

- **MySolidWorks Training**. Opens the Training section at MySolidWorks.com.

- **Introducing SOLIDWORKS (Samples)**. Opens local folders containing sample models.

- **Tutorials (Samples)**. Opens the SOLIDWORKS Tutorials (videos) section at solidworks.com.

- **What's New (Samples)**. List of new changes.

- **3DContentCentral**. Opens 3DContentCentral.com.

- **My Training**. Opens the My Training section at MySolidWorks.com.

- **Certification**. Opens the SOLIDWORKS Certification Program section at solidworks.com.

- **Curriculum**. Opens the Curriculum section at solidworks.com.

When you install the software, if you do not install the Help Files or Example Files, the Tutorials and Samples links are unavailable.

Alerts Tab

The Alerts tab keeps you updated with SOLIDWORKS news.

Sections in the Alerts tab include Critical, Troubleshooting, and Technical.

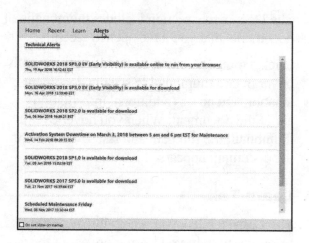

The Critical section does not appear if there are no critical alerts to display.

- **Troubleshooting**. The Troubleshooting section includes troubleshooting messages and recovered documents that used to be on the SOLIDWORKS Recovery tab in the Task Pane.

If the software has a technical problem and an associated troubleshooting message exists, the Welcome dialog box opens to the Troubleshooting section automatically on startup, even if you selected **Do not show at startup** in the dialog box.

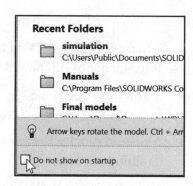

- **Technical Alerts**. The Technical Alerts section opens the contents of the SOLIDWORKS Support Bulletins RSS feed at solidworks.com.

Close the Welcome dialog box.

6) Click **Close** ✕ from the Welcome dialog box. The SOLIDWORKS Graphics window is displayed.

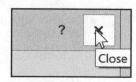

View the SOLIDWORKS Graphics window.

If you do not see this screen, click the SOLIDWORKS Resources  icon on the right side of the Graphics window located in the Task Pane.

7)   **Hover** the mouse pointer over the SOLIDWORKS icon.

8)   **Pin** ✛ the Menu Bar toolbar.

9)   **View** your options from the Menu bar menu: **File**, **View**, **Tools**, and **Help**.

Menu Bar toolbar

The SOLIDWORKS (UI) is designed to make maximum use of the Graphics window. The Menu Bar toolbar contains a set of the most frequently used tool buttons from the Standard toolbar.

The following default tools are available:

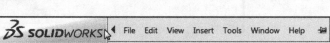

- **Welcome to SOLIDWORKS** ⌂ - Opens the Welcome dialog box, **New** ▱ - Creates a new document; **Open** ▱ - Opens an existing document; **Save** ▱ - Saves an active document; **Print** ▱ - Prints an active document; **Undo** ↰ - Reverses the last action; **Select** ▱ - Selects Sketch entities, components and more; **Rebuild** ▯ - Rebuilds the active part, assembly or drawing; **File Properties** ▤ - Shows the summary information on the active document; and **Options** ⚙ ▾ - Changes system options and Add-Ins for SOLIDWORKS.

Menu Bar menu

Click SOLIDWORKS in the Menu Bar toolbar to display the Menu Bar menu. SOLIDWORKS provides a context-sensitive menu structure. The menu titles remain the same for all three types of documents, but the menu items change depending on which type of document is active.

Example: The Insert menu includes features in part documents, mates in assembly documents, and drawing views in drawing documents. The display of the menu is also dependent on the workflow customization that you have selected. The default menu items for an active document are File, Edit, View, Insert, Tools, Window, Help and Pin.

The Pin ⚲ option displays the Menu bar toolbar and the Menu bar menu as illustrated. Throughout the book, the Menu bar menu and the Menu bar toolbar are referred to as the Menu bar.

Drop-down menu

SOLIDWORKS takes advantage of the familiar Microsoft® Windows user interface. Communicate with SOLIDWORKS through drop-down menus, Context sensitive toolbars, Consolidated toolbars or the CommandManager tabs.

🔆 A command is an instruction that informs SOLIDWORKS to perform a task.

To close a SOLIDWORKS drop-down menu, press the Esc key. You can also click any other part of the SOLIDWORKS Graphics window or click another drop-down menu.

Create a New Part Document

In the next section create a new part document.

Activity: Create a new Part Document.

A part is a 3D model, which consists of features. What are features?

Features are geometry building blocks.

Most features either add or remove material.

Some features do not affect material (Cosmetic Thread).

Features are created either from 2D or 3D sketched profiles or from edges and faces of existing geometry.

Features are individual shapes that combined with other features make up a part or assembly. Some features, such as bosses and cuts, originate as sketches. Other features, such as shells and fillets, modify a feature's geometry.

Features are displayed in the FeatureManager as illustrated (Boss-Extrude1, Cut-Extrude1, Cut-Extrude2, Mirror1, Cut-Extrude3 and CirPattern1).

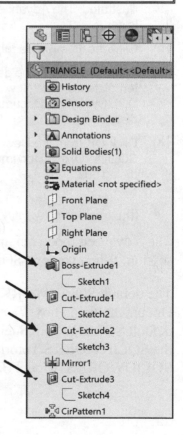

🔆 The first sketch of a part is called the Base Sketch. The Base sketch is the foundation for the 3D model. The book focuses on 2D sketches and 3D features.

There are two modes in the New SOLIDWORKS Document dialog box: Novice and Advanced. The Novice option is the default option with three templates. The Advanced mode contains access to additional templates and tabs that you create in system options. Use the Advanced mode in this book.

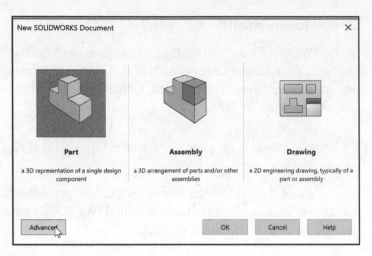

Create a new part.

10) Click **New** from the Menu bar. The New SOLIDWORKS Document dialog box is displayed.

Select the Advanced mode.

11) Click the **Advanced** button as illustrated. The Advanced mode is set.

12) Click the **Templates** tab.

13) Click **Part**. Part is the default template from the New SOLIDWORKS Document dialog box.

14) Click **OK** from the New SOLIDWORKS Document dialog box.

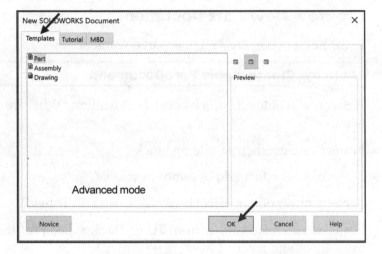

 Illustrations may vary depending on your SOLIDWORKS version and operating system.

The Advanced mode remains selected for all new documents in the current SOLIDWORKS session. When you exit SOLIDWORKS, the Advanced mode setting is saved.

The default SOLIDWORKS installation contains three tabs in the New SOLIDWORKS Document dialog box: *Templates, Tutorial* and *MBD*. The *Templates* tab corresponds to the default SOLIDWORKS templates. The *Tutorial* tab corresponds to the templates utilized in the SOLIDWORKS Tutorials. The *MBD* tab corresponds to the templates utilized in the SOLIDWORKS (Model Based Definition).

Part1 is displayed in the FeatureManager and is the name of the document. Part1 is the default part window name.

The Part Origin 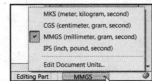 is displayed in blue in the center of the Graphics window. The Origin represents the intersection of the three default reference planes: *Front Plane*, *Top Plane* and *Right Plane*. The positive X-axis is horizontal and points to the right of the Origin in the Front view. The positive Y-axis is vertical and points upward in the Front view. The FeatureManager contains a list of features, reference geometry, and settings utilized in the part.

Edit the document units directly from the Graphics window as illustrated.

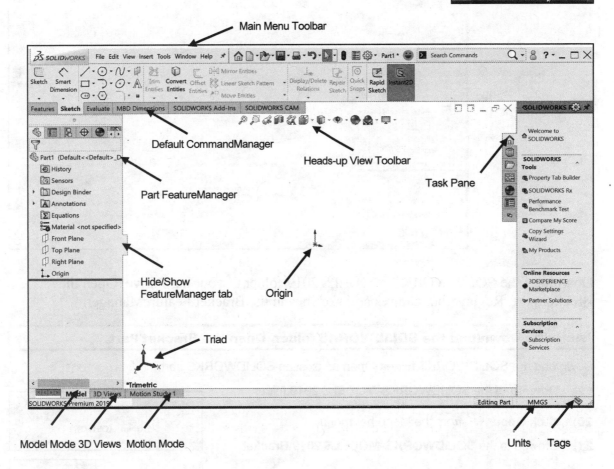

View the Default Sketch Planes.

15) Click the **Front Plane** from the FeatureManager.

16) Click the **Top Plane** from the FeatureManager.

17) Click the **Right Plane** from the FeatureManager.

18) Click the **Origin** from the FeatureManager. The Origin is the intersection of the Front, Top, and Right Planes.

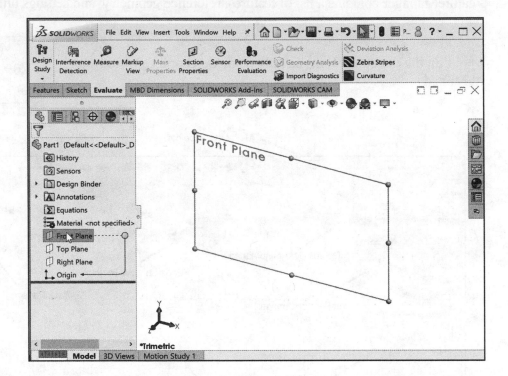

Download the SOLIDWORKS-MODELS 2019 folder to your hard drive. Open the Bracket part. Review the features and sketches in the Bracket FeatureManager.

Activity: Download the SOLIDWORKS folder. Open the Bracket Part.

Download the SOLIDWORKS folder. Open an existing SOLIDWORKS part.

19) **Download** the SOLIDWORKS-MODELS 2019 folder.

20) Click **Open** from the Menu bar menu.

21) Browse to the **SOLIDWORKS-MODELS 2019\Bracket** folder.

22) Double-click the **Bracket** part. The Bracket part is displayed in the Graphics window.

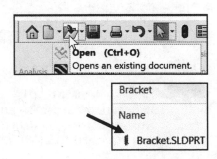

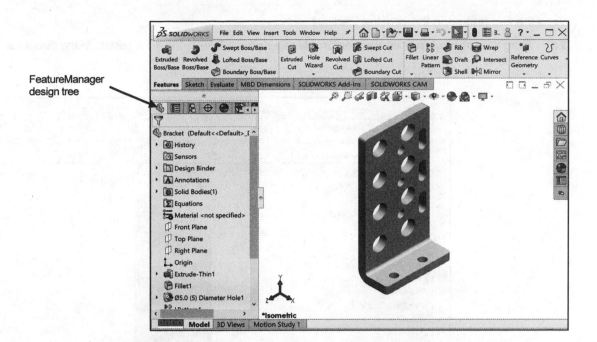

FeatureManager design tree

The FeatureManager design tree is located on the left side of the SOLIDWORKS Graphics window. The FeatureManager provides a summarized view of the active part, assembly, or drawing document. The tree displays the details on how the part, assembly or drawing document was created.

Use the FeatureManager rollback bar to temporarily roll back to an earlier state, to absorbed features, roll forward, roll to previous, or roll to the end of the FeatureManager design tree. You can add new features or edit existing features while the model is in the rolled-back state. You can save models with the rollback bar placed anywhere.

In the next section, review the features in the Bracket FeatureManager using the Rollback bar.

Activity: Use the FeatureManager Rollback Bar option.

Apply the FeatureManager Rollback Bar. Revert to an earlier state in the model.

23) Place the **mouse pointer** over the rollback bar in the FeatureManager design tree as illustrated. The pointer changes to a hand 🖐. Note the provided information on the feature. This is called Dynamic Reference Visualization.

24) Drag the **rollback bar** up the FeatureManager design tree until it is above the features you want rolled back, in this case 10.0 (10) Diameter Hole1.

25) **Release** the mouse button.

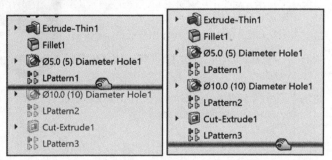

View the first feature in the Bracket Part.

26) Drag the **rollback bar** up the FeatureManager above Fillet1. View the results in the Graphics window.

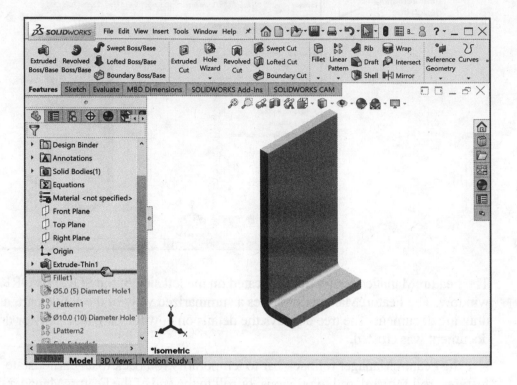

Return to the original Bracket Part FeatureManager.

27) Right-click **Extrude-Thin1** in the FeatureManager. The Pop-up Context toolbar is displayed.

28) Click **Roll to End**. View the results in the Graphics window.

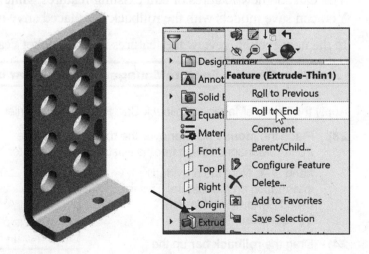

Heads-up View toolbar

SOLIDWORKS provides the user with
numerous view options. One of the most
useful tools is the Heads-up View toolbar displayed
in the Graphics window when a document is active.

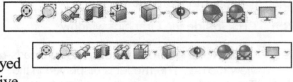

Dynamic Annotation Views 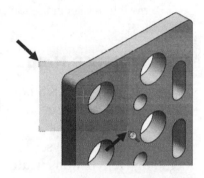 : Only available with SOLIDWORKS MBD (Model Based
Definition). Provides the ability to control how annotations are displayed when you rotate
models.

In the next section, apply the following tools: Zoom to Fit, Zoom to Area, Zoom out, Rotate
and select various view orientations from the Heads-up View toolbar.

Activity: Utilize the Heads-up View toolbar.

Zoom to Fit the model in the Graphics window.

29) Click the **Zoom to Fit** 🔍 icon. The tool fits the model to the
Graphics window.

Zoom to Area on the model in the Graphics window.

30) Click the **Zoom to Area** 🔍 icon. The Zoom to Area 🔍 icon
is displayed.

Zoom in on the top left hole.

31) **Window-select** the top left corner as illustrated. View the
results.

De-select the Zoom to Area tool.

32) Click the **Zoom to Area** 🔍 icon.

Fit the model to the Graphics window.

33) Press the **f** key.

Rotate the model.

34) Hold the **middle mouse button** down. Drag
upward ↻, downward ↻, to the **left** ↻
and to the **right** ↻ to rotate the model in the
Graphics window.

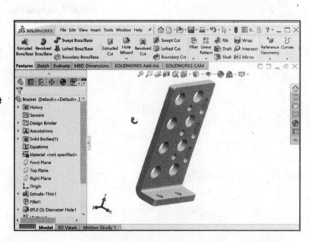

Display a few Standard Views.

35)   Click **inside** the Graphics window.

36)   Click **Front** ⬜ from the drop-down Heads-up view toolbar. The model is displayed in the Front view.

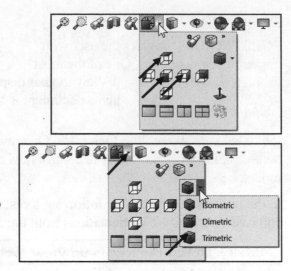

37)   Click **Right** ⬜ from the drop-down Heads-up view toolbar. The model is displayed in the Right view.

38)   Click **Top** ⬜ from the drop-down Heads-up view toolbar. The model is displayed in the Top view.

Display a Trimetric view of the Bracket model.

39)   Click **Trimetric** ⬛ from the drop-down Heads-up view toolbar as illustrated. Note your options. View the results in the Graphics window.

SOLIDWORKS Help

Help in SOLIDWORKS is context-sensitive and in HTML format. Help is accessed in many ways, including Help buttons in all dialog boxes and PropertyManager (or press F1) and Help ⑦ tool on the Standard toolbar for SOLIDWORKS Help.

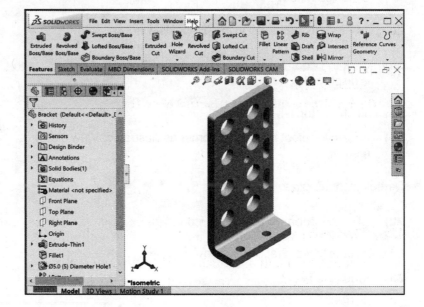

40) Click **Help** from the Menu bar.

41) Click **SOLIDWORKS Help**. The SOLIDWORKS Help Home Page is displayed by default. View your options.

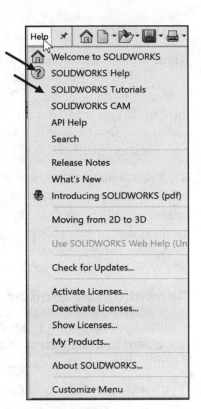

🔅 SOLIDWORKS Web Help is active by default under Help in the Main menu.

Close Help. Return to the SOLIDWORKS Graphics window.

42) **Close** ❌ SOLIDWORKS Home.

SOLIDWORKS Tutorials

Display and explore the SOLIDWORKS tutorials.

43) Click **Help** from the Menu bar.

44) Click **SOLIDWORKS Tutorials**. The SOLIDWORKS Tutorials are displayed. The SOLIDWORKS Tutorials are presented by category.

45) Click the **Getting Started** category. The Getting Started category provides lessons on parts, assemblies, and drawings.

In the next section, close all models, tutorials and view the additional User Interface tools.

Activity: Close all Tutorials and Models.

Close SOLIDWORKS Tutorials and models.

46) **Close** ❌ SOLIDWORKS Tutorials.

47) Click **Window**, **Close All** from the Menu bar menu.

User Interface Tools

The book utilizes additional areas of the SOLIDWORKS User Interface. Explore an overview of these tools in the next section.

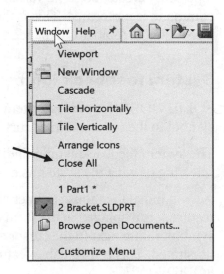

Right-click

Right-click in the Graphics window on a model, or in the FeatureManager on a feature or sketch to display the Context-sensitive toolbar. If you are in the middle of a command, this toolbar displays a list of options specifically related to that command.

Right-click an empty space in the Graphics window of a part or assembly, and a selection context toolbar above the shortcut menu is displayed. This provides easy access to the most commonly used selection tools.

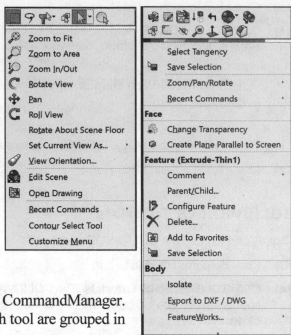

Consolidated toolbar

Similar commands are grouped together in the CommandManager. For example, variations of the Rectangle sketch tool are grouped in a single fly-out button as illustrated.

If you select the Consolidated toolbar button without expanding:

For some commands such as Sketch, the most commonly used command is performed. This command is the first listed and the command shown on the button.

For commands such as rectangle, where you may want to repeatedly create the same variant of the rectangle, the last used command is performed. This is the highlighted command when the Consolidated toolbar is expanded.

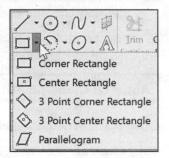

System feedback icon

SOLIDWORKS provides system feedback by attaching a symbol to the mouse pointer cursor.

The system feedback symbol indicates what you are selecting or what the system is expecting you to select.

As you move the mouse pointer across your model, system feedback is displayed in the form of a symbol, riding next to the cursor as illustrated. This is a valuable feature in SOLIDWORKS.

Confirmation Corner

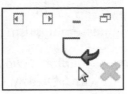

When numerous SOLIDWORKS commands are active, a symbol or a set of symbols is displayed in the upper right-hand corner of the Graphics window. This area is called the Confirmation Corner.

When a sketch is active, the confirmation corner box displays two symbols. The first symbol is the sketch tool icon. The second symbol is a large red X. These two symbols supply a visual reminder that you are in an active sketch. Click the sketch symbol icon to exit the sketch and to save any changes that you made.

When other commands are active, the confirmation corner box provides a green check mark and a large red X. Use the green check mark to execute the current command. Use the large red X to cancel the command.

Confirm changes you make in sketches and tools by using the D keyboard shortcut to move the OK and Cancel buttons to the pointer location in the Graphics window.

Heads-up View toolbar

SOLIDWORKS provides the user with numerous view options from the Standard Views, View and Heads-up View toolbar.

The Heads-up View toolbar is a transparent toolbar that is displayed in the Graphics window when a document is active.

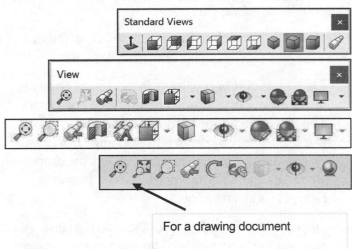

For a drawing document

You can hide, move or modify the Heads-up View toolbar. To modify the Heads-up View toolbar, right-click on a tool and select or deselect the tools that you want to display.

The following views are available. Note: available views are document dependent.

- *Zoom to Fit* : Zooms the model to fit the Graphics window.

- *Zoom to Area* : Zooms to the areas you select with a bounding box.

- *Previous View* : Displays the previous view.

- *Section View* 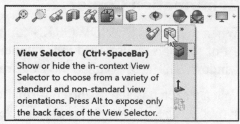 : Displays a cutaway of a part or assembly, using one or more cross section planes.

- *Dynamic Annotation Views* : Only available with SOLIDWORKS MBD. Provides the ability to control how annotations are displayed when you rotate models.

The Orientation dialog has an option to display a view cube (in-context View Selector) with a live model preview. This helps the user to understand how each standard view orientates the model. With the view cube, you can access additional standard views. The views are easy to understand, and they can be accessed simply by selecting a face on the cube.

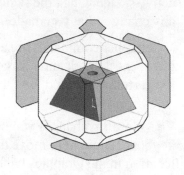

To activate the Orientation dialog box, press (Ctrl + spacebar) or click the View Orientation ˘ icon from the Heads up View toolbar. The active model is displayed in the View Selector in an Isometric orientation (default view).

Click the View Selector icon in the Orientation dialog box to show or hide the in-context View Selector.

💡 Press **Ctrl + spacebar** to activate the View Selector.

💡 Press the **spacebar** to activate the Orientation dialog box.

- *View Orientation box* : Provides the ability to select a view orientation or the number of viewports. The available options are *Top*, *Left*, *Front*, *Right*, *Back*, *Bottom*, *Single view, Two view - Horizontal, Two view - Vertical, Four view.* Click the drop-down arrow to access Axonometric views: Isometric, Dimetric and Trimetric.

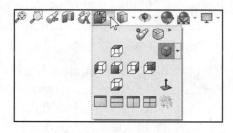

- *Display Style* : Provides the ability to display the style for the active view. The available options are *Wireframe*, *Hidden Lines Visible*, *Hidden Lines Removed*, *Shaded, Shaded With Edges.*

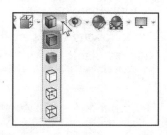

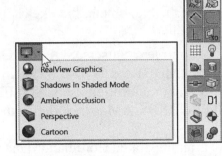

- *Hide/Show Items* 👁 ˇ: Provides the ability to select items to hide or show in the Graphics window. The available items are document dependent. Note the View Center of Mass ✛ icon.

- *Edit Appearance* 🌑: Provides the ability to edit the appearance of entities of the model.

- *Apply Scene* 🌐 ˇ: Provides the ability to apply a scene to an active part or assembly document. View the available options.

- *View Setting* 🖥 ˇ: Provides the ability to select the following settings: *RealView Graphics*, *Shadows In Shaded Mode*, *Ambient Occlusion*, *Perspective* and *Cartoon*.

- *Rotate view* ↻ : Provides the ability to rotate a drawing view. Input Drawing view angle and select the ability to update and rotate center marks with view.

- *3D Drawing View* 🔄: Provides the ability to dynamically manipulate the drawing view in 3D to make a selection.

To display a grid for a part, click Options ⚙ ˇ, Document Properties tab. Click Grid/Snaps, check the Display grid box.

🔅 Add a custom view to the Heads-up View toolbar. Press the space key. The Orientation dialog box is displayed. Click the New View 🗞 tool. The Name View dialog box is displayed. Enter a new named view. Click OK.

Use commands to display information about the triad or to change the position and orientation of the triad. Available commands depend on the triad's context.

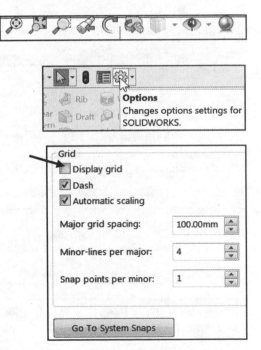

Triad

SOLIDWORKS CommandManager

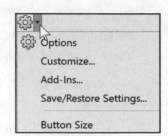

The SOLIDWORKS CommandManager is a Context-sensitive toolbar. By default, it has toolbars embedded in it based on your active document type. When you click a tab below the CommandManager, it updates to display that toolbar. For example, if you click the Sketch tab, the Sketch toolbar is displayed.

For commercial users, SOLIDWORKS Model Based Definition (MBD) and SOLIDWORKS CAM is a separate application. For education users, SOLIDWORKS MBD and SOLIDWORKS CAM is included in the SOLIDWORKS Education Edition.

Below is an illustrated CommandManager for a Part document. Your tabs may vary depending on your SOLIDWORKS applications.

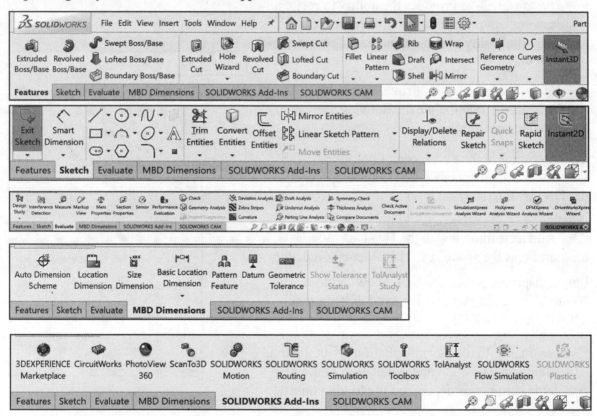

Set button size from the Toolbars tab of the Customize dialog box. To facilitate element selection on touch interfaces such as tablets, you can set up the larger Size buttons and text from the Options menu (Standard toolbar).

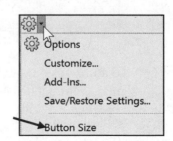

The SOLIDWORKS CommandManager is a Context-sensitive toolbar that automatically updates based on the toolbar you want to access. By default, it has toolbars embedded in it based on your active document type.

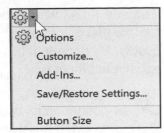

💡 For commercial users, SOLIDWORKS Model Based Definition (MBD) and SOLIDWORKS CAM is a separate application. For education users, SOLIDWORKS MBD and SOLIDWORKS CAM is included in the SOLIDWORKS Education Edition.

Below is an illustrated CommandManager for a Drawing document. Your tabs may vary depending on your SOLIDWORKS applications.

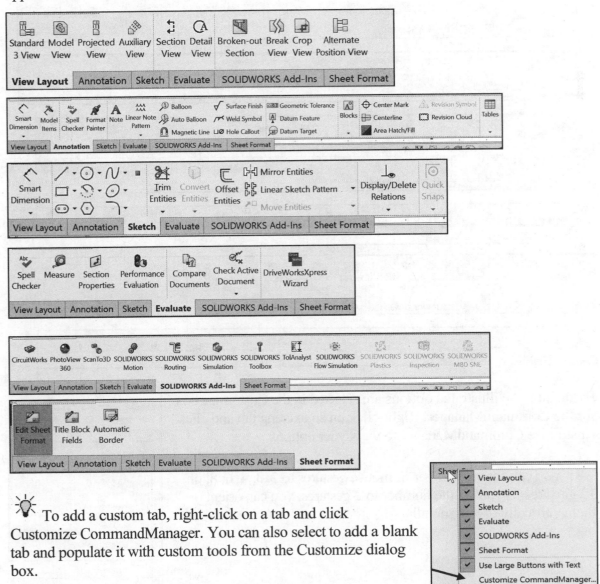

💡 To add a custom tab, right-click on a tab and click Customize CommandManager. You can also select to add a blank tab and populate it with custom tools from the Customize dialog box.

The SOLIDWORKS CommandManager is a Context-sensitive toolbar that automatically updates based on the toolbar you want to access. By default, it has toolbars embedded in it based on your active document type.

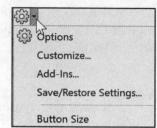

For commercial users, SOLIDWORKS Model Based Definition (MBD) is a separate application. For education users, SOLIDWORKS MBD is included in the SOLIDWORKS Education Edition.

Below is an illustrated CommandManager for an Assembly document. Your tabs may vary depending on your SOLIDWORKS applications.

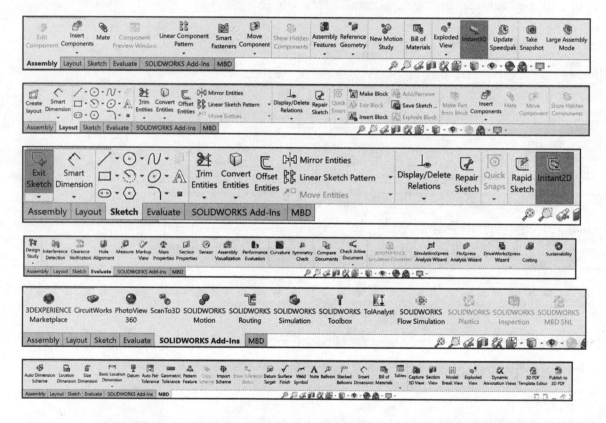

By default, the illustrated options are selected in the Customize box for the CommandManager. Right-click on an existing tab and click Customize CommandManager to view your options.

You can set the number of mouse gestures to 2, 3, 4, 6, 8, or 12 gestures. If you set the number to 2 gestures, you can orient them vertically or horizontally.

Float the CommandManager. Drag the Features, Sketch or any CommandManager tab. Drag the CommandManager anywhere on or outside the SOLIDWORKS window.

To dock the CommandManager, perform one of the following:

While dragging the CommandManager in the SOLIDWORKS window, move the pointer over a docking icon -

◻️ Dock above , ◧ Dock left , ▧ Dock right and click the needed command.

Double-click the floating CommandManager to revert the CommandManager to the last docking position.

Screen shots in the book were made using SOLIDWORKS 2019 SP0 running Windows® 10.

Selection Enhancements

Right-click an empty space in the Graphics window of a part or assembly; a selection context toolbar above the shortcut menu provides easy access to the most commonly used selection tools.

- **Box Selection** ▢. Provides the ability to select entities in parts, assemblies, and drawings by dragging a selection box with the pointer.

- **Lasso Selection** ✐. Provides the ability to select entities by drawing a lasso around the entities.

- **Selection Filters** ⟈. Displays a list of selection filter commands.

- **Select Other** ⟐. Displays the Select Other dialog box.

- **Select** ⬎. Displays a list of selection commands.

- **Magnified Selection** ⟲. Displays the magnifying glass, which gives you a magnified view of a section of a model.

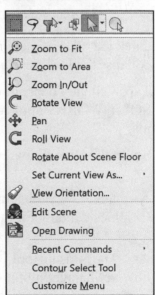

💡 Save space in the CommandManager, right-click in the CommandManager and un-check the Use Large Buttons with Text box. This eliminates the text associated with the tool.

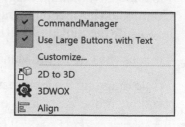

💡 DimXpert provides the ability to graphically check if the model is fully dimensioned and toleranced. DimXpert automatically recognizes manufacturing features. Manufacturing features are not SOLIDWORKS features. Manufacturing features are defined in 1.1.12 of the ASME Y14.5M-1994 Dimensioning and Tolerancing standard. See SOLIDWORKS Help for additional information.

FeatureManager Design Tree

The FeatureManager consists of various tabs:

- *FeatureManager design tree* tab.
- *PropertyManager* tab.
- *ConfigurationManager* tab.
- *DimXpertManager* tab.
- *DisplayManager* tab.
- *CAM FeatureManager tree* tab.

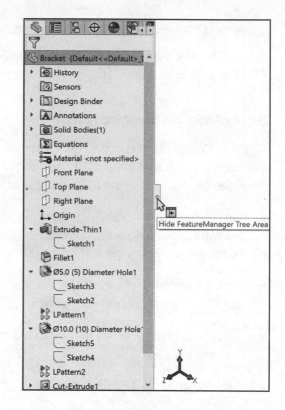

Click the direction arrows to expand or collapse the FeatureManager design tree.

Tabs will vary depending on your SOLIDWORKS applications and Add-ins.

Select the Hide/Show FeatureManager.

Area tab as illustrated to enlarge the Graphics window for modeling.

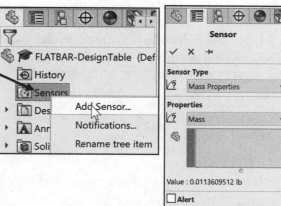

💡 The Sensors tool located in the FeatureManager monitors selected properties in a part or assembly and alerts you when values deviate from the specified limits. There are five sensor types: Simulation Data, Mass properties, Dimensions, Measurement and Costing Data.

Various commands provide the ability to control what is displayed in the FeatureManager design tree.

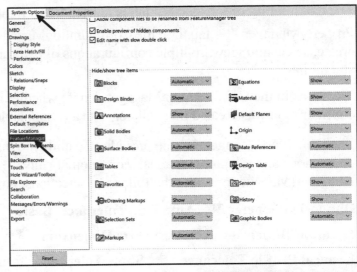

1. Show or Hide FeatureManager items.

Click **Options** ⚙ from the Menu bar. Click **FeatureManager** from the System Options tab. Customize your FeatureManager from the Hide/Show tree Items dialog box.

2. Filter the FeatureManager design tree. Enter information in the filter field. You can filter by *Type of features, Feature names, Sketches, Folders, Mates, User-defined tags* and *Custom properties*.

Tags are keywords you can add to a SOLIDWORKS document to make them easier to filter and to search. The Tags icon is located in the bottom right corner of the Graphics window.

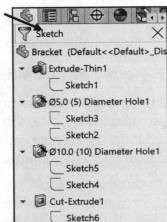

Collapse all items in the FeatureManager, **right-click** and select **Collapse items**, or press the **Shift + C** keys.

The FeatureManager design tree and the Graphics window are dynamically linked. Select sketches, features, drawing views, and construction geometry in either pane.

Split the FeatureManager design tree and either display two FeatureManager instances, or combine the FeatureManager design tree with the ConfigurationManager or PropertyManager.

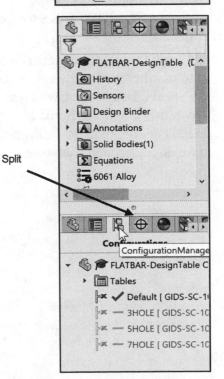

Move between the FeatureManager design tree, PropertyManager, ConfigurationManager, DimXpertManager, DisplayManager and others by selecting the tab at the top of the menu.

Note: Tabs will vary depending on your SOLIDWORKS applications and Add-ins.

The ConfigurationManager 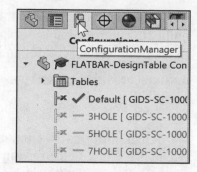 tab is located to the right of the PropertyManager  tab. Use the ConfigurationManager to create, select and view multiple configurations of parts and assemblies.

The icons in the ConfigurationManager denote whether the configuration was created manually or with a design table.

The DimXpertManager ⊕ tab provides the ability to insert dimensions and tolerances manually or automatically. The DimXpertManager provides the following selections: **Auto Dimension Scheme** ⊕, **Auto Pair Tolerance**, **Basic, Location Dimension** ⊢⊡⊣, **Basic Size Dimension** ⬝, **General Profile Tolerance** 📊, **Show Tolerance Status** ⁺ₑ, **Copy Scheme** ⊕, **Import Scheme** ⊕ and **TolAnalyst Study** ᴋᴵ.

Fly-out FeatureManager

The fly-out FeatureManager design tree provides the ability to view and select items in the PropertyManager and the FeatureManager design tree at the same time.

Throughout the book, you will select commands and command options from the drop-down menu, fly-out FeatureManager, Context toolbar, or from a SOLIDWORKS toolbar.

🔆 Another method for accessing a command is to use the accelerator key. Accelerator keys are special key strokes, which activate the drop-down menu options. Some commands in the menu bar and items in the drop-down menus have an underlined character.

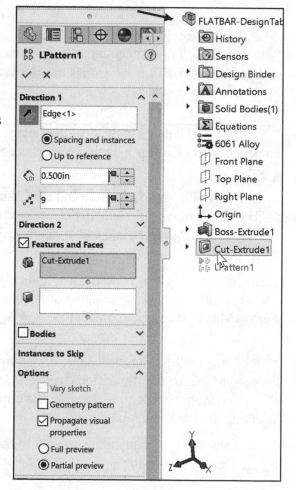

Task Pane

The Task Pane is displayed when a SOLIDWORKS session starts. You can show, hide, and reorder tabs in the Task Pane. You can also set a tab as the default so it appears when you open the Task Pane, pin or unpin to the default location.

The Task Pane contains the following default tabs:

- *SOLIDWORKS Resources* 🏠.

- *Design Library* 📦.

- *File Explorer* 📁.

- *View Palette* 🔲.

- *Appearances, Scenes and Decals* ⬤.

- *Custom Properties* 📋.

- *SOLIDWORKS Forum* 💬.

💡 Additional tabs are displayed with Add-Ins.

Use the **Back** and **Forward** buttons in the Design Library tab and the Appearances, Scenes, and Decals tab of the Task Pane to navigate in folders.

SOLIDWORKS Resources

The basic SOLIDWORKS Resources 🏠 menu displays the following default selections:

- *Welcome to SOLIDWORKS*.

- *SOLIDWORKS Tools*.

- *Online Resources*.

- *Subscription Services*.

Other user interfaces are available during the initial software installation selection: *Machine Design, Mold Design, Consumer Products Design, etc.*

Design Library

The Design Library 🗄 contains reusable parts, assemblies, and other elements including library features.

The Design Library tab contains four default selections. Each default selection contains additional sub categories.

The default selections are:

- *Design Library.*

- *Toolbox.*

- *3D ContentCentral (Internet access required).*

- *SOLIDWORKS Content (Internet access required).*

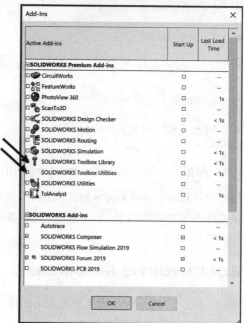

💡 Activate the SOLIDWORKS Toolbox. Click Tools, Add-Ins.., from the Main menu. Check the SOLIDWORKS Toolbox Library and SOLIDWORKS Toolbox Utilities box from the Add-ins dialog box or click SOLIDWORKS Toolbox from the SOLIDWORKS Add-Ins tab.

To access the Design Library folders in a non-network environment, click Add File Location 🗄 and browse to the needed path. Paths may vary depending on your SOLIDWORKS version and window setup. In a network environment, contact your IT department for system details.

File Explorer

File Explorer 🗁 duplicates Windows Explorer from your local computer and displays:

- *Recent Documents.*

- *Samples.*

- *Open in SOLIDWORKS*

- *Desktop.*

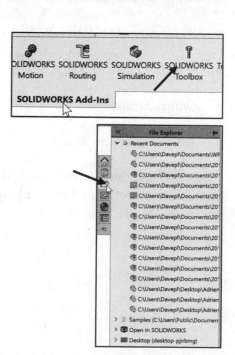

Search

The SOLIDWORKS Search box is displayed in the upper right corner of the SOLIDWORKS Graphics window (Menu Bar toolbar). Enter the text or key words to search.

New search modes have been added to SOLIDWORKS Search as illustrated.

View Palette

The View Palette tool located in the Task Pane provides the ability to insert drawing views of an active document, or click the Browse button to locate the desired document.

Click and drag the view from the View Palette into an active drawing sheet to create a drawing view.

The selected model is FLATBAR-DesignTable in the illustration.

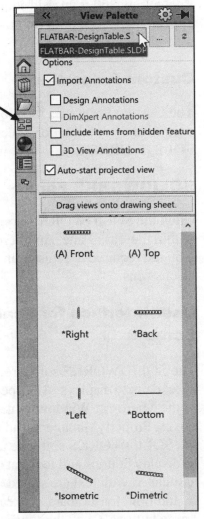

Appearances, Scenes, and Decals

Appearances, Scenes, and Decals 🌐 provide a simplified way to display models in a photo-realistic setting using a library of Appearances, Scenes, and Decals.

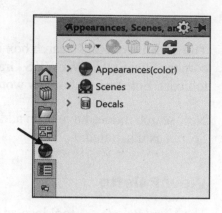

An appearance defines the visual properties of a model, including color and texture. Appearances do not affect physical properties, which are defined by materials.

Scenes provide a visual backdrop behind a model. In SOLIDWORKS they provide reflections on the model. PhotoView 360 is an Add-in. Drag and drop a selected appearance, scene or decal on a feature, surface, part or assembly.

Custom Properties

The Custom Properties 🗒 tool provides the ability to enter custom and configuration specific properties directly into SOLIDWORKS files.

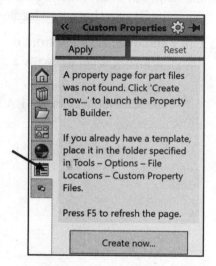

SOLIDWORKS Forum

Click the SOLIDWORKS Forum 🗨 icon to search directly within the Task Pane. An internet connection is required. You are required to register and to log in for postings and discussions.

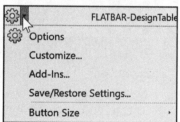

User Interface for Scaling High Resolution Screens

The SOLIDWORKS software supports high-resolution, high-pixel density displays. All aspects of the user interface respond to the Microsoft Windows® display scaling setting. In dialog boxes, PropertyManagers, and the FeatureManager design tree, the SOLIDWORKS software uses your display scaling setting to display buttons and icons at an appropriate size. Icons that are associated with text are scaled to a size appropriate for the text. In addition, for toolbars, you can display Small, Medium, or Large buttons. Click the **Options drop-down arrow** from the Standard Menu bar, and click Button size to size the icons.

Motion Study tab

Motion Studies are graphical simulations of motion for an assembly. Access the MotionManager from the Motion Study tab. The Motion Study tab is located in the bottom left corner of the Graphics window.

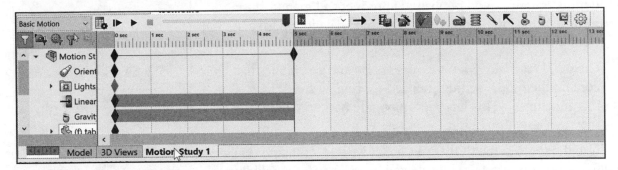

Incorporate visual properties such as lighting and camera perspective. Click the Motion Study tab to view the MotionManager. Click the Model tab to return to the FeatureManager design tree.

The MotionManager displays a timeline-based interface and provides the following selections from the drop-down menu as illustrated:

- *Animation:* Apply Animation to animate the motion of an assembly. Add a motor and insert positions of assembly components at various times using set key points. Use the Animation option to create animations for motion that do **not** require accounting for mass or gravity.

- *Basic Motion:* Apply Basic Motion for approximating the effects of motors, springs, collisions and gravity on assemblies. Basic Motion takes mass into account in calculating motion. Basic Motion computation is relatively fast, so you can use this for creating presentation animations using physics-based simulations. Use the Basic Motion option to create simulations of motion that account for mass, collisions or gravity.

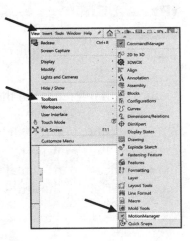

If the Motion Study tab is not displayed in the Graphics window, click **View** ➤ **Toolbars** ➤ **MotionManager** from the Menu bar.

3D Views tab

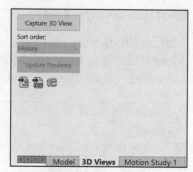

SOLIDWORKS MBD (Model Based Definition) lets you create models without the need for drawings giving you an integrated manufacturing solution. MBD helps companies define, organize, and publish 3D product and manufacturing information (PMI), including 3D model data in industry standard file formats.

Create 3D drawing views of your parts and assemblies that contain the model settings needed for review and manufacturing. This lets users navigate back to those settings as they evaluate the design.

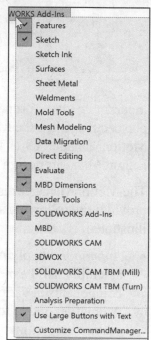

Use the tools in the SOLIDWORKS MBD CommandManager to set up your model with selected configurations, including explodes and abbreviated views, annotations, display states, zoom level, view orientation and section views. Capture those settings so that you and other users can return to them at any time using the 3D view palette.

To access the 3D View palette, click the 3D Views tab at the bottom of the SOLIDWORKS window or the SOLIDWORKS MBD tab in the CommandManager. The Capture 3D View button opens the Capture 3D View PropertyManager, where you specify the 3D view name, and the configuration, display state and annotation view to capture. See SOLIDWORKS help for additional information.

Dynamic Reference Visualization (Parent/Child)

Dynamic Reference Visualization provides the ability to view the parent/child relationships between items in the FeatureManager design tree. When you hover over a feature with references in the FeatureManager design tree, arrows display showing the relationships. If a reference cannot be shown because a feature is not expanded, the arrow points to the feature that contains the reference and the actual reference appears in a text box to the right of the arrow.

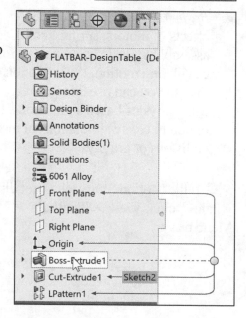

Use Dynamic reference visualization for a part, assembly and ever mates.

To display the Dynamic Reference Visualization, click **View ➤ User Interface ➤ Dynamic Reference Visualization Parent/Child)** from the Main menu bar.

Mouse Movements

A mouse typically has two buttons: a primary button (usually the left button) and a secondary button (usually the right button). Most mice also include a scroll wheel between the buttons to help you scroll through documents and to Zoom in, Zoom out and rotate models in SOLIDWORKS. It is highly recommended that you use a mouse with at least a Primary, Scroll and Secondary button.

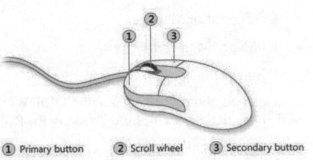

Single-click

To click an item, point to the item on the screen, and then press and release the primary button (usually the left button). Clicking is most often used to select (mark) an item or open a menu. This is sometimes called single-clicking or left-clicking.

Double-click

To double-click an item, point to the item on the screen, and then click twice quickly. If the two clicks are spaced too far apart, they might be interpreted as two individual clicks rather than as one double-click. Double-clicking is most often used to open items on your desktop. For example, you can start a program or open a folder by double-clicking its icon on the desktop.

Right-click

To right-click an item, point to the item on the screen, and then press and release the secondary button (usually the right button). Right-clicking an item usually displays a list of things you can do with the item. Right-click in the open Graphics window or on a command in SOLIDWORKS, and additional pop-up context is displayed.

Scroll wheel

Use the scroll wheel to zoom-in or to zoom-out of the Graphics window in SOLIDWORKS. To zoom-in, roll the wheel backward (toward you). To zoom-out, roll the wheel forward (away from you).

Summary

The SOLIDWORKS 2019 User Interface and CommandManager consist of the following main options: Menu bar toolbar, Menu bar menu, Drop-down menus, Context toolbars, Consolidated fly-out menus, System feedback icons, Confirmation Corner and Heads-up View toolbar.

The default CommandManager Part tabs control the display of the *Features*, *Sketch*, *Evaluate*, *DimXpert* and *SOLIDWORKS Add-Ins* toolbars.

The FeatureManager consists of five default tabs:

- FeatureManager design tree.

- PropertyManager.

- ConfigurationManager.

- DimXpertManager.

- DisplayManager.

You learned about creating a new SOLIDWORKS part and opening an existing SOLIDWORKS part along with using the Rollback bar to view the sketches and features.

You also learned about SOLIDWORKS Help, SOLIDWORKS Tutorials and basic mouse movements to manipulate your SOLIDWORKS model.

Templates are part, drawing and assembly documents which include user-defined parameters. Open a new part, drawing or assembly. Select a template for the new document.

- *Parts*. The Parts default template is located in the C:\ProgramData\SolidWorks\\SOLIDWORKS 2019\templates\Part.prtdot folder.

- *Assemblies*. The Assemblies default template is located in the C:\ProgramData\SolidWorks\\SOLIDWORKS 2019\templates\Assembly.asmdot folder.

- *Drawings*. The Drawings default template is located in the C:\ProgramData\SolidWorks\\SOLIDWORKS 2019\templates\Drawing.drwdot folder.

In Chapter 5, create two Part Templates and two parts for the FLASHIGHT assembly. The Part Templates are:

- PART-IN-ANSI.
- PART-MM-ISO.

A Template is the foundation for a SOLIDWORKS document. Templates are part, drawing, and assembly documents that include user-defined parameters and are the basis for new documents.

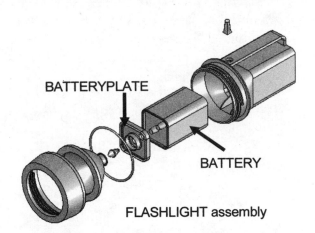

FLASHLIGHT assembly

Create two parts for the FLASHLIGHT assembly:

- BATTERY.
- BATTERYPLATE.

Times in academic integrity cases. Right-clicking on a sketch or feature. Select properties from the Pop-up menu. View who created the sketches and features in a SOLIDWORKS model, and when (date + time).

Notes:

Chapter 5

Introduction to SOLIDWORKS Part Modeling

Below are the desired outcomes and usage competencies based on the completion of Chapter 5.

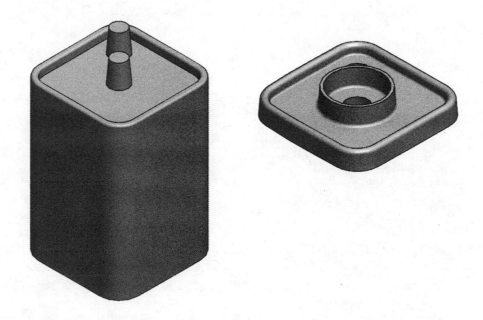

Desired Outcomes:	Usage Competencies:
• Address File Management with file folders.	• Aptitude to create file folders for various Projects and Templates.

• Create two Part Templates:
 	
  o PART-IN-ANSI
  o PART-MM-ISO
	• Skill to address System Options and Document Properties.

• Create two Parts:
 	
  o BATTERY
  o BATTERYPLATE
	• Specific knowledge and understanding of 2D sketching and the following 3D features: Extruded Boss/Base, Extruded Cut, Fillet, and Instant3D.

Notes:

Chapter 5 - Introduction to SOLIDWORKS Part Modeling

Chapter Overview

Provide a comprehensive understanding of System Options, Document Properties, Part templates, File management and more. Create two Part Templates utilized for the parts in this chapter:

- PART-IN-ANSI.

- PART-MM-ISO.

A Template is the foundation for a SOLIDWORKS document. Templates are part, drawing, and assembly documents that include user-defined parameters and are the basis for new documents.

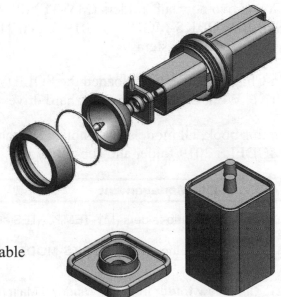

Create two parts for the FLASHLIGHT assembly:

- BATTERY.

- BATTERYPLATE.

On the completion of this chapter, you will be able to:

- Apply and understand System Options and Document Properties.

- Create a new Part Template.

- Open, Save and Close Part documents and Templates.

- Create 2D sketch profiles on the correct Sketch plane.

- Utilize the following Sketch tools: Smart Dimension, Line, Centerline, Circle, Center Rectangle, Convert Entities, Offset Entities and Mirror Entities.

- Apply and edit sketch dimensions.

- Establish Geometric relations, dimensions, and determine the status of the sketch.
  - Under defined, Fully defined and Over defined

- Utilize the Instant3D tool to create an Extruded Boss/Base feature.

- Utilize the Save As, Delete, Edit Feature and Modify tools.

💡 Screen shots and illustrations in the book display the SOLIDWORKS user default setup.

File Management

File management organizes parts, assemblies, drawings, and templates. Why do you require file management? Answer: A top level assembly has hundreds or even thousands of documents that require organization. Utilize folders to organize projects, vendor components, templates, and libraries.

Create two new sub-folders (MY-TEMPLATES and PROJECTS) in the SOLIDWORKS-MODELS 2019 folder. The procedure will be different depending on your operating system.

In Chapter 4, you downloaded the SOLIDWORKS-MODELS 2019 folder to your hard drive. Work directly from your hard drive.

In this book, all models, assemblies and templates are saved to the SOLIDWORKS-MODELS 2019 folder and sub-folders.

Activity: File Management

Create two new sub-folders (MY-TEMPLATES and PROJECTS) under the SOLIDWORKS-MODELS 2019 folder.

1) Double-click the **SOLIDWORKS-MODELS 2019** folder on your hard drive.

2) Click **New folder** from the Windows Main menu. A new folder icon is displayed.

Create the first sub-folder. The procedure will be different depending on your operating system.
3) Enter **MY-TEMPLATES** for the folder name.

Create the second sub-folder.
4) Click **New folder** from the Windows Main menu. A new folder icon is displayed.

5) Enter **PROJECTS** for the second sub-folder name.

Return to the SOLIDWORKS-MODELS 2019 folder.
6) Click the **up arrow** button to return to the SOLIDWORKS-MODELS 2019 folder.

Utilize the MY-TEMPLATES folder and the PROJECTS folder throughout the text. Store the Part template, Assembly template and Drawing template in the MY-TEMPLATES folder. Store the parts, assemblies and drawings that you create in the PROJECTS folder.

Start a SOLIDWORKS Session. Open a new Part.

Create a new part. Click File, New from the Menu bar menu or click New ⬜ from the Menu bar toolbar.

There are two options for new documents: *Novice* and *Advanced*. Select the Advanced option. Select the default Part document from the Templates tab.

Activity: Start a SOLIDWORKS Session. Open a New Part Document.

Start a SOLIDWORKS session and open a new Part document.

7) Click the **SOLIDWORKS 2019** application or click the **SOLIDWORKS desktop** icon. The SOLIDWORKS program window opens.

8) Click **File**, **New** or click **New** ⬜ from the Menu bar toolbar. The New SOLIDWORKS dialog box is displayed.

9) Double-click **Part** from the Templates tab. View the default Part document FeatureManager and the empty Graphics window.

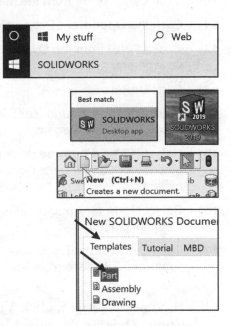

Part Template

The Part Template is the foundation for a SOLIDWORKS part. Part1 displayed in the FeatureManager utilizes the *Part(*.prt, *sldprt)* default template located in the New SOLIDWORKS dialog box.

Document properties contain the default settings for the Part Template. The document properties include the dimensioning standard, units, dimension decimal display, grids, note font, and line styles. There are hundreds of document properties. You will modify the following document properties in this chapter: Dimensioning standard, unit and decimal places.

The Dimensioning (drafting) standard determines the display of dimension text, arrows, symbols, and spacing. Units are the measurement of physical quantities. MMGS (millimeter, gram, second) and IPS (inch, pound, second) are the two most common unit systems specified for engineering parts and drawings.

Document properties are stored with the document. Apply the document properties to the Part Template. Create two Part Templates: PART-IN-ANSI and PART-MM-ISO. Save the Part Templates in the MY-TEMPLATE folder.

System Options are stored in the registry of your computer. The File Locations option controls the file folder location of SOLIDWORKS documents.

Utilize the File Locations option to reference your Part Templates in the MY-TEMPLATES folder. Add the SOLIDWORKS-MODELS 2019\MY-TEMPLATES folder path name to the Document Templates File Locations list.

> **Activity: Create Two Part Templates. Add A New Document Templates Tab.**

Create the **PART-IN-ANSI** Part template. Set drafting standard, units, and precision.

10) Click **Options** ⚙ from the Main menu.

11) Click the **Document Properties** tab from the dialog box.

12) Select **ANSI** from the Overall drafting standard drop-down menu.

13) Click **Units**.

14) Click **IPS** (inch, pound, second), **[MMGS]** for Unit system.

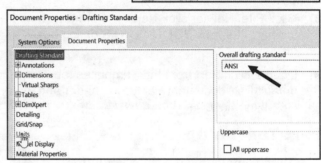

15) Select **.123**, **[.12]** (three decimal places) for Length units Decimal places.

16) Select **None** for Angular units Decimal places.

17) Click **OK** from the Document Properties - Units dialog box.

🔆 To view the Origin, click **View**, **Hide/show**, **Origins** from the Menu bar menu.

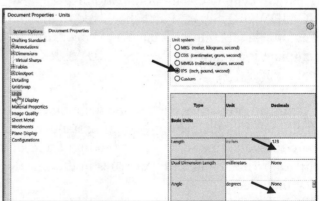

Set the Grid/Snap option.

18) Click **Grid/Snap**. The Document Properties Grid/Snap dialog box is displayed.

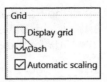

19) Un-check the **Display grid** box.

Return to the SOLIDWORKS Graphics window.

20) Click **OK** from the Document Properties Grid/Snap dialog box.

Save the Part Template. Enter name.

21) Click **Save As** from the Menu bar drop-down menu. The Save As dialog box is displayed.

22) Select **Part Templates (*.prtdot)** from the Save as type box.

23) Select the **SOLIDWORKS-MODELS 2019\MY-TEMPLATES** folder.

24) Enter **PART-IN-ANSI** in the File name box.

25) Click **Save** from the Save As dialog box.

Create the PART-MM-ISO Part Template. Set drafting standard.

26) Click **Options** ⚙ from the Main menu.

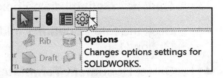

27) Click the **Document Properties** tab from the dialog box. The Document Properties - Drafting Standard dialog box is displayed.

28) Select **ISO** from the Overall drafting standard drop-down box.

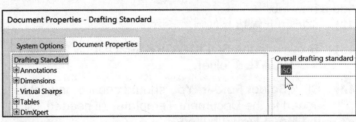

Set units and precision.

29) Click **Units**. The Document Properties - Unit dialog box is displayed.

30) Select **MMGS, (millimeter, gram, second)** for Unit system.

31) Select **.12** (two decimal places) for Length basic units.

32) Select **None** for Angular units Decimal places.

33) Click **OK**.

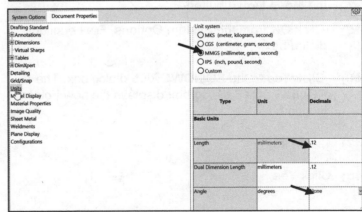

Save the Part Template. Enter name.

34) Click **Save As** from the Menu bar drop-down menu. The Save As dialog box is displayed.

35) Select **Part Templates (*.prtdot)** from the Save as type box.

36) Select the **SOLIDWORKS-MODELS 2019\MY-TEMPLATES** folder.

37) Enter **PART-MM-ISO** in the File name box.

38) Click **Save**.

In the next section, utilize the File Locations option to reference your Templates in the MY-TEMPLATES folder. Add the SOLIDWORKS-MODELS 2019\MY-TEMPLATES folder path name to the Document Templates File Locations list.

Create a New Documents Templates tab. Set folder location.

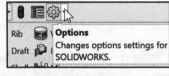

39) Click **Options** ⚙ from the Menu bar. The System Options - General dialog box is displayed.

40) Click the **File Locations** folder from the System Options tab.

41) Select **Document Templates** from Show folders for.

Add a new tab.

42) Click the **Add** button.

43) Browse to the **SOLIDWORKS-MODELS 2019\MY-TEMPLATES** folder.

44) Click **Select Folder**. You should see the new folder added to the Document Templates. If needed, click the **Move Down** button.

45) Click **OK** from the System Options, File Locations dialog box.

46) Click **Yes** to the SOLIDWORKS dialog box. The SOLIDWORKS dialog box displays the new Folder location.

47) Click **Yes**.

48) Click **Yes**.

Close All documents.

49) Click **Window**, **Close All** from the Menu bar.

Display the MY-TEMPLATES folder and templates.

50) Click **New**  from the Menu bar.

51) Click the **MY-TEMPLATES** tab. View the two new Part Templates. Note: Additional templates are displayed.

52) Click **Cancel** from the New SOLIDWORKS Document dialog box.

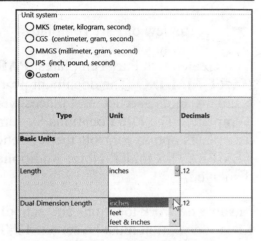

Each folder listed in the System Options, File Locations, Document Templates, Show Folders For option produces a corresponding tab in the New SOLIDWORKS Document dialog box.

The MY-TEMPLATES tab is only visible when the folder contains a SOLIDWORKS Template document. Create the PART-MM-ANSI template as an exercise.

The PART-IN-ANSI Template contains document properties settings for the parts contained in the FLASHLIGHT assembly. Substitute the PART-MM-ISO or PART-MM-ANSI template to create the identical parts in millimeters.

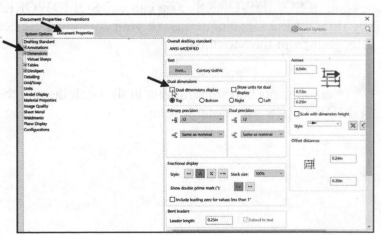

The primary units in this book are IPS (inch, pound, second). The optional secondary units are MMGS (millimeter, gram, second) and are indicated in brackets [ ]. Illustrations are provided in both inches and millimeters.

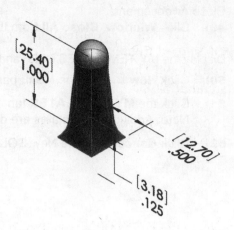

 Additional information on System Options, Document Properties, File Locations, and Templates is located in SOLIDWORKS Help Topics. Keywords: Options (detailing, units), Templates, Files (locations), menus and toolbars (features, sketch).

Review of Part Templates

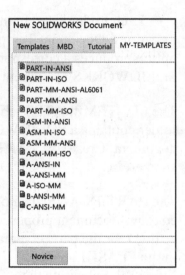

You created two Part Templates: **PART-IN-ANSI** and **PART-MM-ISO**. The document properties Overall drafting standard, units and decimal places were stored in the Part Templates. The File Locations System Option, Document Templates option controls the reference to the MY-TEMPLATES folder. Note: Additional templates are displayed.

In some network locations and school environments, the File Locations option must be set to MY-TEMPLATES for each session of SOLIDWORKS. You can exit SOLIDWORKS at any time during this chapter. Save your document. Select File, Exit from the Menu bar.

Screen shots and illustrations in the book display the SOLIDWORKS user default setup.

BATTERY Part

The BATTERY is a simplified representation of a purchased OEM part. Represent the battery terminals as cylindrical extrusions. The battery dimensions are obtained from the ANSI standard 908D.

A 6-Volt lantern battery weighs approximately 1.38 pounds (0.62kg). Locate the center of gravity closest to the center of the battery.

Create the BATTERY part. Use features to create parts. Features are building blocks that add or remove material.

Utilize the Instant3D tool to create the Extruded Boss/Base 🔲 (Boss-Extrude1) feature. The Extrude Boss/Base feature adds material. The Base feature is the first feature of the part.

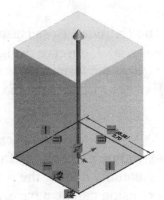

Apply symmetry. Use the Center Rectangle Sketch tool on the Top Plane. The 2D Sketch profile is centered at the Origin.

Extend the profile perpendicular (⊥) to the Top Plane.

Utilize the Fillet 🔲 feature to round the four vertical edges.

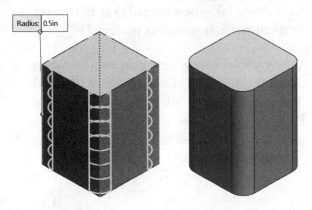

The Extruded Cut 📵 feature removes material from the top face. Utilize the top face for the Sketch plane. Utilize the Offset Entity Sketch tool to create the profile.

Utilize the Fillet feature 📦 to round the top narrow face. Fillet/Round features create a rounded internal or external face on the part. You can fillet all edges of a face, selected sets of faces, selected edges, or edge loops.

The Extruded Boss/Base 📦 feature adds material. Conserve design time. Represent each of the terminals as a cylindrical Extruded Boss (Boss-Extrude2) feature.

🔅 Extrude Features creates a feature by extruding a 3D object from a 2D sketch, essentially adding the third dimension. An extrusion can be a base, a boss (which adds material, often on another extrusion), or a cut (which removes material).

When you create a new part or assembly, the three default Planes (Front, Right and Top) are aligned with specific views. The Plane you select for the Base sketch determines the orientation of the part.

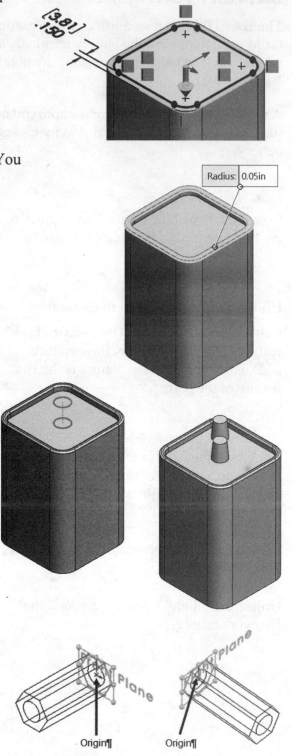

BATTERY Part-Extruded Boss/Base Feature

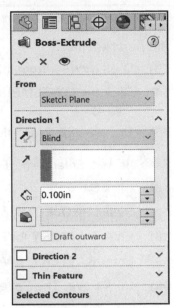

The Extruded Boss/Base feature requires:

- Sketch plane (Top).
- Sketch profile (Rectangle).
  - Geometric relations and dimensions.
- End Condition Depth (Blind) in Direction 1.

Create a new part named BATTERY. Insert an Extruded Boss/Base feature. Extruded features require a Sketch plane. The Sketch plane determines the orientation of the Extruded Base feature. The Sketch plane locates the Sketch profile on any plane or face.

The Top Plane is the Sketch plane. The Sketch profile is a rectangle. The rectangle consists of two horizontal lines and two vertical lines.

Geometric relations and dimensions constrain the sketch in 3D space. The Blind End Condition in Direction 1 requires a depth value to extrude the 2D Sketch profile and to complete the 3D feature.

Alternate between the Features tab and the Sketch tab in the CommandManager to display the available Feature and Sketch tools for the Part document.

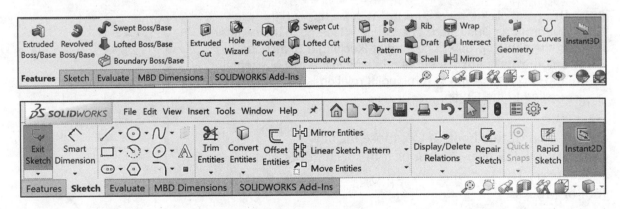

Activity: BATTERY Part - Create the Extruded Base Feature

Create a New part.

53)   Click **New** ⬜ from the Menu bar.

54)   Click the **MY-TEMPLATES** tab. Additional templates are displayed.

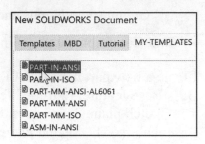

55)   Double-click **PART-IN-ANSI**, [**PART-MM-ISO**].

Save the part. Enter name. Enter description.

56)   Click **Save** 💾.

57)   Select **PROJECTS** for Save in folder.

58)   Enter **BATTERY** for File name.

59)   Enter **BATTERY**, **6-VOLT** for Description.

60)   Click **Save**. The Battery FeatureManager is displayed.

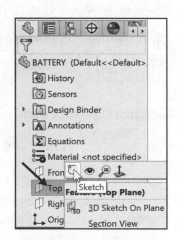

Select the Sketch plane.

61)   Right-click **Top Plane** from the FeatureManager. This is your Sketch plane.

Sketch the 2D Sketch profile centered at the Origin.

62)   Click **Sketch** 📝 from the Context toolbar. The Sketch toolbar is displayed.

63)   Click the **Center Rectangle** ▭ Sketch tool. The Center Rectangle icon is displayed.

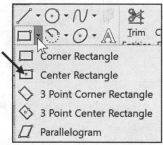

64)   Click the **Origin**. This is your first point.

65)   Drag and click the **second point** in the upper right quadrant as illustrated. The Origin is located in the center of the sketch profile. The Center Rectangle Sketch tool automatically applies equal relations to the two horizontal and two vertical lines. A midpoint relation is automatically applied to the Origin.

💡 A goal of this book is to expose the new SOLIDWORKS user to various tools, techniques and procedures. The text may not always use the most direct tool or process.

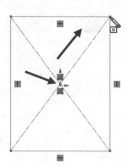

💡 Click **View**, **Hide/Show**, **Sketch Relations** from the Main menu to view sketch relations in the Graphics area.

Dimension the sketch.

66) Click the **Smart Dimension** ✎ Sketch tool.

67) Click the **top horizontal line**. Click a **position** above the horizontal line.

68) Enter **2.700**in, **[68.58]** for width.

69) Click the **Green Check mark** ✔ in the Modify dialog box. Enter **2.700**in, [68.58] for height as illustrated.

70) Click the **Green Check mark** ✔ in the Modify dialog box. The black Sketch status is fully defined.

71) Click **OK** ✔ from the Dimension PropertyManager.

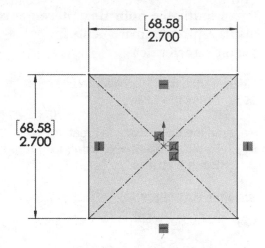

Exit the Sketch.
72) Click **Exit Sketch**.

Insert an Extruded Boss/Base feature. Apply the Instant3D tool. The Instant3D tool provides the ability to drag geometry and dimension manipulator points to resize or to create features directly in the Graphics window.

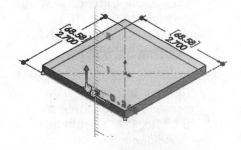

Display an Isometric view. Use the on-screen ruler.

73) Press the **space bar** to display the Orientation dialog box. Click **Isometric view** ▢ .

74) Click the **front horizontal line** as illustrated. A green arrow is displayed.

75) Click and drag the **green/red arrow** upward.

76) Click the on-screen ruler at **4.1**in, [104.14] as illustrated. This is the depth in direction 1. The extrude direction is upwards. Boss-Extrude1 is displayed in the FeatureManager.

Check the Boss-Extrude1 feature depth dimension.

77) Right-click **Boss-Extrude1** from the FeatureManager.

78) Click **Edit Feature** ▢ from the Context toolbar. 4.100in is displayed for depth. Blind is the default End Condition. Note: If you did not select the correct depth, input the depth in the Boss-Extrude1 PropertyManager.

79) Click **OK** ✔ from the Boss-Extrude1 PropertyManager.

☀ Modify the **Spin Box Increments** in System Options to display different increments in the on-screen ruler.

Fit the part to the Graphics window.
80)   Press the **f** key.

Rename the Boss-Extrude1 feature.
81)   Rename **Boss-Extrude1** to **Base Extrude**.

Save the BATTERY.
82)   Click **Save** 💾.

Modify the BATTERY.
83)   Click **Base Extrude** from the FeatureManager. Note: Instant3D is activated by default.

84)   Drag the **manipulator point** upward and click the on-screen ruler to create a **5.000in, [127]** depth as illustrated. Blind is the default End Condition.

Return to the 4.100 depth.

85)   Click the **Undo** ↶ button from the Menu bar. The depth of the model is 4.100in, [104.14]. Blind is the default End Condition. Practice may be needed to select the correct on-screen ruler dimension.

The color of the sketch indicates the sketch status.

- Light Blue - Currently selected.

- Blue - Under defined, requires additional geometric relations and or dimensions.

- Black - Fully defined.

- Red/Yellow - Over defined, requires geometric relations and or dimensions to be deleted or redefined to solve the sketch.

☀ The Instant3D tool is active by default in the Features toolbar located in the CommandManager.

BATTERY Part-Fillet Feature Edge

Fillet features remove sharp edges. Utilize Hidden Lines Visible ⬡ from the Heads-up View toolbar to display hidden edges.

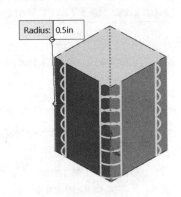

An edge Fillet feature requires:

- A selected edge

- Fillet radius

Select a vertical edge. Select the Fillet feature from the Features toolbar. Enter the Fillet radius. Add the other vertical edges to the Items To Fillet option.

The order of selection for the Fillet feature is not predetermined. Select edges to produce the correct result. The Fillet feature uses the Fillet PropertyManager. The Fillet PropertyManager provides the ability to select either the *Manual* or *FilletXpert* tab.

Each tab has a separate menu and PropertyManager. The Fillet PropertyManager and FilletXpert PropertyManager displays the appropriate selections based on the type of fillet you create.

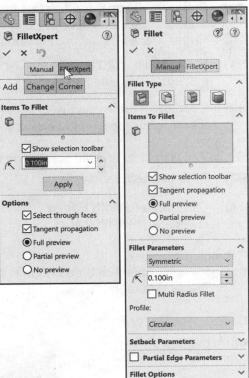

🔆 The FilletXpert automatically manages, organizes and reorders your fillets in the FeatureManager design tree. The FilletXpert PropertyManager provides the ability to add, change or corner fillets in your model. The PropertyManager remembers its last used state. View the SOLIDWORKS tutorials for additional information on fillets.

🔆 Use the Fillet tool to create symmetrical conic shaped fillets for parts, assemblies, and surfaces. You can apply conic shapes to *Constant Size*, *Variable Size* and *Face* fillets.

Activity: BATTERY Part - Fillet Feature Edge

Display hidden edges.

86) Click **Hidden Lines Visible** ⬡ from the Heads-up View toolbar.

Insert a Constant Size Symmetric Fillet feature.

87) Click the **Fillet** ⬜ feature tool. The Fillet PropertyManager is displayed.

88) Click the **Manual** tab. Click the **Constant Size Fillet** button for Fillet Type.

89) Click the **left front vertical edge** as illustrated.

Note the mouse pointer edge ⬠ icon. Edge<1> is displayed in the Items To Fillet box. The fillet option pop-up toolbar is displayed. Options are model dependent.

90) Select the **Connected to start face, 3 Edges icon.** The four selected edges are displayed in the Edges, Faces, Features and Loop box.

91) Enter **.500**in, **[12.7]** for Radius. Accept the default settings (Symmetric type).

92) Click **OK** ✔ from the Fillet PropertyManager. Fillet1 is displayed in the FeatureManager.

Display an Isometric, Shaded with Edges view.

93) Click **Isometric view** 🟦.

94) Click **Shaded With Edges** 🟦 from the Heads-up View toolbar.

Rename the feature.

95) Rename **Fillet1** to **Side Fillets** in the FeatureManager.

Save the BATTERY.

96) Click **Save** 💾.

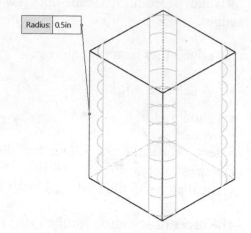

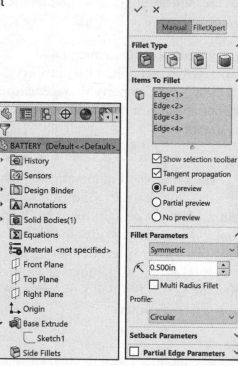

BATTERY Part - Extruded Cut Feature

An Extruded Cut feature removes material. An Extruded Cut feature requires:

- Sketch plane (Top face).

- Sketch profile (Offset Entities).

- End Condition depth (Blind) in Direction 1.

The Offset Entity Sketch tool uses existing geometry, extracts an edge or face and locates the geometry on the current Sketch plane.

Offset the existing Top face for the 2D sketch. Utilize the default Blind End Condition in Direction 1.

Activity: BATTERY Part - Extruded Cut Feature

Select the Sketch plane.

97) Right-click the **Top face** of the BATTERY in the Graphics window. Base Extruded is highlighted in the FeatureManager.

Create a sketch.

98) Click **Sketch** from the Context toolbar. The Sketch toolbar is displayed.

Display the Top face.

99) Press the **space bar** to display the Orientation dialog box.

100) Click **Top view**.

Offset the existing geometry from the boundary of the Sketch plane.

101) Click the **Offset Entities** Sketch tool. The Offset Entities PropertyManager is displayed.

102) Enter .150in, [3.81] for the Offset Distance.

103) If needed check the **Reverse** box. The new Offset yellow profile displays inside the original profile.

104) Click **OK** from the Offset Entities PropertyManager.

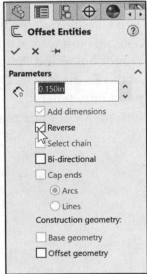

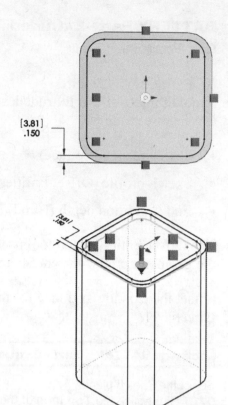

💡 A leading zero is displayed in the spin box. For inch dimensions less than 1, the leading zero is not displayed in the part dimension in the ANSI standard.

Display an Isometric view, with Hidden Lines Removed.

105) Press the **space bar** to display the Orientation dialog box.

106) Click **Isometric view** 📦.

107) Click **Hidden Lines Removed** from the Heads-up View toolbar.

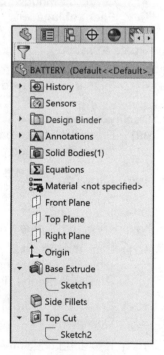

Insert an Extruded Cut feature. As an exercise, use the Instant3D tool to create the Extruded Cut feature. In this section, the Extruded-Cut PropertyManager is used. Note: With the Instant3D tool, you may lose the design intent of the model.

108) Click the **Extruded Cut** 📦 feature tool. The Cut-Extrude PropertyManager is displayed.

109) Enter **.200**in, **[5.08]** for Depth in Direction 1. Accept the default settings.

110) Click **OK** ✔ from the Cut-Extrude PropertyManager. Cut-Extrude1 is displayed in the FeatureManager.

Rename the feature.

111) Rename **Cut-Extrude1** to **Top Cut** in the FeatureManager.

Save the BATTERY

112) Click **Save** 💾.

The Cut-Extrude PropertyManager contains numerous options. The Reverse Direction option determines the direction of the Extrude. The Extruded Cut feature is valid only when the direction arrow points into material to be removed.

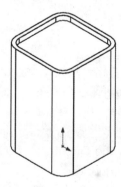

The Flip side to cut option
determines if the cut is to the
inside or outside of the Sketch
profile. The Flip side to cut
arrow points outward. The
Extruded Cut feature occurs on
the outside of the BATTERY.

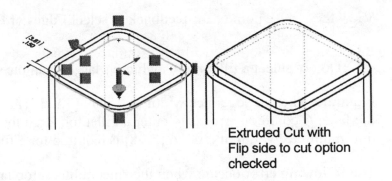

Extruded Cut with
Flip side to cut option
checked

BATTERY Part - Fillet Feature

The Fillet 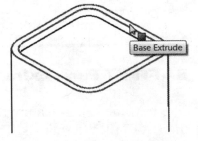 feature tool rounds sharp edges with a constant radius by selecting a face.
A Fillet requires a:

- Selected face

- Fillet radius

Activity: BATTERY Part - Fillet Feature Face

Insert a Constant Size Symmetric Fillet feature on the top face.

113) Click the **Fillet** feature tool. The Fillet PropertyManager
is displayed.

114) Click the **top thin face** as illustrated. Note the face icon
feedback symbol. Face<1> is displayed in the Items To
Fillet box.

115) Click the **Manual** tab.

116) Click the **Constant Size Fillet** button for Fillet Type.

117) Enter **.050**in, **[1.27]** for Radius. Symmetric is selected by
default.

118) Click **OK** from the Fillet PropertyManager. Fillet2 is
displayed in the FeatureManager.

Rename the feature.
119) Rename **Fillet2** to **Top Face Fillet**.

Fit the model to the Graphics window.
120) Press the **f** key.

Save the BATTERY.
121) Click **Save**.

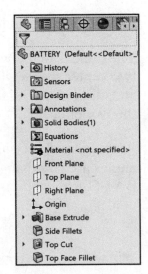

View the mouse pointer for feedback to select Edges or Faces for the fillet.

 Do not select a fillet radius which is larger than the surrounding geometry.

Example: The top edge face width is .150in, [3.81]. The fillet is created on both sides of the face. A common error is to enter a Fillet too large for the existing geometry. A minimum face width of .200in, [5.08] is required for a fillet radius of .100in, [2.54].

The following error occurs when the fillet radius is too large for the existing geometry.

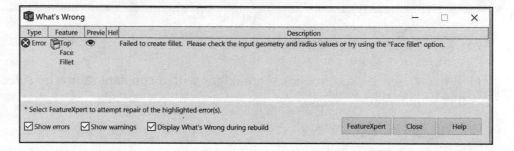

Avoid the fillet rebuild error. Use the FeatureXpert to address a constant radius fillet build error or manually enter a smaller fillet radius size.

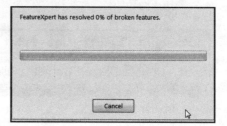

BATTERY Part - Extruded Boss/Base Feature

The Extruded Boss feature requires a truncated cone shape to represent the geometry of the BATTERY terminals. The Draft Angle option creates the tapered shape.

Sketch the first circle on the Top face. Utilize the Ctrl key to copy the first circle.

The dimension between the center points is critical. Dimension the distance between the two center points with an aligned dimension. The dimension text toggles between linear and aligned. An aligned dimension is created when the dimension is positioned between the two circles.

An angular dimension is required between the Right Plane and the centerline. Acute angles are less than 90°. Acute angles are the preferred dimension standard. The overall BATTERY height is a critical dimension. The BATTERY height is 4.500in, [114.3].

Calculate the depth of the extrusion: For inches: 4.500in - (4.100in Base-Extrude height - .200in Offset cut depth) = .600in. The depth of the extrusion is .600in.

For millimeters: 114.3mm - (104.14mm Base-Extrude height - 5.08mm Offset cut depth) = 15.24mm. The depth of the extrusion is 15.24mm.

Activity: BATTERY Part - Extruded Boss Feature

Select the Sketch plane.

122) Right-click the **Top face** of the Top Cut feature in the Graphics window. This is your Sketch plane.

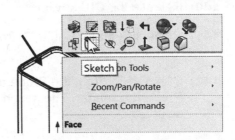

Create the sketch.

123) Click **Sketch** from the Context toolbar. The Sketch toolbar is displayed.

124) Click **Top view** .

Sketch the Close profile.

125) Click the **Circle** Sketch tool. The Circle PropertyManager is displayed.

126) Click the **center point** of the circle coincident to the Origin .

127) Drag and click the **mouse pointer** to the right of the Origin as illustrated.

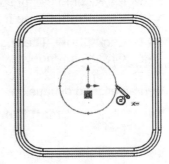

Add a dimension.

128) Click the **Smart Dimension** Sketch tool.

129) Click the **circumference** of the circle.

130) Click a **position** diagonally to the right.

131) Enter **.500**in, [12.7].

132) Click the **Green Check mark** in the Modify dialog box. The black sketch is fully defined.

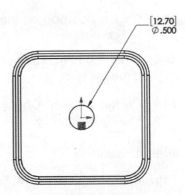

Copy the sketched circle.

133) Right-click **Select** to de-select the Smart Dimension Sketch tool.

134) Hold the **Ctrl** key down.

135) Click and drag the **circumference** of the circle to the upper left quadrant as illustrated.

136) Release the **mouse button**.

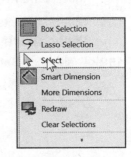

137) Release the **Ctrl** key. The second circle is selected and is displayed in blue. If needed click **OK** 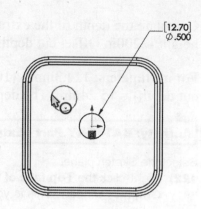 from the Circle PropertyManager.

Add an Equal relation.

138) Hold the **Ctrl** key down.

139) Click the **circumference of the first circle**. The Properties PropertyManager is displayed. Both circles are selected and are displayed in green.

140) Release the **Ctrl** key.

141) Right-click **Make Equal =** from the Context toolbar.

142) Click **OK** 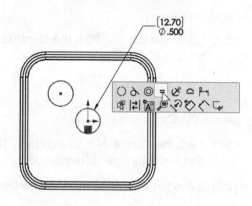 from the Properties PropertyManager. The second circle remains selected.

Show the Right Plane for the dimension reference.

143) Click **Right Plane** from the FeatureManager.

144) Click **Show**. The Right Plane is displayed in the Graphics window.

Add an aligned dimension.

145) Click the **Smart Dimension** Sketch tool.

146) Click the **two center points** of the two circles.

147) Click a **position** off the profile in the upper left corner.

148) Enter **1.000**in, **[25.4]** for the aligned dimension.

149) Click the **Green Check mark** in the Modify dialog box.

Insert a centerline.

150) Click the **Centerline** Sketch tool. The Insert Line PropertyManager is displayed.

151) Sketch a centerline between the **two circle center points** as illustrated.

152) Right-click **Select** to end the line.

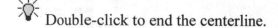 Double-click to end the centerline.

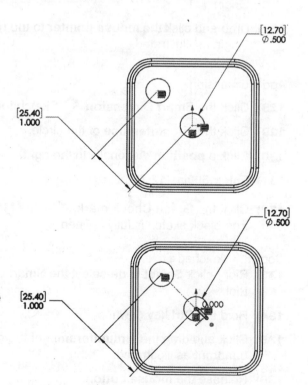

💡 Press the Enter key to accept the value in the Modify dialog box. The Enter key replaces the Green Check mark.

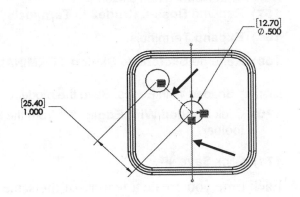

Add an angular dimension.

153) Click the **Smart Dimension** ✎ Sketch tool. Click the **centerline** between the two circles.

154) Click the **Right Plane** (vertical line) in the Graphics window. Note: You can also click Right Plane in the FeatureManager.

155) Click a **position** between the centerline and the Right Plane, off the profile.

156) Enter **45**.

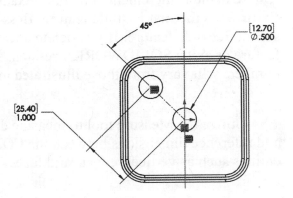

157) Click **OK** ✔ from the Dimension PropertyManager.

Fit the model to the Graphics window.
158) Press the **f** key.

Hide the Right Plane. Save the model.
159) Right-click **Right Plane** in the FeatureManager. Click **Hide** from the Context toolbar.

160) Click **Save** 💾.

💡 Create an angular dimension between three points or two lines. Sketch a centerline/construction line when an additional point or line is required.

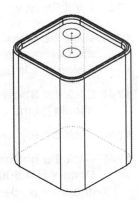

Display an Isometric view. Insert an Extruded Boss feature.

161) Click **Isometric view** 🔳.

162) Click the **Extruded Boss/Base** 🔲 feature tool. The Boss-Extrude PropertyManager is displayed. Blind is the default End Condition Type.

163) Enter **.600**in, **[15.24]** for Depth in Direction 1.

164) Click the **Draft ON/OFF** button.

165) Enter **5deg** in the Draft Angle box.

166) Click **OK** ✔ from the Boss-Extrude PropertyManager. The Boss-Extrude2 feature is displayed in the FeatureManager.

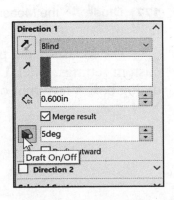

Rename the feature and sketch.

167) Rename **Boss-Extrude2** to **Terminals**.

168) Expand Terminals.

169) Rename **Sketch3** to **Sketch-TERMINALS**.

Display Shaded With Edges. Save the model.

170) Click **Shaded With Edges** 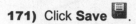 from the Heads-up View toolbar.

171) Click **Save** 💾.

Each time you create a feature of the same feature type, the feature name is incremented by one. Example: Boss-Extrude1 is the first Extrude feature. Boss-Extrude2 is the second Extrude feature. If you delete a feature, rename a feature or exit a SOLIDWORKS session, the feature numbers will vary from those illustrated in the text.

💡 Utilize the Measure tool to measure distances and add reference dimensions between the COM point and entities such as vertices, edges, and faces.

Measure the overall BATTERY height.

172) Click **Front view** 🗗.

173) Click the **Measure** 🔍 tool from the Evaluate tab in the CommandManager. The Measure - BATTERY dialog box is displayed.

174) Click the **Show XYZ Measurements** option. This should be the *only active* option.

175) Click the **top edge** of the battery terminal as illustrated.

176) Click the **bottom edge** of the battery. The overall height, Delta Y is 4.500, [114.3]. Apply the Measure tool to ensure a proper design.

177) Close ✕ the Measure - BATTERY dialog box.

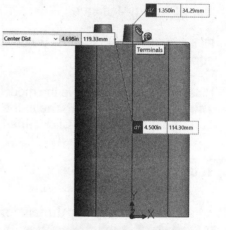

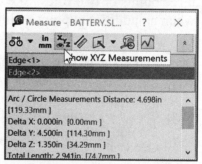

💡 The Measure tool provides the ability to display custom settings. Click **Units/Precision** from the Measure dialog box. View your options. Click **OK**.

The Selection Filter ⬚ option toggles the Selection Filter toolbar. When Selection Filters are activated, the mouse pointer displays the Filter icon ⬚ . The Clear All Filters ⬚ tool removes the current Selection Filters. The Help ⬚ icon displays the SOLIDWORKS Online Users Guide.

Display a Trimetric view.

178) Click **Trimetric view** ⬚ from the Heads-up View toolbar.

Save the BATTERY.

179) Click **Save** ⬚ .

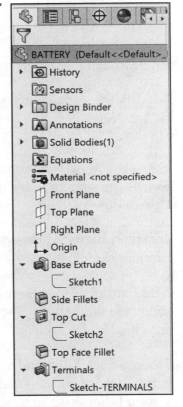

🔍 Additional information on Extruded Boss/Base Extruded Cut and Fillets is located in SOLIDWORKS Help Topics. Keywords: Extruded (Boss/Base, Cut), Fillet (Constant radius fillet), Geometric relations (sketch, equal, midpoint), Sketch (rectangle, circle), Offset Entities and Dimensions (angular).

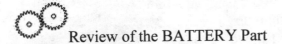 Review of the BATTERY Part

The BATTERY utilized a 2D Sketch profile located on the Top Plane. The 2D Sketch profile utilized the Center Rectangle Sketch tool. The Center Rectangle Sketch tool applied equal geometric relations to the two horizontal and two vertical lines. A midpoint relation was added to the Origin.

The Extruded Boss/Base feature was created using the Instant3D tool. Blind was the default End Condition. The Fillet feature rounded sharp edges. All four edges were selected to combine common geometry into the same Fillet feature. The Fillet feature also rounded the top face. The Sketch Offset Entity created the profile for the Extruded Cut feature.

The Terminals were created with an Extruded Boss feature. You sketched a circular profile and utilized the Ctrl key to copy the sketched geometry.

A centerline was required to locate the two holes with an angular dimension. The Draft Angle option tapered the Extruded Boss feature. All feature names were renamed.

BATTERYPLATE Part

The BATTERYPLATE is a critical plastic part. The BATTERYPLATE:

- Aligns the LENS assembly.

- Creates an electrical connection between the BATTERY and LENS.

Design the BATTERYPLATE. Utilize features from the BATTERY to develop the BATTERYPLATE. The BATTERYPLATE is manufactured as an injection molded plastic part. Build Draft into the Extruded Boss/Base features.

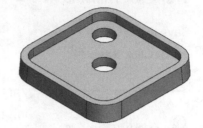

Edit the BATTERY features. Create two holes from the original sketched circles. Apply the Instant3D tool to create an Extruded Cut feature.

Modify the dimensions of the Base feature. Add a 3° draft angle.

A sand pail contains a draft angle. The draft angle assists the sand to leave the pail when the pail is flipped upside down.

Insert an Extruded Boss/Base feature. Offset the center circular sketch.

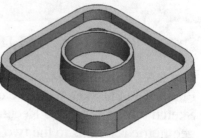

The Extruded Boss/Base feature contains the LENS. Create an inside draft angle. The draft angle assists the LENS into the Holder.

Insert a Face Fillet and a Multi-radius Edge Fillet to remove sharp edges. Plastic parts require smooth edges. Group Fillet features together into a folder.

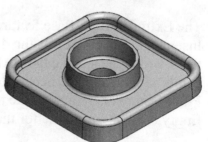

Group fillets together into a folder to locate them quickly. Features listed in the FeatureManager must be continuous in order to be placed as a group into a folder.

Tangent Edges and Origin are displayed for educational purposes.

Save As, Delete, Edit Feature and Modify

Create the BATTERYPLATE part from the BATTERY part.
Utilize the Save As tool from the Menu bar to copy the
BATTERY part to the BATTERYPLATE part.

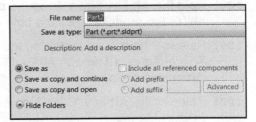

- Utilize the Save As/Save as command to save
  the file in another file format.

- Utilize the Save as copy and continue command
  to save the document to a new file name without
  replacing the active document.

- Utilize the Save as copy and open command to save the document to a new file name
  that becomes the active document. The original document remains open. References
  to the original document are not automatically assigned to the copy.

Reuse existing geometry. Create two holes. Delete the Terminals feature and reuse the
circle sketch. Select the sketch in the FeatureManager. Create an Extruded Cut feature
from the Sketch-TERMINALS using the Instant3D tool. Blind is the default End
Condition. Edit the Bass-Extrude feature. Modify the overall depth. Rebuild the model.

Activity: BATTERYPLATE Part - Save As, Delete, Modify and Edit Feature

Apply the Save As tool. Create and save a new part.

180) Click **Save As** from the drop-down Menu bar.

181) Select **PROJECTS** for Save In folder.

182) Enter **BATTERYPLATE** for File name.

183) Enter **BATTERY PLATE, FOR 6-VOLT** for
Description.

184) Click **Save** from the Save As dialog box. The
BATTERYPLATE FeatureManager is displayed. The
BATTERY part is closed.

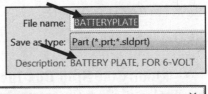

Delete the Terminals feature.

185) Right-click **Terminals** from the FeatureManager.

186) Click **Delete**.

187) Click **Yes** from the Confirm Delete dialog box. Do not
delete the two-circle sketch, Sketch-TERMINALS.

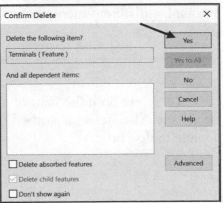

Create an Extruded Cut feature from the Sketch-TERMINALS using Instant3D.

188) Click **Sketch-TERMINALS** from the FeatureManager.

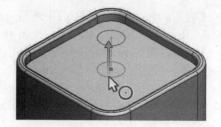

189) Click the **circumference** of the center circle as illustrated. A color arrow is displayed.

190) Hold the **Alt** key down. Drag the **color arrow** downward below the model to create a hole in Direction 1.

191) Release the mouse button on the **vertex** as illustrated. This ensures a Through All End Condition with model dimension changes.

192) Release the **Alt** key. Boss-Extrude1 is displayed in the FeatureManager.

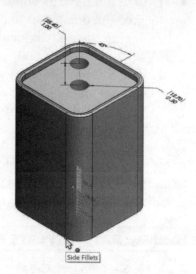

193) Rename the **Boss-Extrude1** feature to **Holes** in the FeatureManager.

Edit the Base Extrude feature.

194) Right-click **Base Extrude** from the FeatureManager.

195) Click **Edit Feature** 🔩 from the Context toolbar. The Base Extrude PropertyManager is displayed.

Modify the overall depth.

196) Enter **.400**in, [**10.16**] for Depth in Direction 1.

197) Click the **Draft ON/OFF** button.

198) Enter **3.00**deg in the Angle box.

199) Click **OK** ✔ from the Base Extrude PropertyManager.

Fit the model to the Graphics window.

200) Press the **f** key.

Save the BATTERYPLATE.

201) Click **Save** 💾.

💡 Modify the **Spin Box Increments** in System Options to display different increments for the Instant3D on-screen ruler.

💡 To delete both the feature and the sketch at the same time, select the Also delete absorbed features check box from the Confirm Delete dialog box.

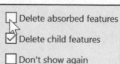

BATTERYPLATE Part - Extruded Boss Feature

The Holder is created with a circular Extruded Boss/Base feature. Utilize the Offset Entities ⊏ Sketch tool to create the second circle. Apply a draft angle of 3° in the Extruded Boss feature.

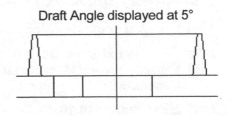

Draft Angle displayed at 5°

When applying the draft angle to the two concentric circles, the outside face tapers inwards and the inside face tapers outwards.

🔆 Plastic parts require a draft angle. Rule of thumb: 1° to 5° is the draft angle. The draft angle is created in the direction of pull from the mold. This is defined by geometry, material selection, mold production and cosmetics. Always verify the draft with the mold designer and manufacturer.

Activity BATTERYPLATE Part - Extruded Boss Feature

Select the Sketch plane.
202) Right-click the **top face** of Top Cut. This is your Sketch plane.

Create the sketch.

203) Click **Sketch** ⌖ from the Context toolbar.

204) Click the **top circular edge** of the center hole. Note: Use the keyboard arrow keys or the middle mouse button to rotate the sketch if needed.

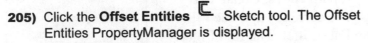

205) Click the **Offset Entities** ⊏ Sketch tool. The Offset Entities PropertyManager is displayed.

206) Enter .300in, [7.62] for Offset Distance. Accept the default settings.

207) Click **OK** ✔ from the Offset Entities PropertyManager.

208) Drag the **dimension** off the model.

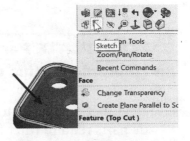

Create the second offset circle.
209) Click the **offset circle** in the Graphics window.

210) Click the **Offset Entities** ⊏ Sketch tool. The Offset Entities PropertyManager is displayed.

211) Enter **.100**in, **[2.54]** for Offset Distance.

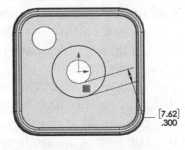

212) Click **OK** ✔ from the Offset Entities PropertyManager. Drag the dimension off the model. Two offset concentric circles define the sketch.

Insert an Extruded Boss/Base feature.

213) Click the **Extruded Boss/Base** 🗐 feature tool. The Boss-Extrude PropertyManager is displayed. Blind is the default End Condition.

214) Enter **.400**in, **[10.16]** for Depth in Direction 1.

215) Click the **Draft ON/OFF** button. Enter **3**deg in the Angle box.

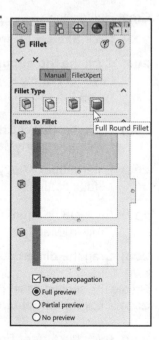

216) Click **OK** ✔ from the Boss-Extrude PropertyManager. The Boss-Extrude1 feature is displayed in the FeatureManager.

Rename the feature.

217) Rename the **Boss-Extrude1** feature to **Holder** in the FeatureManager.

Save the model.

218) Click **Save** 💾.

BATTERYPLATE Part - Fillet Features: Full Round and Multiple Radius Options

Use the Fillet 🗐 feature tool to smooth rough edges in a model. Plastic parts require fillet features on sharp edges. Create two Fillets. Utilize different techniques. The current Top Face Fillet produced a flat face. Delete the Top Face Fillet. The first Fillet feature is a Full Round fillet. Insert a Full Round fillet feature on the top face for a smooth rounded transition.

The second Fillet feature is a Multiple radius fillet. Select a different radius value for each edge in the set. Select the inside and outside edge of the Holder. Select all inside tangent edges of the Top Cut. A Multiple radius fillet is utilized next as an exercise. There are machining instances where the radius must be reduced or enlarged to accommodate tooling. Note: There are other ways to create Fillets.

🔆 Group Fillet features into a Fillet folder. Placing Fillet features into a folder reduces the time spent for your mold designer or toolmaker to look for each Fillet feature in the FeatureManager.

Activity: BATTERYPLATE Part - Fillet Features: Full Round, Multiple Radius Options

---

Delete the Top Edge Fillet.

219) Right-click **Top Face Fillet** from the FeatureManager.

220) Click **Delete**.

221) Click **Yes** to confirm delete.

222) Drag the **Rollback** bar below Top Cut in the FeatureManager.

Create a Full Round Fillet feature.

223) Click **Hidden Lines Visible** from the Heads-up View toolbar.

224) Click the **Fillet** feature tool. The Fillet PropertyManager is displayed.

225) Click the **Manual** tab.

226) Click the **Full Round Fillet button** for Fillet Type.

227) Click the **inside Top Cut face** for Side Face Set 1 as illustrated.

228) Click **inside** the Center Face Set box.

229) Click the **top face** for Center Face Set as illustrated.

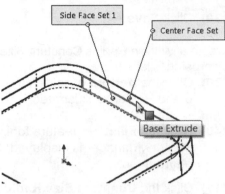

Rotate the part.

230) Press the **Left Arrow** key until you can select the outside Base Extrude face.

231) Click **inside** the Side Face Set 2 box.

232) Click the **outside Base Extrude face** for Side Face Set 2 as illustrated. Accept the default settings.

233) Click **OK** from the Fillet PropertyManager. Fillet1 is displayed in the FeatureManager.

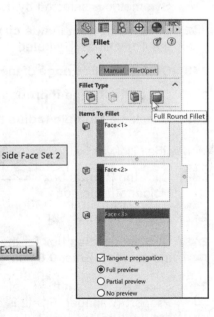

Rename Fillet1.

234) Rename **Fillet1** to **TopFillet**.

Display an Isometric view with Hidden Lines Removed. Save the
BATTERYPLATE.

235) Click **Isometric view** .

236) Click **Hidden Lines Removed** from
the Heads-up View toolbar.

237) Drag the **Rollback bar** to the bottom of
the FeatureManager.

238) Click **Save** .

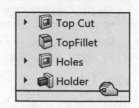

Create a Multiple Radius Constant Size
Symmetric Fillet feature.

239) Click the **bottom outside circular edge**
of the Holder as illustrated.

240) Click the **Fillet** feature tool. The Fillet
PropertyManager is displayed. Click the
Manual tab.

241) Click the **Constant Size Radius** Fillet
button. Enter **.050**in, **[1.27]** for Radius.
Symmetric is selected by default.

242) Click the **bottom inside circular edge** of
the Top Cut as illustrated.

243) Click the **inside edge** of the Top Cut.

244) Check the **Tangent propagation** box.

245) Check the **Multiple radius fillet** box.

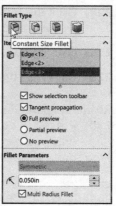

Modify the Fillet values.

246) Click the **Radius** box for the
Holder outside edge.

247) Enter **0.060**in, **[1.52]**.

248) Click the **Radius** box for the Top Cut
inside edge. Enter **0.040**in, **[1.02]**.

249) Click **OK** from the Fillet
PropertyManager. Fillet2 is displayed in
the FeatureManager.

Rename the Fillet2 folder.
250) Rename **Fillet2** to **HolderFillet**.

Display Shaded With Edges.

251) Click **Shaded With Edges** from the
Heads-up View toolbar. View the results
in the Graphics window.

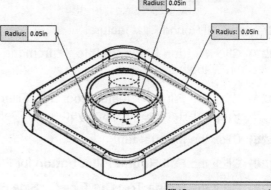

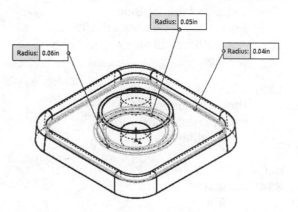

Group the Fillet features into a new folder.

252) Click **TopFillet** from the FeatureManager.

253) Drag the **TopFillet** feature directly above the HolderFillet feature in the FeatureManager.

254) Click **HolderFillet** in the FeatureManager.

255) Hold the **Ctrl** key down.

256) Click **TopFillet** in the FeatureManager.

257) Right-click **Add to New Folder**.

258) Release the **Ctrl** key.

Rename Folder1.

259) Rename **Folder1** to **FilletFolder**.

Save the BATTERYPLATE.

260) Click **Save** .

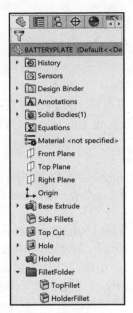

Multi-body Parts and Extruded Boss/Base Feature

A Multi-body part has separate solid bodies within the same part document.

A WRENCH consists of two cylindrical bodies. Each extrusion is a separate body. The oval profile is sketched on the right plane and extruded with the Up to Body option.

The BATTERY consisted of a solid body with one sketched profile. The BATTERY is a single body part.

Additional information on Save, Extruded Boss/Base, Extruded Cut, Fillets, Copy Sketched Geometry, and Multi-body is located in SOLIDWORKS Help.

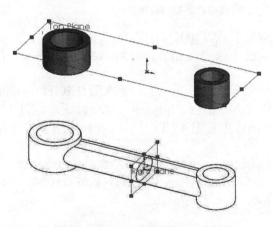

Multi-body part Wrench

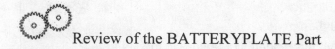

Review of the BATTERYPLATE Part

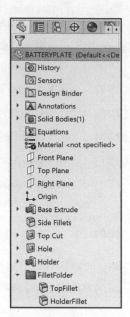

The Save As option was utilized to copy the BATTERY part to the BATTERYPLATE part. You created a hole in the BATTERYPLATE using the Instant3D tool and modified features using the PropertyManager.

The BATTERYPLATE is a plastic part. The Draft Angle option was added in the Base Extrude feature.

The Holder Extruded Boss feature utilized a circular sketch and the Draft Angle option. The Sketch Offset tool created the circular ring profile. Multi radius Edge Fillets and Face Fillets removed sharp edges. Similar Fillet features were grouped together into a folder. Features were renamed in the FeatureManager. The BATTERY and BATTERYPLATE utilized an Extruded Boss/Base feature.

Design Intent is how your part reacts as parameters are modified. Example: If you have a hole in a part that must always be .125≤ from an edge, you would dimension to the edge rather than to another point on the sketch. As the part size is modified, the hole location remains .125≤ from the edge.

Chapter Summary

SOLIDWORKS is a design software application used to model and create 2D and 3D sketches, 3D parts, 3D assemblies and 2D drawings.

You are designing a FLASHLIGHT assembly that is cost effective, serviceable and flexible for future design revisions. The FLASHLIGHT assembly consists of various parts. The BATTERY and BATTERYPLATE parts were modeled in this chapter.

Folders organized your models and templates. The Part Template is the foundation for all parts in the FLASHLIGHT assembly. You created the *PART-IN-ANSI* and *PART-MM-ISO* Part Template.

If you modify a document property from an Overall drafting standard, a modify message is displayed as illustrated.

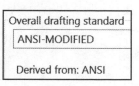

This chapter concentrated on the Extruded Boss/Base feature. The Extruded Boss/Base feature required a Sketch plane, Sketch profile and End Condition (Depth).

The BATTERY and BATTERYPLATE parts incorporated an Extruded Boss/Base feature.

You also addressed the Extruded Cut, Fillet and Instant3D features. You addressed the following Sketch tools: *Smart Dimension, Line, Centerline, Center Rectangle, Circle, Convert Entities, Offset Entities* and *Mirror Entities*.

You created 2D sketches and addressed the three key states of a sketch: *Fully Defined, Over Defined* and *Under Defined*. Note: Always review your FeatureManager for the proper sketch state.

You addressed additional tools that utilized existing geometry: Add Relations, copy, Save As, Edit feature, and more.

Geometric relations were utilized to build symmetry into the sketches. Practice these concepts with the chapter exercises.

Click the 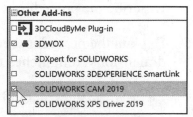 direction arrows in the FeatureManager to expand or collapse the FeatureManager design tree. Your tabs will vary depending on your SOLIDWORKS applications and Add-ins.

For many features (Extruded Boss/Base, Extruded Cut, Simple Hole, Revolved Boss/Base, Revolved Cut, Fillet, Chamfer, Scale, Shell, Rib, Circular Pattern, Linear Pattern, Curve Driven Pattern, Revolved Surface, Extruded Surface, Fillet Surface, Edge Flange and Base Flange), you can enter and modify equations directly in the PropertyManager fields that allow numerical inputs.

Create Equations with Global Variables, functions and file properties without accessing the Equations, Global Variables and Dimensions dialog box.

You can enter equations in:

- Depth fields for Direction 1 and Direction 2

- Draft fields for Direction 1 and Direction 2

- Thickness fields for a Thin Feature with two direction types

- Offset Distance field

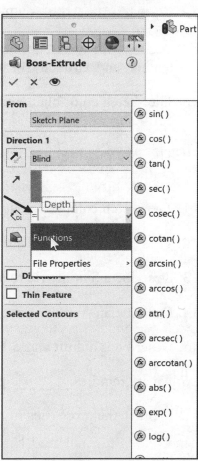

Questions

1. Identify and describe the function of the following features:

    - Extruded Boss/Base

    - Fillet

    - Chamfer

    - Extruded Cut

2. Explain the differences between a Template and a Part document.

3. Explain the steps in starting a SOLIDWORKS session.

4. Describe the procedure to develop a new 2D sketch.

5. Explain the procedure required to change part unit dimensions from inches to millimeters.

6. Identify the three default Reference planes.

7. What is a Base feature? Provide an example.

8. Describe the differences between an Extruded Boss/Base feature, an Extruded Cut feature and a Fillet feature.

9. The sketch color black indicates a sketch is _____ defined.

10. The sketch color blue indicates a sketch is _____ defined.

11. The sketch color red/yellow indicates a sketch is _____ defined.

12. True or False. Folders are utilized to only store part documents.

13. Describe a Symmetric relation.

14. Describe an Angular dimension.

15. Describe is a draft angle. Provide an example.

16. An arc requires _____ points.

17. Identify the properties of a Multi-body part.

18. Identify the name of the following Feature tool icons.

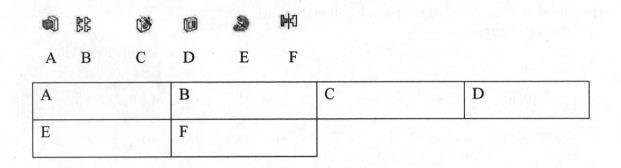

A   B   C   D   E   F

A		B		C		D	
E		F					

19. Identify the name of the following Sketch tool icons.

A   B   C   D   E   F   G   H   I   J

A		B		C		D	
E		F		G		H	
I		J					

Exercises

Exercise 5.1: Identify the Sketch plane for the Boss-Extrude1 feature. View the Origin location. Simplify the number of features.

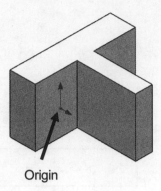

Origin

A: Top Plane

B: Front Plane

C: Right Plane

D: Left Plane

Correct answer _____.

Exercise 5.2: Identify the Sketch plane for the Boss-Extrude1 feature. View the Origin location. Simplify the number of features.

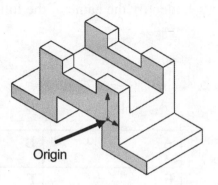

Origin

A: Top Plane

B: Front Plane

C: Right Plane

D: Left Plane

Correct answer _____.

Exercise 5.3: Identify the Sketch plane for the Boss-Extrude1 feature. View the Origin location. Simplify the number of features.

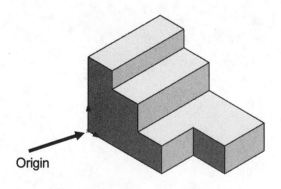

Origin

A: Top Plane

B: Front Plane

C: Right Plane

D: Left Plane

Correct answer _____.

Exercise 5.4: Identify the Sketch plane for the Boss-Extrude1 feature. View the Origin location. Simplify the number of features.

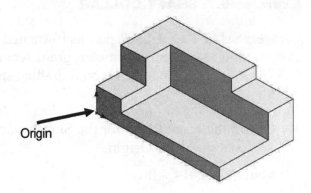

A: Top Plane

B: Front Plane

C: Right Plane

D: Left Plane

Correct answer _____.

Exercise 5.5: Identify the material category for 6061 Alloy.

A: Steel

B: Iron

C: Aluminum Alloys

D: Other Alloys

E: None of the provided

Correct answer _____.

Exercise 5.6: AXLE

Create an AXLE part as illustrated with dual units. Overall drafting standard - ANSI. IPS is the primary unit system.

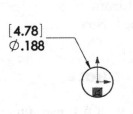

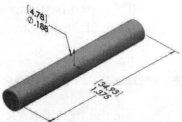

- Utilize the Front Plane for the Sketch plane.

- Utilize the Mid Plane End Condition. The AXLE is symmetric about the Front Plane. Note the location of the Origin.

- Material - 1060 Alloy.

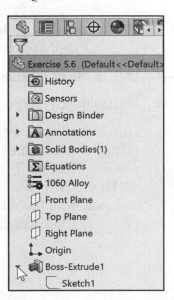

Exercise 5.7: SHAFT COLLAR

Create a SHAFT COLLAR part as illustrated with dual system units: MMGS (millimeter, gram, second) and IPS (inch, pound, second). Overall drafting standard - ANSI.

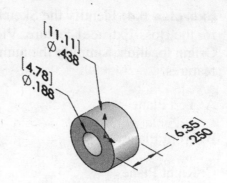

- Utilize the Front Plane for the Sketch plane. Note the location of the Origin.

- Material - 2014 Alloy.

Exercise 5.8: FLAT BAR - 3 HOLE

Create the FLAT BAR - 3 HOLE part as illustrated with dual system units: MMGS (millimeter, gram, second) and IPS (inch, pound, second). Overall drafting standard - ANSI.

- Utilize the Front Plane for the Sketch plane.

- Utilize the Centerpoint Straight Slot Sketch tool.

- Utilize a Linear Pattern feature for the three holes. The FLAT BAR - 3 HOLE part is stamped from 0.060in, [1.5mm] Stainless Steel (ferritic).

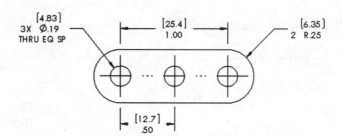

Exercise 5.9: FLAT BAR - 9 HOLE

Create the the FLAT BAR - 9 HOLE part. Overall drafting standard - ANSI.

- The dimensions for hole spacing, height and end arcs are the same as the FLAT BAR - 3 HOLE part.

- Utilize the Front Plane for the Sketch plane.

- Utilize the Centerpoint Straight Slot Sketch tool.

- Utilize the Linear Pattern feature to create the hole pattern. The FLAT BAR - 9 HOLE part is stamped from 0.060in, [1.5mm] 1060 Alloy.

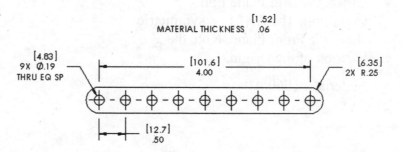

Exercise 5.10

Create the illustrated part. Note the location of the Origin. Overall drafting standard - ANSI.

- Calculate the overall mass of the illustrated model. Apply the Mass Properties tool.

- Think about the steps that you would take to build the model.

- Review the provided information carefully.

- Units are represented in the IPS (inch, pound, second) system.

- A = 3.50in, B = .70in

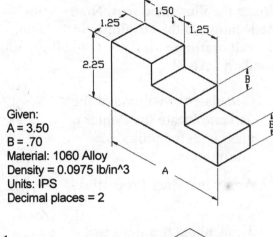

Given:
A = 3.50
B = .70
Material: 1060 Alloy
Density = 0.0975 lb/in^3
Units: IPS
Decimal places = 2

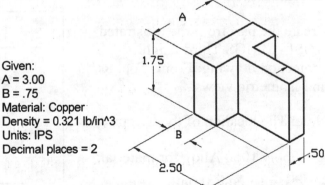

Origin

Exercise 5.11

Create the illustrated part. Note the location of the Origin. Overall drafting standard - ANSI.

- Calculate the overall mass of the illustrated model. Apply the Mass Properties tool.

- Think about the steps that you would take to build the model.

- Review the provided information carefully.

- Units are represented in the IPS (inch, pound, second) system.

- A = 3.00in, B = .75in

Given:
A = 3.00
B = .75
Material: Copper
Density = 0.321 lb/in^3
Units: IPS
Decimal places = 2

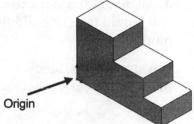

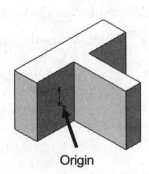

Origin

Exercise 5.12

Create the illustrated part. Note the location of the Origin. Overall drafting standard - ANSI.

- Calculate the volume of the part and locate the Center of mass with the provided information.

- Apply the Mass Properties tool.

- Think about the steps that you would take to build the model.

- Review the provided information carefully.

Given:
A = 3.30
B = 2.00
Material: 2014 Alloy
Density = .101 lb/in^3
Units: IPS
Decimal places = 2

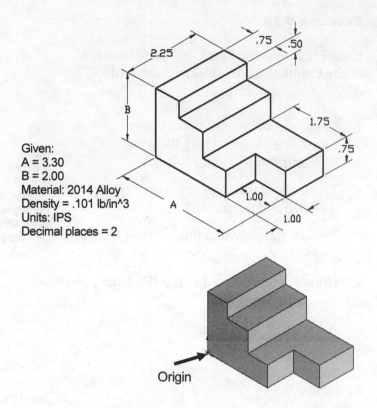

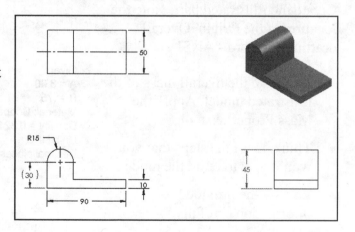

Origin

Exercise 5.13

Create the part from the illustrated ANSI - MMGS Third Angle Projection drawing: Front, Top, Right and Isometric views.

Note: The location of the Origin.

- Apply 1060 Alloy for material.

- Calculate the Volume of the part and locate the Center of mass.

- Think about the steps that you would take to build the model. The part is symmetric about the Front Plane.

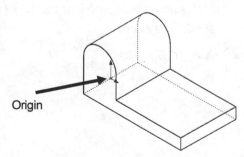

Origin

Exercise 5.14

Create the part from the illustrated ANSI - MMGS Third Angle Projection drawing: Front, Top, Right and Isometric views.

Note: The location of the Origin.

- Apply the Hole Wizard feature.

- Apply 1060 Alloy for material.

- The part is symmetric about the Front Plane.

- Calculate the Volume of the part and locate the Center of mass.

- Think about the steps that you would take to build the model.

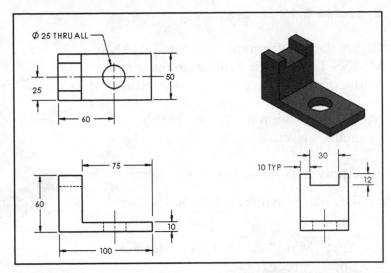

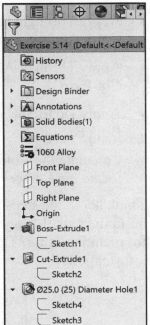

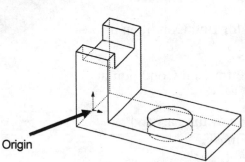

Origin

Exercise 5.15

Create the part from the illustrated A-ANSI - MMGS Third Angle Projection drawing below: Front, Top, Right and Isometric view.

Note: The location of the Origin (Shown in an Isometric view).

- Apply Cast Alloy steel for material.

- The part is symmetric about the Front Plane.

- Apply Mid Plane for End Condition in Boss-Extrude1.

- Apply Through All for End Condition in Cut-Extrude1.

- Apply Through All for End Condition in Cut-Extrude2.

- Apply Up to Surface for End Condition in Boss Extrude2

- Calculate the Volume of the part and locate the Center of mass.

Think about the steps that you would take to build the model. Do you need the Right view for manufacturing? Does it add any important information?

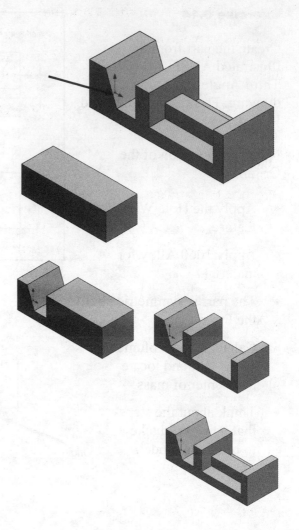

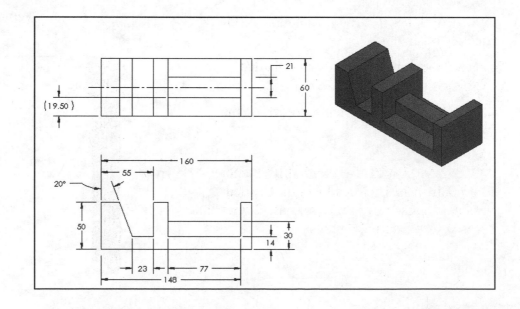

Exercise 5.16: Cosmetic Threads

Apply a Cosmetic thread: ¼-20-1 UNC 2A. A cosmetic thread represents the inner diameter of a thread on a boss or the outer diameter of a thread.

Copy and open the Cosmetic thread part from the Chapter 5 Homework folder.

Create a Cosmetic thread.

Click the bottom edge of the part as illustrated.

Click Insert, Annotations, Cosmetic Thread from the Menu bar menu. View the Cosmetic Thread PropertyManager. Edge<1> is displayed.

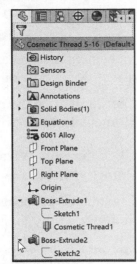

Select ANSI Inch.

Select Machine Threads.

Select ¼-20 for Size.

Blind for End Condition.

Enter 1.00 for depth.

Select 2A for Thread class.

Click OK ✔ from the Cosmetic Thread FeatureManager.

Expand the FeatureManager.

View the Cosmetic Thread feature.

If needed, right-click the Annotations folder, click Details.

Check the Cosmetic threads and Shaded cosmetic threads box.

Click OK. View the cosmetic thread on the model.

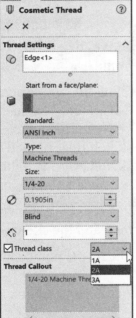

🔆 ¼-20-1 UNC 2A - ¼ inch drill diameter - 20 threads/inch, 1in thread length, Unified National Coarse thread series, Class 2 (General Thread), A - External threads.

Exercise 5.17: 3D SKETCH/HOLE WIZARD

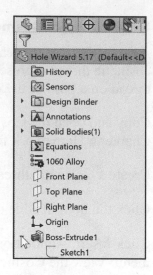

Create a part using the Hole Wizard feature. Apply the 3D sketch placement method as illustrated in the FeatureManager. Insert and dimension a hole on a cylindrical face.

Copy and open the Hole Wizard 5.17 model from the Chapter 5 Homework folder.

Click the Hole Wizard 🔘 Features tool. The Hole Specification PropertyManager is displayed.

Create a **Counterbore, ANSI Inch, Socket Head Cap Screw fastener Type**. Size - ¼. Fit - **Normal**. End Condition - **Through All** with a **.100in** Head clearance.

Click the **Positions** Tab.

Click the **3D Sketch** button. The Hole Position PropertyManager is displayed. SOLIDWORKS displays an active 3D interface with the

Point ⁺XY⤝ tool. When the Point tool is active, wherever you click, you will create a point.

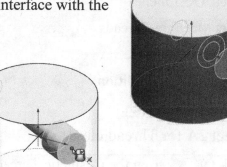

Click the cylindrical face of the model as illustrated. The selected face is displayed in orange. This indicates that an OnSurface sketch relation will be created between the sketch point and the cylindrical face. The hole is displayed in the model.

Insert a **.25in** dimension between the top face and the sketch point.

Locate the point angularly around the cylinder.

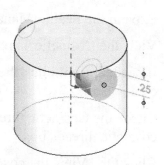

Apply construction geometry. Display the **Temporary Axes**.

Click the **Line** ✏ Sketch tool. Note: 3D Sketch is still activated.

Ctrl + Click the top flat face of the model. This moves the red space handle origin to the selected face. This also constrains any new sketch entities to the top flat face. Note

the mouse pointer ⤓▪⤝ icon.

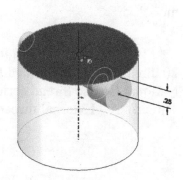

Move the **mouse pointer** near the center of the activated top flat face as illustrated. View the small black circle. The circle indicates that the end point of the line will pick up a Coincident relation.

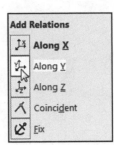
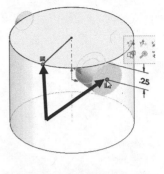

Click the **center point** of the circle.

Sketch a line so it picks up the **Along Z sketch relation**. The cursor displays the relation to be applied. **This is a very important step.** Create an **AlongZ sketch relation** if needed.

Deselect the Line Sketch tool.
Right-click **Select**.

Create an **Along Y** sketch relation between the center point of the hole on the cylindrical face and the endpoint of the sketched line. The sketch is fully defined.

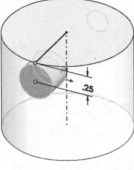
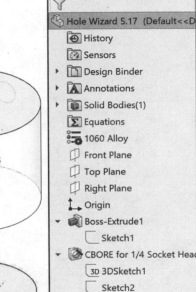

Click **OK** from the PropertyManager.

Click **OK** to return to the Part FeatureManager.

Expand the FeatureManager and view the results. The two sketches are fully defined. One sketch is the hole profile, the other sketch is to define the position of the feature.

Close the model.

When creating a Hole Wizard hole, you no longer have to preselect a planar face to create a 2D sketch. For a 3D sketch, request a 3D sketch as illustrated.

Slots have been added to the Hole Wizard. You can create regular slots as well and counterbore and countersink slots. You also have options for position and orientation of the slot. If you have hardware already mated in place, the mates will not be broken if you switch from a hole to a slot.

Notes:

Chapter 6

Revolved Boss/Base Features

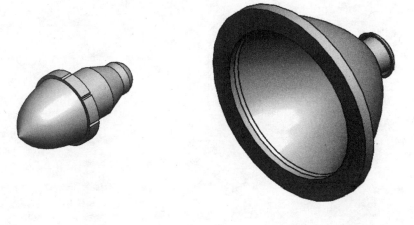

Below are the desired outcomes and usage competencies based on the completion of Chapter 6.

Desired Outcomes:	Usage Competencies:

• Two FLASHLIGHT parts:
  ○ LENS
  ○ BULB
	• Specific knowledge and understanding of the following Features: Extruded Boss/Base, Extruded Cut, Revolved Base, Revolved Boss Thin, Revolved Thin Cut, Dome, Shell, Hole Wizard, and Circular Pattern.

• Insert the following Geometric relations: Equal, Coincident, Symmetric, Intersection and Perpendicular.	• Ability to insert multiple Geometric relations to a model. • Ability to apply Design Intent in Sketches, Features, Parts and Assemblies.

Notes:

Chapter 6 - Revolved Boss/Base Features

Chapter Overview

Provide a comprehensive understanding of
the Revolved Boss/Base feature. Create two
parts for the FLASHLIGHT assembly:

* LENS and BULB.

A Revolved Boss/Base feature requires a 2D
sketch profile and a centerline. Utilize sketch
geometry and sketch tools to create the
following features:

* Revolved Base.

* Revolved Boss.

* Revolved Boss Thin.

* Revolved Cut.

Utilize existing faces to create the following features:

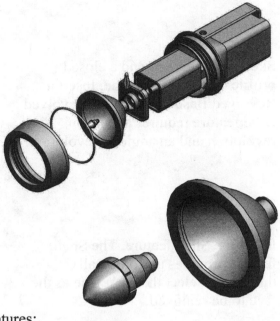

BULB　　　　　LENS

* Shell.

* Dome.

* Hole Wizard.

Utilize the Extruded Cut feature (seed feature) to create the Circular Pattern feature. After
completing the activities in this chapter, you will be able to:

* Utilize the following Sketch tools: Circle, Line, 3 Point Arc, Centerpoint Arc, Spline,
  Mirror, Offset Entities, Trim and Convert Entities.

* Insert the following Geometric relations: Equal, Coincident, Symmetric, Intersection
  and Perpendicular.

* Apply Transparent Optical Properties to the LENS.

* Create and edit the following features: Extruded Boss/Base, Extruded Cut, Revolved
  Base, Revolved Boss, Revolved Boss Thin, Revolved Thin Cut, Shell, Hole Wizard,
  Dome and Circular Pattern.

* Create two parts for the FLASHLIGHT assembly: LENS and BULB.

LENS Part

Create the LENS. The LENS is a purchased part.

The LENS utilizes a Revolved Base feature.

Sketch a centerline and a closed profile on the Right Plane. Insert a Revolved Base feature. The Revolved Base feature requires an axis of revolution and an angle of revolution.

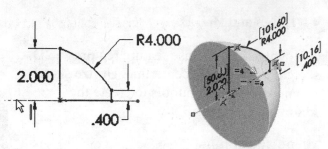

Insert the Shell feature. The Shell feature provides uniform wall thickness. Select the front face as the face to be removed.

Utilize the Convert Entities sketch tool to extract the back circular edge for the sketched profile. Insert an Extruded Boss feature from the back of the LENS.

Sketch a single profile. Insert a Revolved Thin feature to connect the LENS to the BATTERYPLATE. The Revolved Thin feature requires a thickness.

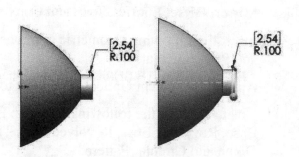

Insert a Counterbore Hole feature using the Hole Wizard feature.

The BULB is located inside the Counterbore Hole.

Insert the front Lens Cover with an Extruded Boss/Base feature. The sketch profile for the Extruded Boss/Base is sketched on the Front Plane. Add a transparent Lens Shield with the Extruded Boss/Base feature.

LENS Part-Revolved Boss/Base Feature

Create the LENS with a Revolved Base feature. The solid Revolved Base feature requires:

- Sketch plane (Right).
- Sketch profile.
- Centerline.
- Angle of Revolution (360°).

The profile lines reference the Top and Front Planes. Create the curve of the LENS with a 3-point arc.

Activity: Create the LENS Part

Create the new part.

1) Click **New** ⬜ from the Menu bar.

2) Click the **MY-TEMPLATES** tab.

3) Double-click **PART-IN-ANSI, [PART-MM-ISO]**.

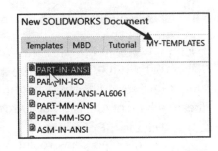

Save the part. Enter name. Enter description.

4) Click **Save** 💾.

5) Select **PROJECTS** for Save in folder.

6) Enter **LENS** for File name.

7) Enter **LENS WITH SHIELD** for Description.

8) Click **Save**. The LENS FeatureManager is displayed.

View the Front Plane.

9)   Right click **Front Plane** from the FeatureManager.

10)   Click **Show** 👁 from the Context toolbar. Hide unwanted planes in the FeatureManager if needed.

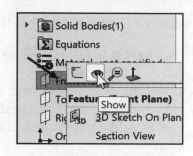

Create a sketch.

11)   Right-click **Right Plane** from the FeatureManager.

12)   Click **Sketch** ⬚ from the Context toolbar. The Sketch toolbar is displayed.

13)   Click the **Centerline** 〃 Sketch tool. The Insert Line PropertyManager is displayed.

14)   Sketch a **horizontal centerline** collinear to the Top Plane, through the Origin ↳ as illustrated.

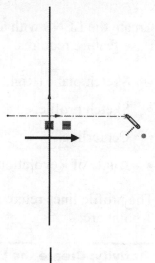

Sketch the profile. Create three lines.

15)   Click the **Line** ✏ Sketch tool. The Insert Line PropertyManager is displayed.

16)   Sketch a **vertical line** collinear to the Front Plane coincident with the Origin.

17)   Sketch a **horizontal line** coincident with the Top Plane.

18)   Sketch a **vertical line** approximately 1/3 the length of the first line.

19)   Right-click **End Chain**.

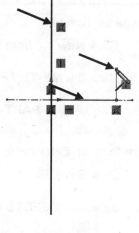

Create a 3 Point Arc. A 3 Point Arc requires three points.

20)   Click the **3 Point Arc** ⌒ Sketch tool from the Consolidated Centerpoint Arc toolbar.

21)   Click the **top point** on the left vertical line. This is your first point. Drag the **mouse pointer** to the right.

22)   Click the **top point** on the right vertical line. This is your second point.

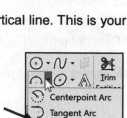

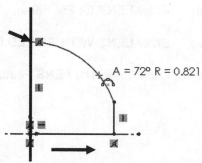

A = 72° R = 0.821

23)　Drag the **mouse pointer** upward.

24)　Click a **position** on the arc.

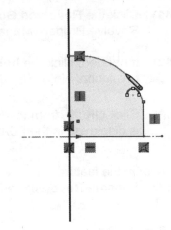

Add an Equal relation.

25)　Right-click **Select** to deselect the sketch tool.

26)　Click the **left vertical** line.

27)　Hold the **Ctrl** key down.

28)　Click the **horizontal** line. The Properties PropertyManager is displayed. The selected entities are displayed in the Selected Entities box.

29)　Release the **Ctrl** key.

30)　Right-click **Make Equal** from the Context toolbar.

31)　Click **OK** ✔ from the Properties PropertyManager.

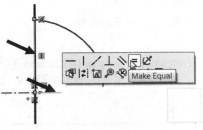

Add dimensions.

32)　Click the **Smart Dimension** ✎ Sketch tool.

33)　Click the **left vertical** line.

34)　Click a **position** to the left of the profile.

35)　Enter **2.000**in, [**50.8**].

36)　Click the **right vertical** line.

37)　Click a **position** to the right of the profile.

38)　Enter **.400**in, [**10.16**]. Click the **arc**.

39)　Click a **position** to the right of the profile.

40)　Enter **4.000**in, [**101.6**]. The black sketch is fully defined.

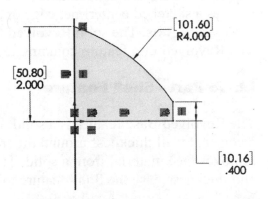

💡 Utilize **Tools, Sketch Tools, Check Sketch for Feature** option to determine if a sketch is valid for a specific feature and to understand what is wrong with a sketch.

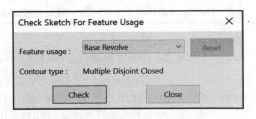

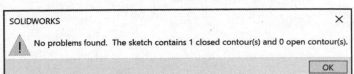

Insert the Revolved Base feature.

41) Click the **Revolved Boss/Base** 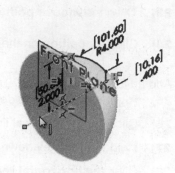 feature tool. The Revolve PropertyManager is displayed.

42) If needed, click the **horizontal centerline** for the axis of revolution. Note: The direction arrow points clockwise.

43) Click **OK** ✔ from the Revolve PropertyManager. Revolve1 is displayed in the FeatureManager.

Rename the feature.
44) Rename **Revolve1** to **BaseRevolve**.

Save the model.
45) Click **Save** 💾.

Display the axis of revolution.
46) Click **View**, **Hide/Show**, check **Temporary Axes** from the Menu bar.

Revolve features contain an axis of revolution. The axis of revolution utilizes a sketched centerline, edge or an existing feature/sketch or a Temporary Axis. The solid Revolved feature contains a closed profile. The Revolved thin feature contains an open or closed profile.

LENS Part - Shell Feature

The Revolved Base feature is a solid. Utilize the Shell feature to create a constant wall thickness around the front face. The Shell feature removes face material from a solid. The Shell feature requires a face and thickness. Use the Shell feature to create thin-walled parts.

Activity: LENS Part - Shell Feature

Insert the Shell feature.
47) Click the **front face** of the BaseRevolve feature.

48) Click the **Shell** 🔲 feature tool. The Shell1 PropertyManager is displayed. Face<1> is displayed in the Faces to Remove box.

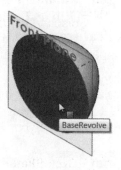

49) Enter .250in, [6.35] for Thickness.

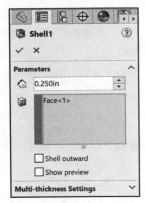

Display the Shell feature.

50) Click **OK** ✔ from the Shell1 PropertyManager. Shell1 is displayed in the FeatureManager.

51) Right-click **Front Plane** from the FeatureManager. Click **Hide** from the Context toolbar.

Rename the feature.
52) Rename **Shell1** to **LensShell**.

Click **Save** 💾.

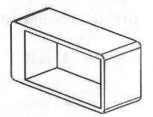

To insert rounded corners inside a shelled part, apply the Fillet feature before the Shell feature. Select the Multi-thickness option to apply different thicknesses.

Extruded Boss/Base Feature and Convert Entities Sketch tool

Create the LensNeck. The LensNeck houses the BULB base and is connected to the BATTERYPLATE. Use the Extruded Boss/Base feature. The back face of the Revolved Base feature is the Sketch plane.

Utilize the Convert Entities Sketch tool to extract the back circular face to the Sketch plane. The new curve develops an On Edge relation. Modify the back face, and the extracted curve updates to reflect the change. No sketch dimensions are required.

Activity: Extruded Boss Feature and Convert Entities Sketch tool

Rotate the LENS.
53) **Rotate** the LENS with the middle mouse button to display the back face as illustrated. The Rotate ↻ icon is displayed.

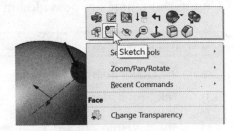

Sketch the profile.
54) Right-click the **back face** for the Sketch plane. BaseRevolve is highlighted in the FeatureManager. This is your Sketch plane.

55) Click **Sketch** ✏️ from the Context toolbar. The Sketch toolbar is displayed.

56)   Click the **Convert Entities**  Sketch tool. The Convert Entities PropertyManager is displayed.

Insert an Extruded Boss/Base feature.

57)   Click the **Extruded Boss/Base** feature tool. The Boss-Extrude PropertyManager is displayed.

58)   Enter **.400**in, **[10.16]** for Depth in Direction 1. Accept the default settings.

59)   Click **OK** ✔ from the Boss-Extrude PropertyManager. Boss-Extrude1 is displayed in the FeatureManager.

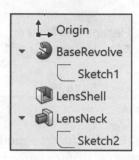

Display an Isometric view. Rename the feature. Save the model.

60)   Click **Isometric view** ⬛.

61)   Rename **Boss-Extrude1** to **LensNeck**.

62)   Click **Save** 💾.

LENS Part-Hole Wizard

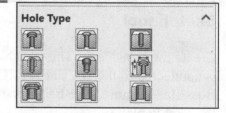

The LENS requires a Counterbore hole. Apply the Hole Wizard 🕳 feature. The Hole Wizard feature assists in creating complex and simple holes. The Hole Wizard Hole type categories are *Counterbore, Countersink, Hole, Straight Tap, Tapered Tap, Legacy Hole* (holes created before SOLIDWORKS 2000), *Counterbore Slot, Countersink Slot* and *Slot*.

Specify the user parameters for the custom Counterbore hole. The parameters are *Description, Standard, Screw Type, Hole Size, Fit, Counterbore diameter, Counterbore depth* and *End Condition*. Select the face or plane to locate the hole profile.

Insert a Coincident relation to position the hole center point. Dimensions for the Counterbore hole are provided in both inches and millimeters.

Activity: LENS Part - Hole Wizard Counterbore Hole Feature

Create the Counterbore hole.

63)   Click **Front view** ▱.

64)   Click the **Hole Wizard** 🕳 feature tool. The Hole Specification PropertyManager is displayed. Type is the default tab.

65) Click the **Counterbore** icon as illustrated.

Note: For a metric hole, skip the next few steps.

For inch Counterbore hole:

66) Select **ANSI Inch** for Standard. Select **Hex Bolt** for Type.

67) Select ½ for Size. Check the **Show custom sizing** box.

68) Click inside the **Counterbore Diameter** value box.

69) Enter **.600**in. Click inside the **Counterbore Depth** value box.

70) Enter **.200**in. Select **Through All** for End Condition.

71) Click the **Position** tab.

72) Click the small **inside back face** of the LensShell feature as illustrated. Do not select the Origin. LensShell is highlighted in the FeatureManager. The Point tool icon is displayed.

73) Click the **origin**. A Coincident relation is displayed.

Deselect the Point tool.

74) Right-click **Select** in the Graphics window.

Note: For an inch hole, skip the next few steps to address millimeter.

For millimeter Counterbore hole:

75) Select **ANSI Metric** for Standard.

76) Enter **Hex Bolt** for Type. Select **M5** for Size.

77) Click **Through All** for End Condition.

78) Check the **Show custom sizing** box.

79) Click inside the **Counterbore Diameter** value box.

80) Enter **15.24**. Click inside the **Counterbore Depth** value box.

81) Enter **5**. Click the **Position** tab. Click the small **inside back face** of the LensShell feature as illustrated. Do not select the Origin. LensShell is highlighted in the FeatureManager. The Point tool icon is displayed.

82) Click the **origin**. A Coincident relation is displayed.

Deselect the Point tool.

83) Right-click **Select** in the Graphics window.

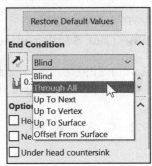

Return to the Type tab.
84) Click the **Type** tab.

Add the new hole type to your Favorites list.
85) **Expand** the Favorites box.

86) Click the **Add or Update Favorite** ✦ icon.

87) Enter **CBORE FOR BULB**.

88) Click **OK** from the Add or Update a Favorite
dialog box.

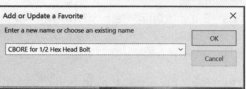

89) Click **OK** ✔ from the Hole Specification
PropertyManager.

Expand the Hole feature.
90) **Expand** the CBORE feature in the
FeatureManager. Note: Sketch3 and Sketch4.

Display the Section view.
91) Click **Right Plane** from the FeatureManager.

92) Click **Section view** 🗐 from the Heads-up View toolbar.

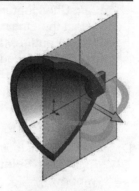

93) Click **OK** ✔ from the Section View PropertyManager.

94) Click **Isometric view** 🔲.

Display the Full view.
95) Click **Section view** 🗐 from the Heads-up View toolbar.

Rename the feature and save the model.
96) Rename **CBORE for ½ Hex Head Bolt1** to **BulbHole**.

97) Click **Save** 💾.

LENS Part - Revolved Boss Thin Feature

Create a Revolved Boss Thin feature. Rotate an open sketched profile around an axis.
The sketch profile must be open and cannot cross the axis. A Revolved Boss Thin feature
requires:

- Sketch plane (Right Plane).

- Sketch profile (Center point arc).

- Axis of Revolution (Temporary axis).

- Angle of Rotation (360).

- Thickness .100in [2.54].

A Revolved feature produces silhouette edges in 2D views. A silhouette edge represents the extent of a cylindrical or curved face.

Select the Temporary Axis for Axis of Revolution. Select the Revolved Boss feature. Enter .100in [2.54] for Thickness in the Revolve PropertyManager. Enter 360° for Angle of Revolution.

Activity: LENS Part - Revolved Boss Thin Feature

Create a sketch.
98) Right-click **Right Plane** from the FeatureManager.

99) Click **Sketch** from the Context toolbar.

100) Click **Right view**.

101) **Zoom in** on the LensNeck.

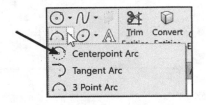

102) Click the **Centerpoint Arc** Sketch tool.

103) Click the **top horizontal edge** of the LensNeck. Do not select the midpoint of the silhouette edge.

104) Click the **top right corner** of the LensNeck.

105) Drag the **mouse pointer** counterclockwise to the left.

106) Click a **position** directly above the first point as illustrated.

Add a dimension.
107) Click the **Smart Dimension** Sketch tool.

108) Click the **arc**.

109) Click a **position** to the right of the profile.

110) Enter .100in, [2.54]. The Sketch is fully defined.

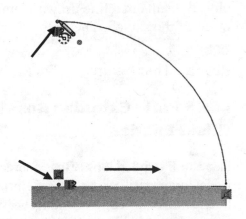

Insert a Revolved Thin feature.

111) Click the **Revolved Boss/Base** 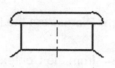 feature tool. The Revolve PropertyManager is displayed.

112) Select **Mid-Plane** for Revolve Type in the Thin Feature box.

113) Enter **.050**in, **[1.27]** for Direction1 Thickness.

114) Click the **Temporary Axis** for Axis of Revolution.

115) Click **OK** ✔ from the Revolve PropertyManager.

Rename the feature.
116) Rename **Revolve-Thin1** to **LensConnector**.

Fit the model to the Graphics window and save.
117) Press the **f** key.

118) Click **Save** 💾.

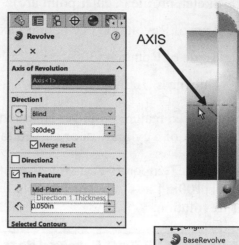

AXIS

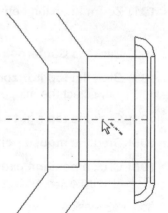

 A Revolved sketch that remains open results in a Revolve

Thin feature ——⊢. A Revolved sketch that is automatically closed results in a line drawn from the start point to the end point of the sketch. The sketch is closed and results in a non-

Revolve Thin feature ——⊢.

LENS Part - Extruded Boss/Base Feature and Offset Entities

Use the Extruded Boss/Base feature tool to create the front LensCover. Utilize the Offset Entities Sketch tool to offset the outside circular edge of the Revolved feature. The Sketch plane for the Extruded Boss feature is the front circular face.

The Offset Entities Sketch tool requires an Offset Distance and direction. Utilize the Bi-direction option to create a circular sketch in both directions. The extrude direction is away from the Front Plane.

Activity: LENS Part - Extruded Boss Feature and Offset Entities

Create the sketch.

119) Click **Isometric view** 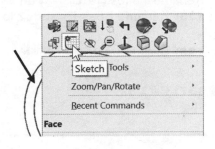 from the Heads-up View toolbar.

120) Click **Hidden Lines Removed** from the Heads-up View toolbar.

121) Right-click the **front circular face** for the Sketch plane.

122) Click **Sketch** from the Context toolbar.

123) Click **Front view**.

Offset the selected edge.

124) Click the **outside circular edge** of the LENS in the Graphics window.

125) Click the **Offset Entities** Sketch tool. The Offset Entities PropertyManager is displayed.

126) Check the **Bi-directional** box.

127) Enter **.250**in, [6.35] for Offset Distance. Accept the default settings.

128) Click **OK** from the Offset Entities PropertyManager.

Display an Isometric view with Shaded With edges.

129) Click **Isometric view** 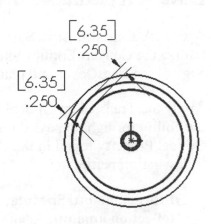 from the Heads-up View toolbar.

130) Click **Shaded With Edges** from the Heads-up View toolbar.

Insert an Extruded Boss/Base feature.

131) Click the **Extruded Boss/Base** feature tool. The Boss-Extrude PropertyManager is displayed.

132) Enter **.250**in, [6.35] for Depth in Direction 1. Accept the default settings.

133) Click **OK** from the Boss-Extrude PropertyManager. 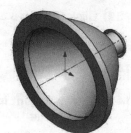 Boss-Extrude2 is displayed in the PropertyManager.

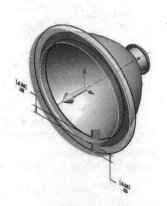

Verify the position of the extruded feature.
134) Click the **Top view** ⬜. View the extruded feature.

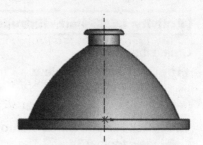

Rename the feature. Display an Isometric view. Save the model.
135) Rename **Boss-Extrude2** to **LensCover**.

136) Click **Isometric view** 🔲.

137) Click **Save** 💾.

LENS Part - Extruded Boss Feature and Transparency

Apply the Extruded Boss/Base feature to create the LensShield. Utilize the Convert Entities Sketch tool to extract the inside circular edge of the LensCover and place it on the Front plane.

Apply the Transparent Optical property to the LensShield to control the ability for light to pass through the surface. Transparency is an Optical Property found in the Color PropertyManager. Control the following properties:

- **Diffuse amount, Specular amount, Specular spread, Reflection amount, Transparent amount and Luminous intensity.**

Activity: LENS Part - Extruded Boss Feature and Transparency

Create the sketch.
138) Right-click **Front Plane** from the FeatureManager. This is your Sketch plane.

139) Click **Sketch** 📝 from the Context toolbar. The Sketch toolbar is displayed.

140) Click **Isometric view** 🔲.

141) Click the **front inner circular edge** of the LensCover (Boss-Extrude2) as illustrated.

142) Click the **Convert Entities** Sketch tool. The circle is projected onto the Front Plane.

Insert an Extruded Boss feature.

143) Click the **Extruded Boss/Base** feature tool. The Boss-Extrude FeatureManager is displayed.

144) Enter **.100**in, **[2.54]** for Depth in Direction 1.

145) Click **OK** ✓ from the Boss-Extrude PropertyManager. Boss-Extrude3 is displayed in the FeatureManager.

Rename the feature. Save the model.
146) Rename **Boss-Extrude3** to **LensShield**.

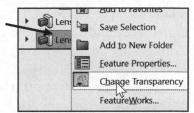

147) Click **Save** 🖫.

Change Transparency of the LensShield.
148) Right-click **LensShield** in the FeatureManager.

149) Click **Change Transparency**. View the results.

Hide the axis of revolution.
150) Click **View**, **Hide/Show**, uncheck **Temporary Axes** from the Menu bar.

Save the model.
151) Click **Save** 🖫.

🔍 Additional information on Revolved Boss/Base, Shell, Hole Wizard and Appearance is located in SOLIDWORKS Help Topics. Keywords: Revolved (features), Shells, Hole Wizard (Counterbore) and Appearances.

Refer to Help, SOLIDWORKS Tutorials, Revolve and Swept for additional information.

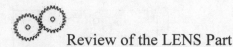 Review of the LENS Part

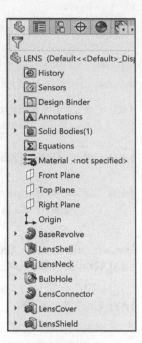

The LENS feature utilized a Revolved Base feature. A Revolved feature required an axis, profile and an angle of revolution. The Shell feature created a uniform wall thickness.

You utilized the Convert Entities Sketch tool to create the Extruded Boss feature for the LensNeck. The Counterbore hole was created using the Hole Wizard feature.

The Revolved Thin feature utilized a single 3 Point Arc. Geometric relations were added to the silhouette edge to define the arc. The LensCover and LensShield utilized existing geometry to Offset and Convert the geometry to the sketch.

The Color and Optics PropertyManager determined the LensShield transparency.

BULB Part

The BULB fits inside the LENS. Use the Revolved feature as the Base feature for the BULB.

Insert the Revolved Base ⊌ feature from a sketched profile on the Right Plane.

Insert a Revolved Boss ⊌ feature using a Spline sketched profile. The profile utilizes a complex curve called a Spline (Non-Uniform Rational B-Spline or NURB). Draw Splines with control points.

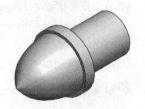

Insert a Revolved Cut Thin ⍔ feature at the base of the BULB. A Revolved Cut Thin feature removes material by rotating an open sketch profile about an axis.

Insert a Dome ⊖ feature at the base of the BULB. A Dome feature creates spherical or elliptical shaped geometry. Use the Dome feature to create the Connector feature of the BULB. The Dome feature requires a face and a height value.

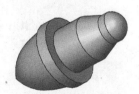

Insert a Circular Pattern ⛶ feature from an Extruded Cut feature.

BULB Part - Revolved Base Feature

Create the new part, BULB. The BULB utilizes a solid Revolved Base feature.

The solid Revolved Base feature requires a:

- Sketch plane (Right Plane).

- Sketch profile (Lines).

- Axis of Revolution (Centerline).

- Angle of Rotation (360°).

Utilize the centerline to create a diameter dimension for the profile. The flange of the BULB is located inside the Counterbore hole of the LENS. Align the bottom of the flange with the Front Plane. The Front Plane mates against the Counterbore face.

Activity: BULB Part - Revolved Base Feature

Create a New part.

152) Click **New** ⬜ from the Menu bar.

153) Click the **MY-TEMPLATES** tab. Additional templates are displayed.

154) Double-click **PART-IN-ANSI**, [**PART-MM-ISO**].

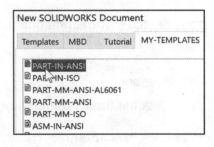

Save the part. Enter name. Enter description.

155) Click **Save** 🖫.

156) Select **PROJECTS** for Save in folder.

157) Enter **BULB** for File name.

158) Enter **BULB FOR LENS** for Description.

159) Click **Save**. The BULB FeatureManager is displayed.

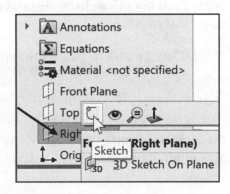

Select the Sketch plane.

160) Right-click **Right Plane** from the FeatureManager. This is your Sketch plane.

Create the sketch.

161) Click **Sketch** ⌐ from the Context toolbar. The Sketch toolbar is displayed.

Sketch a centerline.

162) Click the **Centerline** 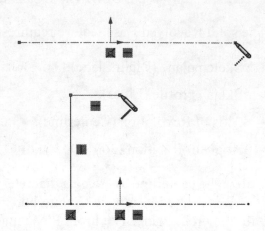 Sketch tool. The Insert Line PropertyManager is displayed.

163) Sketch a **horizontal centerline** through the Origin.

Create six profile lines.

164) Click the **Line** Sketch tool. The Insert Line PropertyManager is displayed.

165) Sketch a **vertical line** to the left of the Front Plane.

166) Sketch a **horizontal line** with the endpoint coincident to the Front Plane.

167) Sketch a short **vertical line** towards the centerline, collinear with the Front Plane.

168) Sketch a **horizontal line** to the right.

169) Sketch a **vertical line** with the endpoint collinear with the centerline.

170) Sketch a **horizontal line** to the first point to close the profile.

Add dimensions.

171) Click the **Smart Dimension** Sketch tool.

172) Click the **centerline**. (Note: Select the centerline).

173) Click the **top right horizontal line** as illustrated.

174) Click a **position** below the centerline and to the right. Enter **.400**in, **[10.016]**.

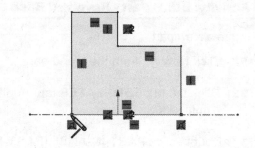

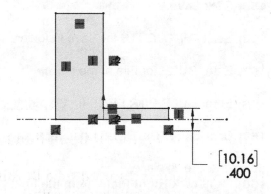

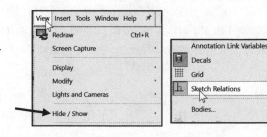

$$\begin{bmatrix}10.16\end{bmatrix}$$
.400

Click **View, Hide/Show, Sketch Relations** from the Menu bar to display the relations of the model in the Graphics window.

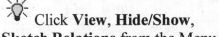

Click the **centerline**. (Note: Select the centerline).

175) Click the **top left horizontal line**.

176) Click a **position** below the centerline and to the left.

177) Enter .590in, [**14.99**]. Click the **top left horizontal line**.

178) Click a **position** above the profile.

179) Enter .100in, [**2.54**]. Click the **top right horizontal line**.

180) Click a **position** above the profile.

181) Enter .500in, [**12.7**].

Fit the model to the Graphics window.
182) Press the **f** key.

Insert a Revolved Base feature.

183) Click the **Revolved Boss/Base** 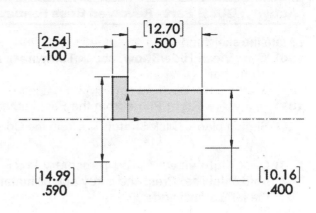 feature tool. The Revolve PropertyManager is displayed. Accept the default settings. Click **OK** ✔ from the Revolve PropertyManager. Revolve1 is displayed in the FeatureManager.

Display an Isometric view. Save the model.
184) Click **Isometric view** .

185) Click **Save** .

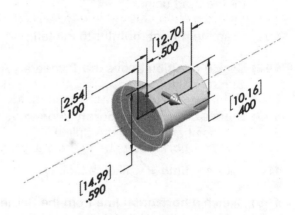

BULB Part - Revolved Boss Feature and Spline Sketch Tool

The BULB requires a second solid Revolved feature. The profile utilizes a complex curve called a Spline (Non-Uniform Rational B-Spline or NURB). Draw Splines with control points. Adjust the shape of the curve by dragging the control points.

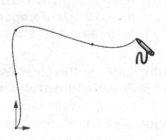

🔅 For additional flexibility, deactivate the Snaps option in Document Properties for this model.

Activity: BULB Part - Revolved Boss Feature and Spline Sketch Tool

Create the sketch.

186) Click **View**, **Hide/Show**, check **Temporary Axes** from the Menu bar.

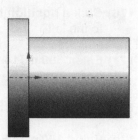

187) Right-click **Right Plane** from the FeatureManager for the Sketch plane. Click **Sketch** 🗔 from the Context toolbar.

188) Click **Right view** ⬜. The Temporary Axis is displayed as a horizontal line. Press the **z** key approximately five times to view the left vertical edge.

Sketch the profile.

189) Click the **Spline** 𝒩 Sketch tool.

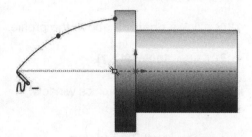

190) Click the **left vertical edge** of the Base feature for the Start point.

191) Drag the **mouse pointer** to the left.

192) Click a **position** above the Temporary Axis for the Control point.

193) Double-click the **Temporary Axis** to create the End point and to end the Spline.

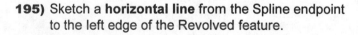

194) Click the **Line** ✏ Sketch tool.

195) Sketch a **horizontal line** from the Spline endpoint to the left edge of the Revolved feature.

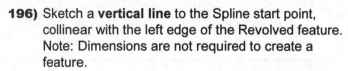

196) Sketch a **vertical line** to the Spline start point, collinear with the left edge of the Revolved feature. Note: Dimensions are not required to create a feature.

De-select the Line Sketch tool. Insert a Revolved Boss feature.

197) Right-click **Select** to de-select the Line Sketch tool. Click the **Temporary Axis** from the Graphics window as illustrated.

198) Click the **Revolved Boss/Base** 🍥 feature tool. The Revolve PropertyManager is displayed. Accept the default settings.

199) Click **OK** ✔ from the Revolve PropertyManager. Revolve2 is displayed in the FeatureManager.

200) Click **Isometric view** .

201) Click **Save** .

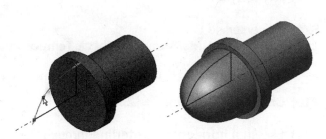

The points of the Spline dictate the shape of the Spline. Edit the control points in the sketch to produce different shapes for the Revolved Boss feature.

BULB Part - Revolved Cut Thin Feature

A Revolved Cut Thin feature removes material by rotating an open sketch profile around an axis. Sketch an open profile on the Right Plane. Add a Coincident relation to the silhouette and vertical edge. Insert dimensions.

💡 Sketch a Centerline to create a diameter dimension for a revolved profile. The Temporary axis does not produce a diameter dimension.

Note: If lines snap to grid intersections, uncheck Tools, Sketch Settings, Enable Snapping for the next activity.

Activity: BULB Part - Revolved Cut Thin Feature

Create the sketch. De-select the Line Sketch tool.
202) Right-click **Right Plane** from the FeatureManager.

203) Click **Sketch** from the Context toolbar.

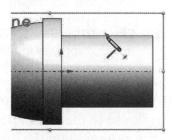

204) Click **Right view** from the Heads-up View toolbar.

205) Click the **Line** Sketch tool. Click the **midpoint** of the top silhouette edge.

206) Sketch a **line** downward and to the right as illustrated.

207) Sketch a **horizontal line** to the right vertical edge.

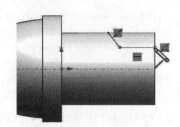

208) Right-click **Select** to de-select the Line Sketch tool.

If needed, add a Coincident relation.
209) Click the **endpoint** of the line.

210) Hold the **Ctrl** key down.

211) Click the right **vertical edge**.

212) Release the **Ctrl** key.

213) Click **Coincident** ✗ .

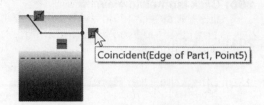

Coincident(Edge of Part1, Point5)

214) Click **OK** ✔ from the Properties
PropertyManager.

Sketch a centerline.
215) Click **View**, **Hide/Show**, un-check **Temporary Axes** from the Menu bar.

216) Click the **Centerline** ⌇ Sketch tool.

217) Sketch a **horizontal centerline** through the Origin.

Add dimensions.
218) Click the **Smart Dimension** ✎ Sketch tool.

219) Click the **horizontal centerline**.

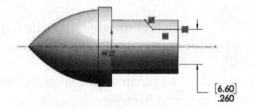

220) Click the **short horizontal line**.

221) Click a **position** below the profile to create a diameter dimension.

222) Enter **.260**in, **[6.6]**. Click the **short horizontal line**.

223) Click a **position** above the profile to create a horizontal dimension.

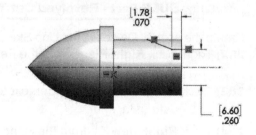

224) Enter **.070**in, **[1.78]**. The Sketch is fully defined and is displayed in black.

💡 For Revolved features, the ∅ symbol is not displayed in the part. The ∅ symbol is displayed when inserted into the drawing.

Insert the Revolved Cut Thin feature.
225) De-select the Smart Dimension Sketch tool. Right-click **Select**.

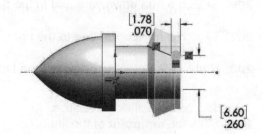

226) Click the **centerline** in the Graphics window.

227) Click the **Revolved Cut** 🗐 feature tool. The Cut-Revolve PropertyManager is displayed.

228) Click **No** to the Warning Message, "Would you like the sketch to be automatically closed?" The Cut-Revolve PropertyManager is displayed.

229) Check the **Thin Feature** box.

230) Enter **.150**in, **[3.81]** for Thickness.

231) Click the **Reverse Direction** box.

232) Click **OK** ✔ from the Cut-Revolve PropertyManager. Cut-Revolve-Thin1 is displayed in the FeatureManager.

Save the model.
233) Click **Save** 🖫.

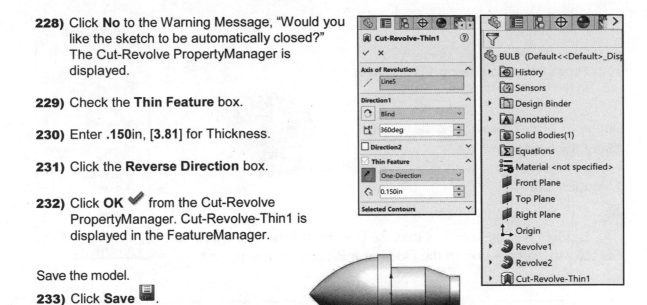

BULB Part - Dome Feature

The Dome 🗇 feature creates spherical or elliptical shaped geometry. Use the Dome feature to create the Connector feature of the BULB. The Dome feature requires a face and a height/distance value.

Activity: BULB Part - Dome Feature

Insert the Dome feature.
234) Click the **back circular face** of Revolve1. Revolve1 is highlighted in the FeatureManager.

235) Click the **Dome** 🗇 feature tool. The Dome PropertyManager is displayed. Face1 is displayed in the Parameters box. If Dome is not displayed in your CommandManager, click **Insert**, **Features**, **Dome** from the Main menu.

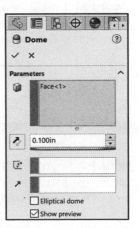

236) Enter **.100**in, **[2.54]** for Distance.

237) Click **OK** ✔ from the Dome
PropertyManager. Dome1 is displayed
in the FeatureManager.

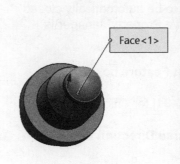

Display an Isometric view. Save the model.
238) Click **Isometric view** 🔲.

239) Click **Save** 💾.

💡 Before creating sketches that use Geometric relations, check the
Enable Snapping option in the Document Properties dialog box.

BULB Part - Circular Pattern Feature

A Pattern feature creates one or more instances of a feature or a
group of features. The Circular Pattern feature places the
instances around an axis of revolution.

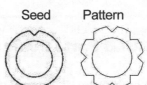

The Circular Pattern 🔧 feature requires a seed feature. The
seed feature is the first feature in the pattern. The seed feature in
this section is a V-shaped Extruded Cut feature.

Activity: BULB Part - Circular Pattern Feature

Create the Seed Cut feature.
240) Right-click the **front face** of the Base feature, Revolve1 in the
Graphics window for the Sketch plane as illustrated. Revolve1
is highlighted in the FeatureManager.

241) Click **Sketch** 📐 from the Context toolbar. The Sketch toolbar
is displayed.

242) Click the **outside circular edge** of the BULB.

243) Click the **Convert Entities** 📋 Sketch tool.

244) Click **Front view** 🔲 from the Heads-up View toolbar.

245) **Zoom in** on the top half of the BULB.

Sketch a centerline.

246) Click the **Centerline** ✐ Sketch tool.

247) Sketch a **vertical centerline** coincident with the top and bottom circular edges and coincident with the Right plane.

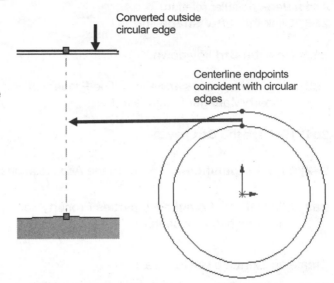

Converted outside circular edge

Centerline endpoints coincident with circular edges

Sketch a V-shaped line.

248) Click **Tools**, **Sketch Tools**, **Dynamic Mirror** from the Menu bar. The Mirror PropertyManager is displayed.

249) Click the **centerline** from the Graphics window.

250) Click the **Line** ✐ Sketch tool.

251) Click the **midpoint** of the centerline.

252) Click the coincident **outside circle edge to the left** of the centerline.

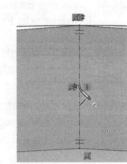

Deactivate the Dynamic Mirror tool.

253) Click **Tools**, **Sketch Tools**, **Dynamic Mirror** from the Menu bar.

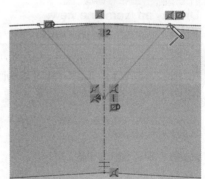

Trim unwanted geometry.

254) Click the **Trim Entities** ✂ Sketch tool. The Trim PropertyManager is displayed.

255) Click **Power trim** ✀ from the Options box.

256) Click a **position** in the Graphics window and drag the mouse pointer until it intersects the circle circumference.

257) Click **OK** ✔ from the Trim PropertyManager.

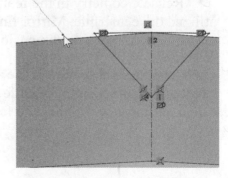

Add a Perpendicular relation.

258) Click the **left V shape** line.

259) Hold the **Ctrl** key down.

260) Click the **right V shape** line. The Properties PropertyManager is displayed.

261) Release the **Ctrl** key.

262) Click **Perpendicular** ⊥ from the Add Relations box.

263) Click **OK** ✔ from the Properties PropertyManager. The sketch is fully defined.

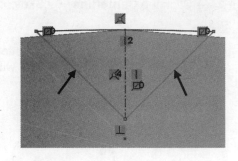

Create an Extruded Cut feature.

264) Click the **Extruded Cut** 🔲 feature tool. The Cut-Extrude PropertyManager is displayed.

265) Click **Through All** for End Condition in Direction 1. Accept the default settings.

266) Click **OK** ✔ from the Cut-Extrude PropertyManager. The Cut-Extrude1 feature is displayed in the FeatureManager.

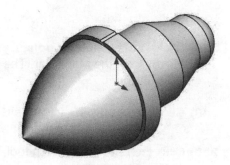

Display an Isometric view.

267) Click **Isometric view** 🎲.

Fit the drawing to the Graphics window.

268) Press the **f** key.

Save the model.

269) Click **Save** 💾.

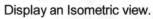

💡 Reuse Geometry in the feature. The Cut-Extrude1 feature utilized the centerline, Mirror Entity, and geometric relations to create a sketch with no dimensions.

The Cut-Extrude1 feature is the seed feature for the pattern. Create four copies of the seed feature. A copy of a feature is called an Instance. Modify the four copies Instances to eight.

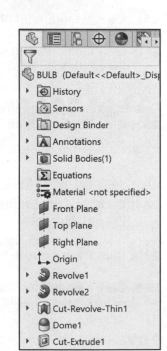

Insert the Circular Pattern feature.

270) Click the **Cut-Extrude1** feature from the FeatureManager.

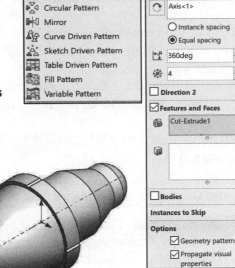

271) Click the **Circular Pattern** 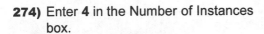 feature tool. The Circular Pattern PropertyManager is displayed. Cut-Extrude1 is displayed in the Features to Pattern box.

272) Click **View**, **Hide/Show**, check **Temporary Axes** from the Menu bar.

273) Click the **Temporary Axis** from the Graphics window. Axis<1> is displayed in the Pattern Axis box.

274) Enter **4** in the Number of Instances box.

275) Check the **Equal spacing** box. Check the **Geometry pattern** box. Accept the default settings.

276) Click **OK** ✔ from the Circular Pattern PropertyManager. CirPattern1 is displayed in the FeatureManager.

Edit the Circular Pattern feature.

277) Right-click **CirPattern1** from the FeatureManager.

278) Click **Edit Feature** from the Context toolbar. The CirPattern1 PropertyManager is displayed.

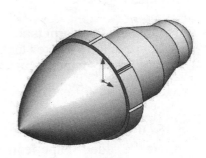

279) Enter **8** in the Number of Instances box.

280) Click **OK** ✔ from the CirPattern1 PropertyManager.

Rename the feature

281) Rename **Cut-Extrude1** to **SeedCut**.

Hide the reference geometry.

282) Click **View**, **Hide/Show**, uncheck **Temporary Axes** from the Menu bar.

Save the model.

283) Click **Save** 💾.

Rename the seed feature of a pattern to locate it quickly for future assembly.

Customizing Toolbars and Short Cut Keys

The default toolbars contain numerous icons that represent basic functions. Additional features and functions are available that are not displayed on the default toolbars.

You have utilized the z key for Zoom In/Out, the f key for Zoom to Fit and Ctrl-C/Ctrl-V to Copy/Paste. Short Cut keys save time.

Assign a key to execute a SOLIDWORKS function. Create a Short Cut key for the Temporary Axis.

Activity: Customizing Toolbars and Short Cut Keys

Customize the toolbar.
284) Click **Tools**, **Customize** from the Menu bar. The Customize dialog box is displayed.

Place the Freeform icon on the Features toolbar.
285) Click the **Commands** tab.

286) Click **Features** from the category text box.

287) Drag the **Freeform** feature icon into the Features toolbar. The Freeform feature option is displayed.

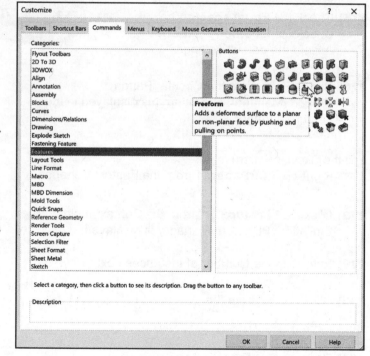

 SOLIDWORKS does not support the older Shape feature.

Customize the keyboard for the Temporary Axes.

288) Click the **Keyboard** tab from the Customize dialog box.

289) Select **View** for Categories.

290) Select **Temporary Axes** for Commands.

291) Enter **R** for Shortcut(s) key.

292) Click **OK**.

293) Press the **R** key to toggle the display of the Temporary Axes.

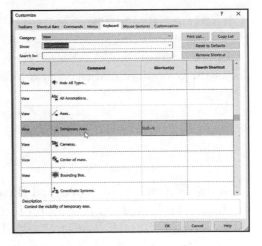

Test the proposed Short cut key before you customize your keyboard. Refer to the default Keyboard Short cut table in the Appendix.

Additional information on Revolved Boss/Base, Spline, Circular Pattern, Dome, Line/Arc Sketching is located in SOLIDWORKS Help Topics. Keywords: Revolved (features), Spline, Pattern (Circular) and Dome.

Review of the BULB Part

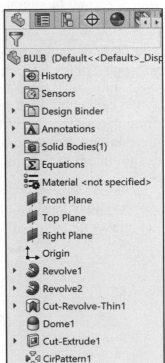

The Revolved Base feature utilized a sketched profile on the Right plane and a centerline. The Revolved Boss feature utilized a Spline sketched profile. A Spline is a complex curve.

You created the Revolved Cut Thin feature at the base of the BULB to remove material. A centerline was inserted to add a diameter dimension. The Dome feature was inserted on the back face of the BULB. The Circular Pattern feature was created from an Extruded Cut feature. The Extruded Cut feature utilized existing geometry and required no dimensions.

Toolbars and keyboards were customized to save time. Always verify that a Short cut key is not predefined in SOLIDWORKS.

Chapter Summary

You are designing a FLASHLIGHT assembly. You created two parts in this chapter:

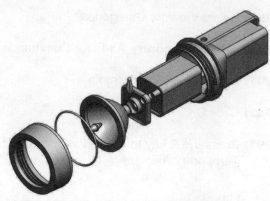

- LENS

- BULB

Both parts utilized the Revolved Base feature. The Revolved feature required a Sketch plane, Sketch profile, Axis of revolution and an Angle of rotation.

You created and edited the following features: Extruded Boss/Base, Extruded Cut, Revolved Base, Revolved Boss, Revolved Boss Thin, Revolved Thin Cut, Shell, Hole Wizard, Dome and Circular Pattern.

You applied transparent optical properties to the LENS part and established the following Geometric relations: Equal, Coincident, Symmetric, Intersection, and Perpendicular.

The other parts for the FLASHLIGHT assembly are addressed in the later chapters.

When defining a sketch, first add geometric relations then dimensions. This keeps the user from having too many unnecessary dimensions. This also helps to show the design intent of the model.

💡 Screen shots and illustrations in the book display the SOLIDWORKS user default setup.

Questions

1. Identify and describe the function of the following features:

   - Revolved Base

   - Revolved Boss

   - Revolved Cut

   - Revolved Cut Thin

   - Dome

2. Describe a Symmetric relation.

3. When is the Trim Entity Sketch tool used?

4. Explain the function of the Shell feature.

5. A Center point arc requires _____ points.

6. Describe the Hole Wizard feature.

7. What is a Spline?

8. Identify the required information for a Circular Pattern feature.

9. Name the Pull down menu that lists the Temporary Axis.

10. Describe the procedure to Show/Hide a Plane.

11. Describe the differences between Offset Entities and Convert Entities.

12. Identify the type of line required to utilize Mirror Entities.

13. Identify the geometric relation automatically created between Mirror Entities.

14. True or False. Select the arc center point to dimension an arc to its max condition.

15. True or False. The Transparency Optical Property is located in the Features toolbar.

16. Additional information on the Revolve Boss/Base and Revolve Cut features is located in _____.

Exercise 6.1: BUSHING

Create the illustrated BUSHING part on the Front Plane.

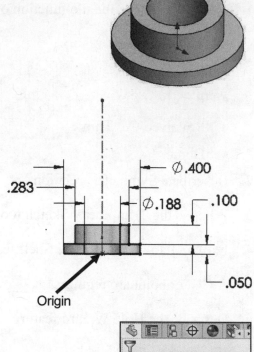

- Note the location of the Origin (center) and the provided dimensions.

- Apply the ANSI overall drafting standard.

- Use the IPS unit system.

- Apply Brass as a material. Sketch1 should be fully defined.

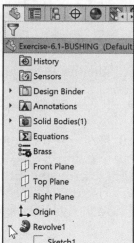

Exercise 6.2: PIN

Create the illustrated PIN part as illustrated on the Front Plane.

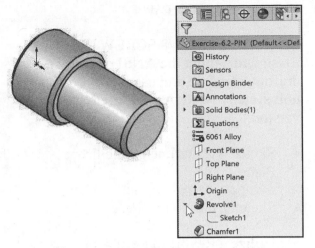

- Note the location of the Origin.

- Apply the ANSI overall drafting standard.

- Use the IPS unit system.

- Apply 6061 Alloy as the material.

- Calculate the total mass of the part. Apply the SOLIDWORKS Mass Properties tool under the Evaluate tab in the CommandManager.

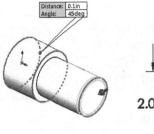

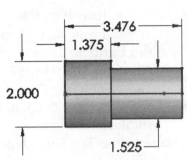

Exercise 6.3: Mounting Nut

Create the illustrated Mounting Nut part as illustrated on the Front Plane.

- Apply the ANSI overall drafting standard and use the MMGS unit system.

- Obtain a metric nut and measure the dimensions.

- Use these dimensions to create the part.

- Apply Plain Carbon Steel as the material.

- Add chamfers.

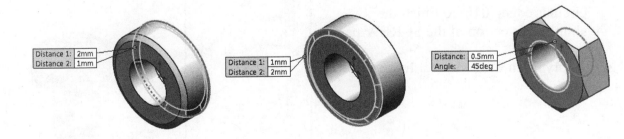

Exercise 6.4: SCREW

Create the 10-24x3/8 SCREW as illustrated. Apply the ANSI overall drafting standard and use the IPS unit system. Note: A simplified version.

- Sketch a centerline on the Front Plane.

- Sketch a closed profile.

- Utilize a Revolved Feature.

- Edit the Revolved Base Sketch. Use the Tangent Arc tool with Trim Entities. Enter an Arc dimension of .304in, [7.72].

- Utilize an Extruded Cut feature with the Mid Plane option to create the Top Cut. Depth = .050in.

- The Top Cut is sketched on the Front Plane.

- Utilize the Convert Entities Sketch tool to extract the left edge of the profile.

- Utilize the Circular Pattern feature and the Temporary Axis to create four Top Cuts.

- Use the Fillet Feature on the top circular edge, .01in to finish the simplified version of the SCREW part.

- Apply Plain Carbon Steel for material.

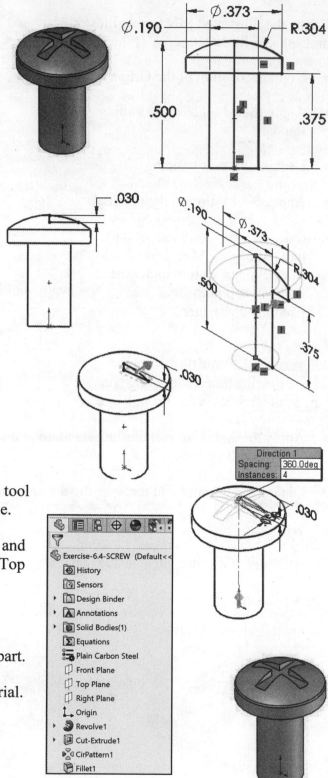

Exercise 6.5:

Create the illustrated part. Apply the ANSI overall drafting standard. All edges of the model are not located on perpendicular planes. Think about the steps required to build the model. Insert two features: Boss-Extrude1 and Extruded Cut 1.

Note the location of the Origin.

Select the Right Plane as the Sketch plane. Apply construction geometry. Insert the required geometric relations and dimensions for Sketch1.

☀ There are numerous ways to build the models in this chapter. Optimize your time.

Given:
A = 3.00, B = 1.00
Material: 6061 Alloy
Density = .097 lb/in^3
Units: IPS
Decimal places = 2

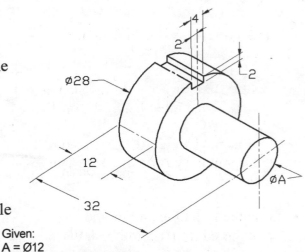

Exercise 6.6:

Create the illustrated part. Apply the ANSI overall drafting standard. Unit system - MMGS.

Think about the required steps to build this part. Insert a Revolved Base feature and an Extruded Cut feature. Note the location of the Origin. Select the Front Plane as the Sketch plane.

Apply the Centerline Sketch tool for the Revolve1 feature.

Insert the required geometric relations and dimensions for Sketch1. Sketch1 is the profile for the Revolve1 feature.

Given:
A = Ø12
Material: Cast Alloy Steel
Density = .0073 g/mm^3
Units: MMGS

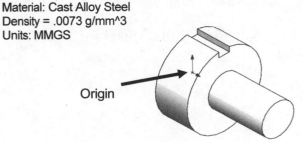

Exercise 6.7:

Create the illustrated part. Calculate the total mass of the part. Apply the SOLIDWORKS Mass Properties tool under the Evaluate tab in the CommandManager.

- Apply the ANSI overall drafting standard. Think about the required steps to build this part. Insert four features: Boss-Extrude1, Cut-Extrude1, Cut-Extrude2 and Fillet.

- Note the location of the Origin.

- Add 6061 Alloy for material.

- Select the Right Plane as the Sketch plane.

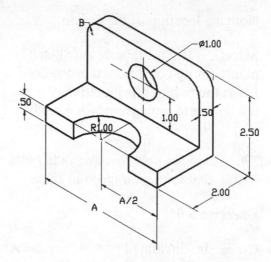

Given:
A = 4.00
B = R.50
Material: 6061 Alloy
Density = .0975 lb/in^3
Units: IPS
Decimal places = 2

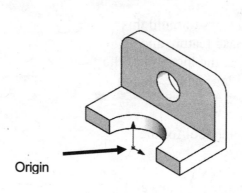

Origin

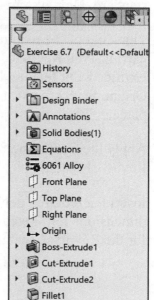

Exercise 6.8:

Create the Hole-Block ANSI - IPS part using the Hole Wizard feature as illustrated. Create the Hole-Block part on the Front Plane.

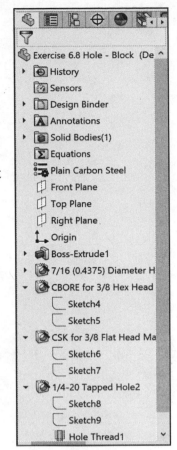

The Hole Wizard 🔵 feature creates either a 2D or 3D sketch for the placement of the hole in the FeatureManager.

You can consecutively place multiple holes of the same type. The Hole Wizard creates 2D sketches for holes unless you select a non-planar face or click the 3D Sketch button in the Hole Position PropertyManager.

Hole Wizard creates two sketches: 1.) A sketch of the revolved cut profile of the selected hole type and 2.) A sketch of the center placement for the profile. Both sketches should be fully defined.

Create a rectangular prism 2 inches wide by 5 inches long by 2 inches high on the top surface of the prism. Place four holes, 1 inch apart.

- Hole #1: Simple Hole Type: Fractional Drill Size, 7/16 diameter, End Condition: Blind, 0.75 inch deep.
- Hole #2: Counterbore hole Type: for 3/8 inch diameter Hex bolt, End Condition: Through All.
- Hole #3: Countersink hole Type: for 3/8 inch diameter Flat head screw, 1.5 inch deep.
- Hole #4: Tapped hole Type, Size ¼-20, End Condition: Blind -1.0 inch deep.

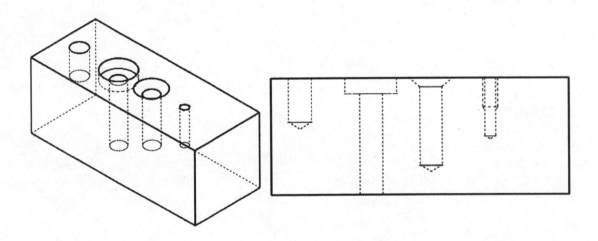

Notes:

Chapter 7

Swept, Lofted, Rib, Mirror and Additional Features

Below are the desired outcomes and usage competencies based on the completion of Chapter 7.

Project Desired Outcomes:	**Usage Competencies:**

• Create four FLASHLIGHT parts:
 	
  o O-RING
 	
  o SWITCH
 	
  o LENSCAP
 	
  o HOUSING
	• Specific knowledge and understanding of the following Features: Extruded Boss/Base, Extruded Cut, Swept Base, Swept Boss, Lofted Base, Lofted Boss, Mirror, Draft, Dome, Rib and Linear Pattern.

• Establish Geometric relations: Pierce, Tangent, Equal, Intersection, Coincident and Midpoint.	• Ability to apply multiple Geometric relations to a model. • Skill to apply Design Intent to 2D Sketches, 3D Features, Parts and Assemblies.

Notes:

Chapter 7 - Swept, Lofted, Rib, Mirror and Additional Features

Chapter Overview

Create four new parts for the
FLASHLIGHT assembly:

- O-RING

- SWITCH

- LENSCAP

- HOUSING

Chapter 7 introduces the Swept and Loft
feature tools. The O-RING utilizes a
simple Swept Base feature. The SWITCH
utilizes the Lofted Base feature. The
LENSCAP and HOUSING utilize the
Swept Boss and Lofted Boss feature.

A simple Swept Base feature requires a *path* and *profile*. Sketch the path and enter the
circular profile diameter for the ORING. The profile follows the path to create the
following Swept features:

- Swept Base

- Swept Boss

The Lofted feature requires a minimum of two profiles sketched on different planes. The
profiles are blended together to create the following Lofted features:

- Lofted Base

- Lofted Boss

Utilize existing features to create the Rib, Linear Pattern, and Mirror features. Utilize
existing faces to create the Draft and Dome feature. The LENSCAP and HOUSING
combines the Extruded Boss/Base, Extruded Cut, Revolved Thin Cut, Shell and Circular
Pattern with the Swept and Lofted feature.

After completing the activities in this chapter, you will be able to:

- Utilize the following Sketch tools: Point, Centerline, Convert Entities, Trim Entities
  and Sketch Fillet.

- Establish the following Geometric relations: Pierce, Tangent, Equal, Intersection,
  Coincident and Midpoint.

- Create the following features: Swept Boss/Base, Lofted Boss/Base, Mirror, Draft, Rib, Dome and Linear Pattern.

- Review the Extruded Boss/Base, Extruded Cut, Revolve Cut Thin, Shell and Circular Pattern features.

- Suppress and Un-suppress various features.

- Reuse geometry from sketches, features and other parts to develop new geometry.

- Apply the Performance Evaluation tool.

O-RING Part - Swept Base Feature

The O-RING part is positioned between the LENSCAP and the LENS.

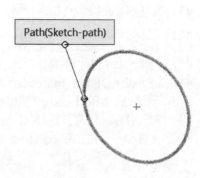

A Swept Boss/Base feature can be simple or complex. The O-RING is classified as a simple Swept Boss/Base feature. A complex Swept Boss/Base feature utilizes 3D curves and Guide Curves.

Create the O-RING with the Swept Base ![swept icon] feature. The Swept Base feature requires a sketch path and a circular profile diameter.

Utilize the PART-IN-ANSI Template created in a previous chapter for inch units. Utilize the PART-MM-ISO Template for millimeter units. Millimeter dimensions are provided in brackets [x].

☼ For non-circular sketch profiles, create the sketch profile on a perpendicular plane to the path and use the pierce relation to locate the profile on the path.

Activity: Create the O-RING Part

Create a New part. Enter name. Enter description.

1)   Click **New** ![new icon] from the Menu bar.

2)   Click the **MY-TEMPLATES** tab.

3)   Double-click **PART-IN-ANSI**, **[PART-MM-ISO]**.

4)   Click **Save**.

5)   Enter **O-RING** for File name.

6)   Enter **O-RING FOR LENS** for Description.

7)   Click **Save**. The O-RING FeatureManager is displayed.

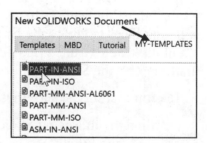

Create the Swept path.

8) Right-click **Front Plane** from the FeatureManager for the Sketch plane. This is your Sketch plane.

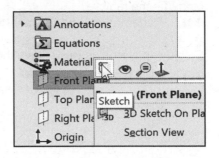

9) Click **Sketch** from the Context toolbar. The Sketch toolbar is displayed.

10) Click the **Circle** Sketch tool. The Circle PropertyManager is displayed.

11) Sketch a circle centered at the **Origin**.

Add a dimension.

12) Click the **Smart Dimension** Sketch tool.

13) Click the **circumference** of the circle.

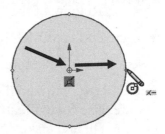

14) Click a **position** off the profile.

15) Enter **4.350**in, [**110.49**] as illustrated.

Close the sketch. Rename the sketch.

16) **Rebuild** the model. Sketch1 is displayed in the FeatureManager.

17) Rename **Sketch1** to **Sketch-path**.

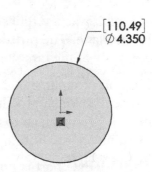

Insert the Swept feature.

18) Click the **Swept Boss/Base** feature tool. The Sweep PropertyManager is displayed.

19) Select **Circular Profile**.

20) **Expand** O-RING from the fly-out FeatureManager.

21) Click **Sketch-path** from the fly-out FeatureManager. Sketch-path is displayed.

22) Enter **.125in** for Diameter.

23) Click **OK** from the Sweep PropertyManager. Sweep1 is displayed in the FeatureManager.

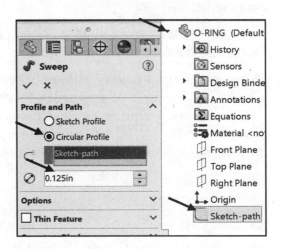

Rename the feature.

24)   Rename **Sweep1** to **Base-Sweep**.

Display an Isometric view. Save the model.

25)   Click **Isometric view** 📦.

26)   Click **Save** 💾.

Review of the O-RING Part

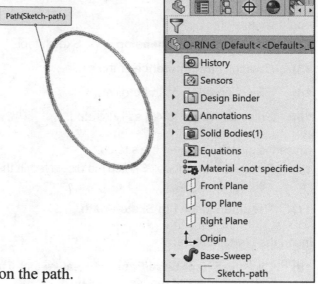

You created the O-RING with the Swept Base 〰️ feature.

The O-RING utilized the Swept feature. The Swept feature required a sketch path and a sketch profile.

The path was a circle sketched on the Front Plane. The profile was the diameter of the circular profile.

For non-circular sketch profiles, create the sketch profile on a perpendicular plane to the path and use a pierce relation to locate the profile on the path.

The O-RING part is positioned between the LENSCAP and the LENS.

A Swept feature can be simple or complex. The O-RING is classified as a simple Swept feature.

A complex Swept feature utilizes 3D curves and Guide Curves. An example of a complex Swept is a violin body. The violin body requires Guide Curves to control the Swept geometry. Without Guide Curves the profile follows the straight path to produce a rectangular shape.

SWITCH Part - Lofted Base Feature

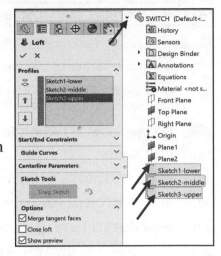

The SWITCH is a purchased part. The SWITCH is a complex assembly. Create the outside casing of the SWITCH as a simplified part.

Create the SWITCH with the Lofted Base ⬇ feature.

The orientation of the SWITCH is based on the position in the assembly. The SWITCH is comprised of three cross section profiles. Sketch each profile on a different plane.

The first plane is the Top Plane. Create two reference planes parallel to the Top Plane.

Sketch one profile on each plane. The design intent of the sketch is to reduce the number of dimensions.

Utilize symmetry, construction geometry and Geometric relations to control three sketches with one dimension.

Insert a Lofted feature. Select the profiles to create the Loft feature.

Insert the Dome feature to the top face of the Loft and modify the Loft Base feature. Modify the dimensions to complete the SWITCH.

You can specify an elliptical dome for cylindrical or conical models or a continuous dome for polygonal models.

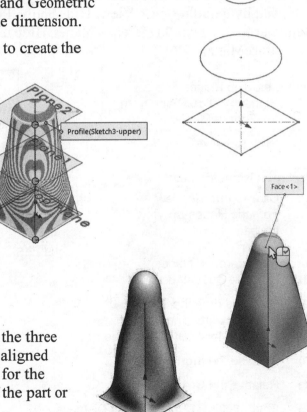

🔆 Tangent Edges and Origin are displayed for educational purposes.

When you create a new part or assembly, the three default planes (Front, Right and Top) are aligned with specific views. The plane you select for the Base sketch determines the orientation of the part or assembly.

Activity: SWITCH Part - Loft Base Feature

Create a New part. Enter name. Enter description.

27) Click **New** ⬜ from the Menu bar.

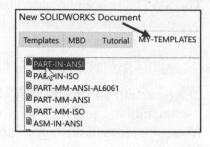

28) Click the **MY-TEMPLATES** tab. Additional templates are displayed.

29) Double-click **PART-IN-ANSI**, **[PART-MM-ISO]**.

30) Click **Save**.

31) Select **PROJECTS** for the Save in folder.

32) Enter **SWITCH** for File name.

33) Enter **BUTTON STYLE** for Description.

34) Click **Save**. The SWITCH FeatureManager is displayed.

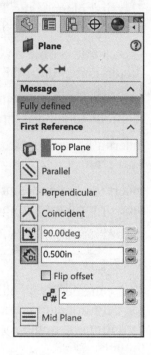

💡 If you upgrade from an older version, your planes may be displayed by default. Click View, Hide Show, Planes from the Menu bar to display all reference planes. Hide unwanted planes in the FeatureManager.

Display the Top Plane.

35) Right-click **Top Plane** from the FeatureManager.

36) Click **Show** 👁 .

Display an Isometric view.

37) Click **Isometric view** 🧊 from the Heads-up View toolbar.

> Hold the Ctrl key down. Drag the Top Plane upward. Pick an edge, not the handles.

Insert two Reference planes.

38) Hold the **Ctrl** key down.

39) Click and drag the **Top Plane** upward. The Plane PropertyManager is displayed.

40) Release the **mouse button**.

41) Release the **Ctrl** key.

42) Enter **.500**in, **[12.7]** for Offset Distance.

43) Enter **2** for # of Planes to Create.

44) Click **OK** ✔ from the Plane PropertyManager. Plane1 and Plane2 are displayed in the FeatureManager.

45) Click **Front view** 🗗 to display Plane1 and Plane2 offset from the Top plane.

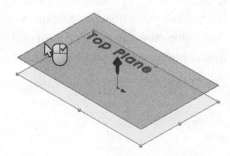

46) Click **Top view** 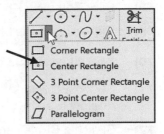.

Insert Sketch1. Sketch1 is a square on the Top Plane centered about the Origin.

47) Right-click **Top Plane** from the FeatureManager.

48) Click **Sketch** from the Context toolbar.

49) Click the **Center Rectangle** tool from the Consolidated Sketch toolbar.

50) Click the **Origin**.

51) Click a **point** in the upper right hand of the window as illustrated. The Center Rectangle Sketch tool automatically inserts a midpoint relation about the Origin.

52) Click **OK** from the Rectangle PropertyManager.

Add an Equal relation between the four edges. If needed, click the Add Relations tool.

53) Click the **left vertical line**.

54) Hold the **Ctrl** key down.

55) Click the **top horizontal line**.

56) Click the **right vertical line**.

57) Click the **bottom horizontal line**.

58) Release the **Ctrl** key.

59) Click **Equal** .

Add a dimension.

60) Click the **Smart Dimension** Sketch tool.

61) Click the **top horizontal line**.

62) Click a **position** above the profile.

63) Enter **.500**in, **[12.7]**.

Close, fit and rename the sketch.

64) **Rebuild** the model.

65) **Fit** the sketch to the Graphics window.

66) Rename **Sketch1** to **Sketch1-lower**.

Save the model.

67) Click **Save** .

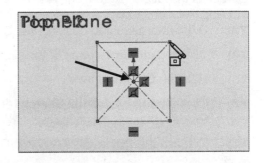

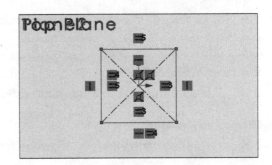

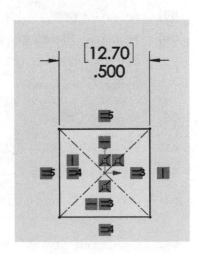

Insert Sketch2 on Plane1.

68) Click **Top view** .

69) Right-click **Plane1** from the FeatureManager.

70) Click **Sketch** from the Context toolbar.

71) Click the **Circle** Sketch tool. The Circle PropertyManager is displayed.

72) Sketch a **Circle** centered at the Origin as illustrated.

De-select the Circle Sketch tool. Add a Tangent relation. If needed, use the Add Relations tool.

73) Right-click **Select** to deselect the Circle Sketch tool.

74) Click the **circumference** of the circle.

75) Hold the **Ctrl** key down.

76) Click the **top horizontal Sketch1-lower line**. The Properties PropertyManager is displayed.

77) Release the **Ctrl** key.

78) Click **Tangent** from the Properties PropertyManager.

79) Click **OK** from the Properties PropertyManager.

Close and rename the sketch.

80) Click **Exit Sketch**.

81) Rename **Sketch2** to **Sketch2-middle**.

Display an Isometric view.

82) Click **Isometric view** . View the results.

Insert Sketch3 on Plane2.

83) Click **Top view** .

84) Right-click **Plane2** from the FeatureManager.

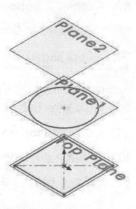

85) Click **Sketch** from the Context toolbar. The Sketch toolbar is displayed.

86) Click the **Centerline** Sketch tool.

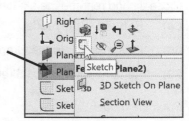

87) Sketch a **centerline** coincident with the Origin and the upper right corner point as illustrated.

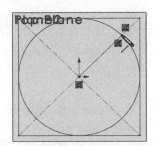

88) Click the **Point** ▪ Sketch tool.

89) Click the **midpoint** of the right diagonal centerline. The Point PropertyManager is displayed.

90) Click **Circle** ⊙ from the Sketch toolbar.

91) Sketch a **Circle** centered at the Origin to the midpoint of the diagonal centerline.

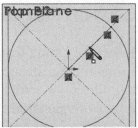

Close and rename the sketch.
92) Click **Exit Sketch**.

93) Rename **Sketch3** to **Sketch3-upper**.

Insert the Lofted Base feature.

94) Click the **Lofted Boss/Base** ⬇ feature tool. The Loft PropertyManager is displayed.

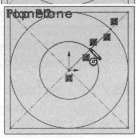

95) Click **Isometric view** 🧊 from the Heads-up View toolbar.

96) Right-click in the **Profiles box**.

97) Click **Clear Selections**.

98) Click **Sketch1-lower** from the fly-out FeatureManager.

99) Click **Sketch2-middle** from the fly-out FeatureManager.

100) Click **Sketch3-upper** from the fly-out FeatureManager. The selected sketch entities are displayed in the Profiles box.

101) Click **OK** ✔ from the Loft PropertyManager. Loft1 is displayed in the FeatureManager.

Rename the feature.
102) Rename **Loft1** to **Base Loft**.

Hide the planes.
103) Click **View**, **Hide/Show**, uncheck **Planes** from the Menu bar.

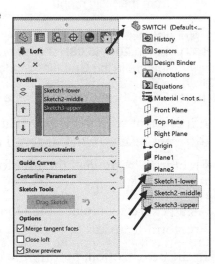

Save the part.
104) Click **Save** 💾.

The system displays a preview curve and loft as you select the profiles. Use the Up button and Down button in the Loft PropertyManager to rearrange the order of the profiles.

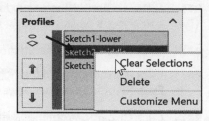

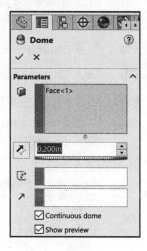 Redefine incorrect selections efficiently. Right-click in the Graphics window and click Clear Selections to remove selected profiles. Select the correct profiles.

SWITCH Part - Dome Feature

Insert the Dome feature on the top face of the Lofted Base feature. The Dome feature forms the top surface of the SWITCH. Note: You can specify an elliptical dome for cylindrical or conical models and a continuous dome for polygonal models. A continuous dome's shape slopes upwards evenly on all sides.

Activity: SWITCH Part - Dome Feature

Insert the Dome feature.
105) Click the **top face** of the Base Loft feature in the Graphics window.

106) Click the **Dome** ⬭ feature tool. The Dome PropertyManager is displayed. Face<1> is displayed in the Faces to Dome box. Note: If the Dome feature is not displayed in the Feature CommandManager, click Insert, Features, Dome from the Main menu.

Enter distance.
107) Enter **.20**in, [**5.08**] for Distance. View the dome feature in the Graphics window.

108) Click **OK** ✔ from the Dome PropertyManager.

Experiment with the Dome feature to display different results. Click Insert, Features, Freeform ⬭ to view the Freeform PropertyManager.

As an exercise, replace the Dome feature with a Freeform feature.

In the next section, modify the offset distance between the Top Plane and Plane1.

Modify the Loft Base feature.

109) **Expand** the Base Loft feature.

110) Right-click on **Annotations** in the FeatureManager.

111) Click **Show Feature Dimensions**.

112) Click on the Plane1 offset dimension, **.500**in, [**12.700**].

113) Enter **.125**in, [**3.180**]. Click **Rebuild**.

114) Click **OK** ✔ from the Dimension PropertyManager.

Hide Feature dimensions.

115) Right-click on **Annotation** in the FeatureManager.

116) Un-check **Show FeatureDimensions**.

Display Performance Evaluation.

117) Click **Performance Evaluation** 🐞 from the Evaluate tab in the CommandManager. View the results.

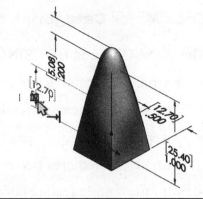

Performance Evaluation — ☐ ✕

Print...	Copy	Refresh	Close

SWITCH
Features 7, Solids 1, Surfaces 0
Total rebuild time in seconds: 0.11

Feature Order	Time %	Time(s)
Dome1	71.56	0.08
Base Loft	28.44	0.03
Plane1	0.00	0.00
Plane2	0.00	0.00
Sketch1-lower	0.00	0.00
Sketch2-middle	0.00	0.00
Sketch3-upper	0.00	0.00

Close SWITCH Performance Evaluation.

118) Click **Close** from the Performance Evaluation dialog box.

119) Click **Save** 💾.

The Dome feature created the top for the SWITCH. The Performance Evaluation report displays the rebuild time for the Dome feature and the other SWITCH features. As feature geometry becomes more complex, the rebuild time increases.

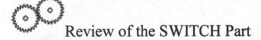 Review of the SWITCH Part

The SWITCH utilized the Lofted Base and Dome feature. The Lofted Base feature required three planes: Sketch1-lower, Sketch2-middle and Sketch3-upper. A profile was sketched on each plane. The three profiles were combined to create the Lofted Base feature.

The Dome feature created the final feature for the SWITCH.

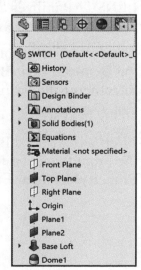

The SWITCH utilized a simple Lofted Base feature. Lofts become more complex with additional Guide Curves. Complex Lofts can contain hundreds of profiles.

Four Major Categories of Solid Features

The LENSCAP and HOUSING combine the four major categories of solid features:

- Extrude: Requires one profile.
- Revolve: Requires one profile and axis of revolution.
- Swept: Requires one profile and one path sketched on different planes.
- Lofted: Requires two or more profiles sketched on different planes.

Identify the simple features of the LENSCAP and HOUSING. Extrude and Revolve are simple features. Only a single sketch profile is required. Swept and Loft are more complex features. Two or more sketches are required.

Example: The O-RING was created as a Swept Base feature.

Could the O-RING utilize an Extrude feature?

Answer: No. Extruding a circular profile produces a cylinder.

Can the O-RING utilize a Revolve feature? Answer: Yes. Revolving a circular profile about a centerline creates the O-RING.

Revolved feature   Sweep feature

A Swept feature is required if the O-RING contained a non-circular path. Example: A Revolved feature does not work with an elliptical path or a more complex curve as in a paper clip. Combine the four major features and additional features to create the LENSCAP and HOUSING.

LENSCAP Part

The LENSCAP is a plastic part used to position the LENS to the HOUSING. The LENSCAP utilizes an Extruded Boss/Base, Extruded Cut, Extruded Thin Cut, Shell, Revolved Cut and Swept features.

The design intent for the LENSCAP requires that the Draft Angle be incorporated into the Extruded Boss/Base and Revolved Cut feature. Create the Revolved Cut feature by referencing the Extrude Base feature geometry. If the Draft angle changes, the Revolved Cut also changes.

Insert an Extruded Boss/Base 🗇 feature with a circular profile on the Front Plane. Use a Draft option in the Boss-Extrude PropertyManager. Enter 5° for Draft angle.

Insert an Extruded Cut 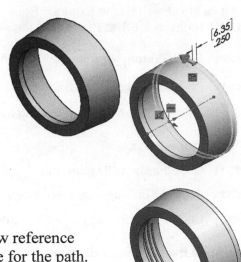 feature. The Extruded Cut feature should be equal to the diameter of the LENS Revolved Base feature.

Insert a Shell feature. Use the Shell feature for a constant wall thickness.

Insert a Revolved Cut feature on the back face. Sketch a single line on the Silhouette edge of the Extruded Base.
Utilize the Thin Feature option in the Cut-Revolve PropertyManager.

Utilize a Swept feature for the thread. Insert a new reference plane for the start of the thread. Insert a Helical Curve for the path. Sketch a trapezoid for the profile.

LENSCAP Part - Extruded Boss/Base, Extruded Cut and Shell Features

Create the LENSCAP. The first feature is a Boss-Extrude feature. Select the Front Plane for the Sketch plane. Sketch a circle centered at the Origin for the profile. Utilize a Draft angle of 5°.

Create an Extruded Cut feature on the front face of the Base feature. The diameter of the Extruded Cut equals the diameter of the Revolved Base feature of the LENS. The Shell feature removes the front and back face from the solid LENSCAP.

Activity: LENSCAP Part - Extruded Base, Extruded Cut and Shell Features

---

Create a New part. Enter name. Enter description.

120) Click **New** from the Menu bar.

121) Click the **MY-TEMPLATES** tab.

122) Double-click **PART-IN-ANSI**, [**PART-MM-ISO**].

123) Click **Save**.

124) Select **PROJECTS** for the Save in folder.

125) Enter **LENSCAP** for File name.

126) Enter **LENSCAP for 6V FLASHLIGHT** for Description.

127) Click **Save**. The LENSCAP FeatureManager is displayed.

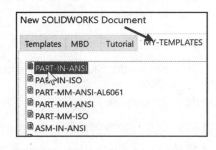

Create the sketch for the Extruded Base feature.

128) Right-click **Front Plane** from the FeatureManager.

129) Click **Sketch** from the Context toolbar.

130) Click the **Circle** Sketch tool. The Circle PropertyManager is displayed. Sketch a **circle** centered at the Origin.

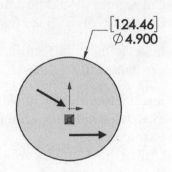

Add a dimension.

131) Click the **Smart Dimension** Sketch tool.

132) Click the **circumference** of the circle.

133) Click a **position** off the profile. Enter **4.900**in, **[124.46]**.

Insert an Extruded Boss/Base feature.

134) Click the **Extruded Boss/Base** feature tool. The Boss-Extrude PropertyManager is displayed. Blind is the default End Condition in Direction 1.

135) Click the **Reverse Direction** box.

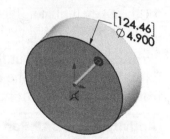

136) Enter **1.725**in, **[43.82]** for Depth in Direction 1.

137) Click the **Draft On/Off** button.

138) Enter **5**deg for Angle. Click the **Draft outward** box.

139) Click **OK** from the Boss-Extrude PropertyManager. Boss-Extrude1 is displayed in the FeatureManager.

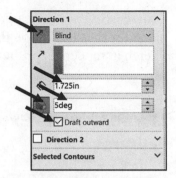

140) Rename **Boss-Extrude1** to **Base Extrude**.

141) Click **Save**.

Create the sketch for the Extruded Cut feature.

142) Right-click the **front face** for the Sketch plane.

143) Click **Sketch** from the Context toolbar. The Sketch toolbar is displayed.

144) Click the **Circle** Sketch tool. The Circle PropertyManager is displayed. Sketch a **circle** centered at the Origin.

Add a dimension.

145) Click the **Smart Dimension** Sketch tool.

146) Click the **circumference** of the circle.

147) Click a **position** off the profile.

148) Enter **3.875**in, **[98.43]**.

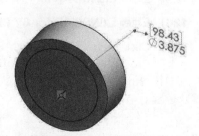

Insert an Extruded Cut feature.

149) Click the **Extruded Cut** feature tool. The Cut-Extrude PropertyManager is displayed. Blind is the default End Condition.

150) Enter **.275**in, [**6.99**] for Depth in Direction 1.

151) Click the **Draft On/Off** button.

152) Enter **5**deg for Angle. Accept the default settings.

153) Click **OK** from the Cut-Extrude PropertyManager. Cut-Extrude1 is displayed in the FeatureManager.

154) Rename **Cut-Extrude1** to **Front-Cut**.

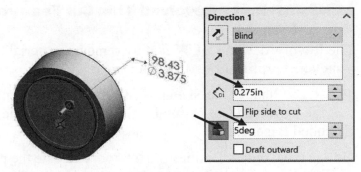

Insert the Shell feature.

155) Click the **Shell** feature tool. The Shell1 PropertyManager is displayed.

156) Click the **front face** of the Front-Cut as illustrated.

157) **Rotate** the model to view the back face. Click the **back face** of the Base Extrude.

158) Enter **.150**in, [**3.81**] for Thickness.

159) Click **OK** from the Shell1 PropertyManager. Shell1 is displayed in the FeatureManager.

Display an Isometric view.

160) Press the **space bar** to display the Orientation dialog box.

161) Click **Isometric view**.

Display the inside of the Shell.

162) Click **Right view**.

163) Click **Hidden Lines Visible** from the Heads-up View toolbar.

164) Click **Save**.

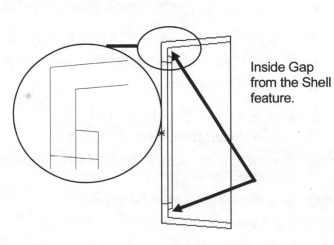

Inside Gap from the Shell feature.

Use the inside gap created by the Shell feature to seat the O-RING in the assembly.

LENSCAP Part - Revolved Thin Cut Feature

The Revolved Thin Cut  feature removes material by rotating a sketched profile around a centerline.

The Right Plane is the Sketch plane. The design intent requires that the Revolved Cut maintains the same Draft angle as the Extruded Base feature.

Utilize the Convert Entities Sketch tool to create the profile. Small thin cuts are utilized in plastic parts. Utilize the Revolved Thin Cut feature for cylindrical geometry in the next activity.

Utilize a Swept Cut for non-cylindrical geometry. The semi-circular Swept Cut profile is explored in the project exercises.

Sweep Cut Example

Activity: LENSCAP Part - Revolved Thin Cut Feature

Create the sketch.

165) Right-click **Right Plane** from the FeatureManager. This is your Sketch plane.

166) Click **Sketch** from the Context toolbar. The Sketch toolbar is displayed.

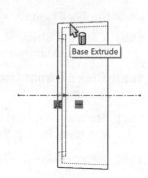

Sketch a centerline.

167) Click the **Centerline** Sketch tool. The Insert Line PropertyManager is displayed.

168) Sketch a **horizontal centerline** through the Origin.

Create the profile.

169) Right-click **Select** to deselect the Centerline Sketch tool.

170) Click the **top silhouette outside edge** of Base Extrude as illustrated.

171) Click the **Convert Entities** Sketch tool.

172) Click and drag the **left endpoint 2/3** towards the right endpoint.

173) Release the **mouse button**.

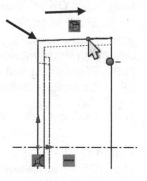

Add a dimension.

174) Click the **Smart Dimension** Sketch tool.

175) Click the **line**. The aligned dimension arrows are parallel to the profile line.

176) Drag the **text upward** and to the right.

177) Enter **.250**in, [**6.35**].

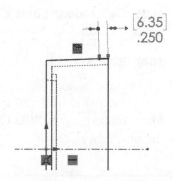

[6.35]
.250

Insert a Revolved Cut feature.

178) Click **Revolved Cut**  from the Features toolbar. Do not close the Sketch. The warning message states: "The sketch is currently open."

179) Click **No**. The Cut-Revolve PropertyManager is displayed.

180) Click the **Reverse Direction** box in the Thin Feature box. The arrow points counterclockwise.

181) Enter **.050**in, **[1.27]** for Direction 1 Thickness.

182) Click **OK**  from the Cut-Revolve PropertyManager. Cut-Revolve-Thin1 is displayed in the FeatureManager.

Display the Revolved Thin Cut feature.
183) **Rotate** the part to view the back face.

184) Click **Isometric view** 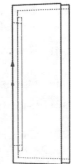 .

185) Click **Shaded With Edges** .

Rename and save the feature.
186) Rename **Cut-Revolve-Thin1** to **BackCut**.

187) Click **Save**.

LENSCAP Part - Thread, Swept Feature and Helix/Spiral Curve

Utilize the Swept feature  to create the required threads. The thread requires a spiral path. This path is called the ThreadPath. The thread requires a Sketched profile. This cross-section profile is called the ThreadSection.

The plastic thread on the LENSCAP requires a smooth lead in. The thread is not flush with the back face. Use an Offset plane to start the thread. There are numerous steps required to create a thread:

- Create a new plane for the start of the thread.

- Create the Thread path. Utilize Convert Entities and click Helix/Spiral from the Curves drop-down menu.

- Create a large thread cross section profile for improved visibility.

- Insert the Swept feature.

- Reduce the size of the thread cross section.

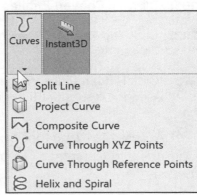

Activity: LENSCAP Part - Thread, Swept Feature, and Helix/Spiral Curve

Create the offset plane.

188) **Rotate** ↻ and **Zoom to Area** 🔍 on the back face of the LENSCAP.

189) Click the **narrow back face** of the Base Extrude feature. Note the mouse feedback icon.

190) Click **Insert**, **Reference Geometry, Plane** from the Menu bar. The Plane PropertyManager is displayed.

191) Enter **.450**in, **[11.43]** for Distance.

192) Click the **Flip offset** box.

193) Click **OK** ✔ from the Plane PropertyManager. Plane1 is displayed in the FeatureManager.

194) Rename **Plane1** to **ThreadPlane**.

Display the Isometric view with Hidden Lines Removed.

195) Click **Isometric view** 🔲.

196) Click **Hidden Lines Removed** 🔲.

Save the model.

197) Click **Save** 💾.

Utilize the Convert Entities Sketch tool to extract the back circular edge of the LENSCAP to the ThreadPlane.

Create the Thread path.

198) Right-click **ThreadPlane** from the FeatureManager.

199) Click **Sketch** 🔲 from the Context toolbar.

200) Click the **back inside circular edge** of the Shell as illustrated.

201) Click the **Convert Entities** 🔲 Sketch tool.

202) Click **Top view** 🔲. The circular edge is displayed on the ThreadPlane.

203) Click **Hidden Lines Visible** 🔲 from the Heads-up View toolbar. View the results.

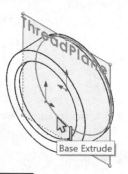

💡 Access the Plane tool from the Consolidated Reference Geometry toolbar.

Insert the Helix/Spiral curve path.

204) Click **Insert, Curve, Helix/Spiral** from the Menu bar. The Helix/Spiral PropertyManager is displayed.

205) Enter .250in, [6.35] for Pitch.

206) Check the **Reverse direction** box.

207) Enter **2.5** for Revolutions.

208) Enter 0deg for Starting angle. The Helix start point and end point are Coincident with the Top Plane.

209) Click the **Clockwise** box.

210) Click the **Taper Helix** box.

211) Enter 5deg for Angle.

212) Uncheck the **Taper outward** box.

213) Click **OK** ✔ from the Helix/Spiral PropertyManager.

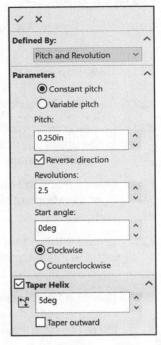

Rename the feature.

214) Rename **Helix/Spiral1** to **ThreadPath**.

Save the model.

215) Click **Save** 💾.

The Helix tapers with the inside wall of the LENSCAP. Position the Helix within the wall thickness to prevent errors in the Swept.

Sketch the profile on the Top Plane. Position the profile to the Top right of the LENSCAP in order to pierce to the ThreadPath in the correct location.

If required, hide the ThreadPlane.

216) Right-click **ThreadPlane** from the FeatureManager.

217) Click **Hide** from the Context toolbar.

218) Click **Hidden Lines Removed** ⬡ from the Heads-up View toolbar.

Select the Plane for the Thread.

219) Right-click **Top Plane** from the FeatureManager.

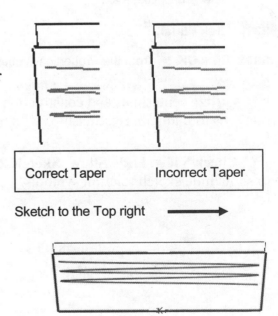

Correct Taper   Incorrect Taper

Sketch to the Top right ➡

Sketch the profile.

220) Click **Sketch**  from the Context toolbar.

221) Click **Top view** .

222) Click the **Centerline** Sketch tool.

223) Create a short **vertical centerline** off to the upper top area of the ThreadPath feature.

224) Create a second **centerline** horizontal from the Midpoint to the left of the vertical line.

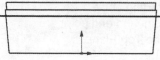

225) Create a third centerline coincident with the left horizontal endpoint. Drag the **centerline upward** until it is approximately the same size as the right vertical line.

226) Create a fourth **centerline** coincident with the left horizontal endpoint. Drag the **centerline** downward until it is approximately the same size as the left vertical line as illustrated.

Add an Equal relation.

227) Right-click **Select** to de-select the Centerline Sketch tool.

228) Click the **right vertical centerline**.

229) Hold the **Ctrl** key down.

230) Click the **two left vertical centerlines**.

231) Release the **Ctrl** key. The selected sketch entities are displayed in the Selected Entities box.

232) Click **Equal** ＝

233) Click **OK** ✔ from the Properties PropertyManager.

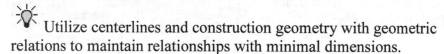

Utilize centerlines and construction geometry with geometric relations to maintain relationships with minimal dimensions.

Check **View, Hide/Show, Sketch Relations** from the Menu bar to show/hide sketch relation symbols.

Add a dimension.

234) Click the **Smart Dimension** ⌖ Sketch tool.

235) Click the two **left vertical endpoints**. Click a **position** to the left.

236) Enter .500in, [12.7].

Sketch the profile. The profile is a trapezoid.

237) Click the **Line** ✎ Sketch tool.

238) Click the **endpoints** of the vertical centerlines to create the trapezoid as illustrated.

239) Right-click **Select** to de-select the Line Sketch tool.

Add an Equal relation.
240) Click the **left vertical line**.

241) Hold the **Ctrl** key down.

242) Click the **top** and **bottom lines** of the trapezoid.

243) Release the **Ctrl** key.

244) Click **Equal** = .

245) Click **OK** ✔ from the Properties PropertyManager.

Click and drag the sketch to a position above the top right corner of the LENSCAP.

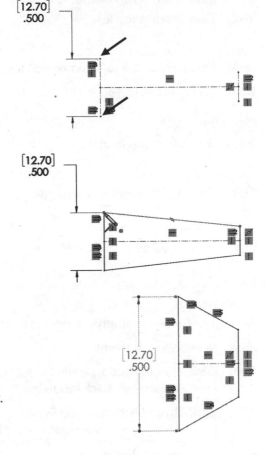

Add a Pierce relation.
246) Click the **left midpoint** of the trapezoid.

247) Hold the **Ctrl** key down.

248) Click the **starting left back edge** of the ThreadPath.

249) Release the **Ctrl** key.

250) Click **Pierce** 𝄢 from the Add Relations box. The sketch is fully defined.

251) Click **OK** ✔ from the Properties PropertyManager.

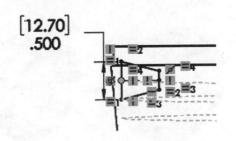

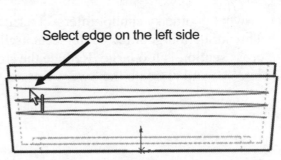

Select edge on the left side

Display the sketch in an Isometric view.

252) Click **Isometric view** .

Modify the dimension.

253) Double-click the **.500** dimension text.

254) Enter **.125**in, **[3.18]**.

Close the sketch.

255) **Rebuild** the model.

Rename the sketch.

256) Rename **Sketch#** to **ThreadSection**.

Save the model.

257) Click **Save** .

Insert the Swept feature.

258) Click the **Swept Boss/Base** feature tool. The Sweep PropertyManager is displayed.

259) Select **Sketch Profile**.

260) Click **ThreadSection** in the fly-out FeatureManager. Click inside the **Path** box.

261) Click **ThreadPath** from the fly-out FeatureManager. ThreadPath is displayed.

262) Click **OK** from the Sweep PropertyManager. Sweep1 is displayed in the FeatureManager.

Rename the sketch. Display Shaded With Edges.

263) Rename **Sweep1** to **Thread**. Click **Shaded With Edges** from the Heads-up View toolbar.

Save the model.

264) Click **Save** .

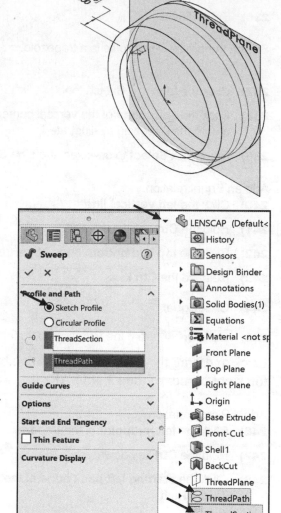

Swept geometry cannot intersect itself. If the ThreadSection geometry intersects itself, the cross section is too large. Reduce the cross section size and recreate the Swept feature.

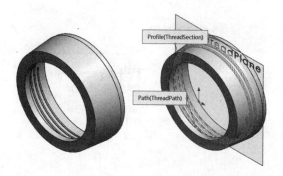

The Thread profile is composed of the following: ThreadSection and ThreadPath.

The ThreadPath contains the circular sketch and the helical curve.

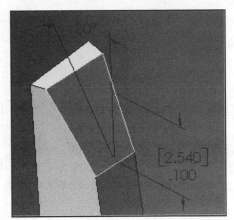

Most threads require a beveled edge or smooth edge for the thread part start point. A 30° Chamfer feature can be utilized on the starting edge of the trapezoid face. This action is left as an exercise.

Create continuous Swept features in a single step. Pierce the cross section profile at the start of the swept path for a continuous Swept feature.

Un-suppress the Pattern feature to resolve both the Pattern feature and the seed feature at the same time.

The LENSCAP is complete. Review the LENSCAP before moving onto the last part of the FLASHLIGHT.

Additional information on Extruded Base/Boss, Extruded Cut, Swept, Helix/Spiral, Circular Pattern and Reference Planes are found in SOLIDWORKS Help Topics.

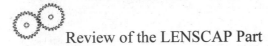

Review of the LENSCAP Part

The LENSCAP utilized the Extruded Base feature with the Draft Angle option. The Extruded Cut feature created an opening for the LENS. You utilized the Shell feature with constant wall thickness to remove the front and back faces.

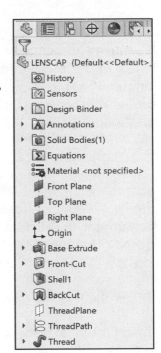

The Revolved Thin Cut feature created the back cut with a single line. The line utilized the Convert Entities tool to maintain the same draft angle as the Extruded Base feature.

You utilized a Swept feature with a Helical Curve and Thread profile to create the thread.

HOUSING Part

The HOUSING is a plastic part utilized to contain the BATTERY and to support the LENS. The HOUSING utilizes an Extruded Boss/Base, Lofted Boss, Extruded Cut, Draft, Swept, Rib, Mirror and Linear Pattern features.

Insert an Extruded Boss/Base
(Boss-Extrude1)  feature centered at the Origin.

Insert a Lofted Boss ⬇ feature. The first profile is the
converted circular edge of the Extruded Base. The
second profile is a sketch on the BatteryLoftPlane.

Insert the second Extruded Boss/Base (Boss-Extrude2) 🔲
feature. The sketch is a converted edge from the Loft Boss.
The depth is determined from the height of the BATTERY.

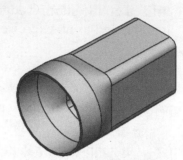

Insert a Shell 🔲 feature to create a thin walled part.

Insert the third Extruded Boss (Boss-Extrude3) 🔲 feature.
Create a solid circular ring on the back circular face of the
Boss-Extrude1 feature. Insert the Draft feature to add a draft
angle to the circular face of the HOUSING. The design
intent for the Boss-Extrude1 feature requires you to maintain
the same LENSCAP draft angle.

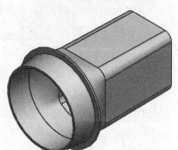

Insert a Swept 〽 feature for the Thread. Insert a Swept
feature for the Handle. Reuse the Thread profile from the
LENSCAP part. Insert an Extruded Cut to create the hole for
the SWITCH.

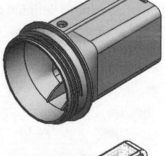

Insert the Rib 🔲 feature on the back face of the HOUSING.

Insert a Linear Pattern 🔲 feature to create a row of Ribs.

Insert a Rib 🔲 feature along the bottom of the HOUSING.

Utilize the Mirror 🔲 feature to create the second Rib.

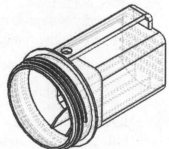

💡 Reuse geometry between parts. The LENSCAP thread is the same as the HOUSING thread. Copy the ThreadSection from the LENSCAP to the HOUSING.

💡 Reuse geometry between features. The Linear Pattern and Mirror Pattern utilized existing features.

Activity: HOUSING Part - Extruded Base Feature

Create the New part. Enter name. Enter description.

265) Click **New** ⬚ from the Menu bar.

266) Click the **MY-TEMPLATES** tab.

267) Double-click **PART-IN-ANSI**, **[PART-MM-ISO]**.

268) Click **Save**. Select **PROJECTS** for the Save in folder.

269) Enter **HOUSING** for File name. Enter **HOUSING FOR 6VOLT FLASHLIGHT** for Description.

270) Click **Save**. The HOUSING FeatureManager is displayed.

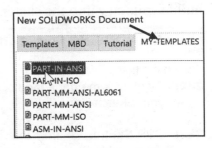

Create the sketch.
271) Right-click **Front Plane** from the FeatureManager. This is your Sketch plane.

272) Click **Sketch** 🗒 from the Context toolbar.

273) Click the **Circle** ⊙ Sketch tool. The Circle PropertyManager is displayed.

274) Sketch a circle centered at the **Origin** ↳ as illustrated.

Add a dimension.
275) Click the **Smart Dimension** ✐ Sketch tool.

276) Click the **circumference**. Enter **4.375**in, **[111.13]**.

Insert an Extruded Boss/Base feature.
277) Click the **Extruded Boss/Base** 🗐 feature tool. The Boss-Extrude PropertyManager is displayed.

278) Enter **1.300**, **[33.02]** for Depth in Direction 1. Accept the default settings. Click **OK** ✔ from the Boss-Extrude1 PropertyManager.

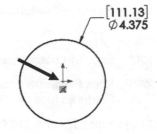

279) Click **Isometric view** 🔲. Note the location of the Origin.

Rename the feature and save the model.
280) Rename **Boss-Extrude1** to **Base Extrude**.

281) Click **Save** 💾.

HOUSING Part - Lofted Boss Feature

The Lofted Boss feature is composed of two profiles. The first sketch is named Sketch-Circle. The second sketch is named Sketch-Square.

Create the first profile from the back face of the Extruded feature. Utilize the Convert Entities sketch tool to extract the circular geometry to the back face.

Create the second profile on an Offset Plane. The FLASHLIGHT components must remain aligned to a common centerline. Insert dimensions that reference the Origin and build symmetry into the sketch. Utilize the Mirror Entities Sketch tool.

Activity: HOUSING Part - Lofted Boss Feature

Create the first profile.

282) Right-click the **back face** of the Base Extrude feature. This is your Sketch plane.

283) Click **Sketch** from the Context toolbar. The Sketch toolbar is displayed.

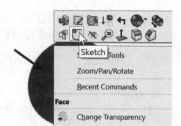

284) Click the **Convert Entities** Sketch tool to extract the face to the Sketch plane.

Close and rename the sketch.

285) Right-click **Exit Sketch**.

286) Rename **Sketch2** to **SketchCircle**.

Create an offset plane.

287) Click the **back face** of the Base Extrude feature.

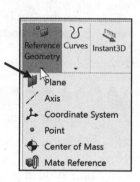

288) Click **Plane** from the Consolidated Reference Geometry Features toolbar. The Plane PropertyManager is displayed.

289) Enter **1.300**in, **[33.02]** for Distance.

290) Click **Top view** to verify the Plane position.

291) Click **OK** from the Plane PropertyManager. Plane1 is displayed in the FeatureManager.

Rename Plane1, rebuild and save the model.

292) Rename **Plane1** to **BatteryLoftPlane**.

293) **Rebuild** the model.

294) Click **Save**.

Create the second profile.

295) Right-click **BatteryLoftPlane** in the FeatureManager.

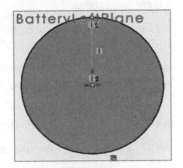

296) Click **Sketch** from the Context toolbar. The Sketch toolbar is displayed.

297) Click **Back view**.

298) Click the **circumference** of the circle.

299) Click the **Convert Entities** Sketch tool.

300) Click the **Centerline** Sketch tool.

301) Sketch a **vertical centerline** coincident to the Origin and to the top edge of the circle as illustrated.

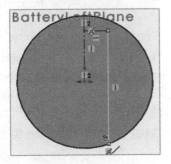

302) Click the **Line** Sketch tool.

303) Sketch a **horizontal line** to the right side of the centerline as illustrated.

304) Sketch a **vertical line** down to the circumference.

305) Click the **Sketch Fillet** Sketch tool. The Sketch Fillet PropertyManager is displayed.

306) **Clear** the Entities to Fillet box.

307) Click the **horizontal line** in the Graphics window.

308) Click the **vertical line** in the Graphics window.

309) Enter **.1**in [**2.54**] for Radius.

310) Click **OK** from the Sketch Fillet PropertyManager. View the Sketch Fillet in the Graphics window.

311) Click **OK** from the PropertyManager.

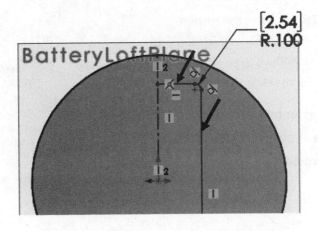

Mirror the profile.

312) Click the **Mirror Entities** 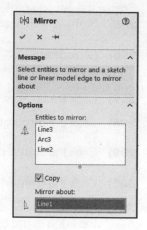 Sketch tool. The Mirror PropertyManager is displayed.

313) Click the **horizontal line**, **fillet** and **vertical line**. The selected entities are displayed in the Entities to mirror box.

314) Click inside the **Mirror about** box.

315) Click the **centerline** from the Graphics window.

316) Click **OK** ✔ from the Mirror PropertyManager.

Trim unwanted geometry.

317) Click the **Trim Entities** ✄ Sketch tool. The Trim PropertyManager is displayed.

318) Click **PowerTrim** ⊞ from the Options box.

319) Click a **position** to the far right of the circle.

320) Drag the **mouse pointer** to intersect the circle.

321) Perform the same **actions** on the left side of the circle.

322) Click **OK** ✔ from the Trim PropertyManager.

Add dimensions.

323) Click the **Smart Dimension** ✎ Sketch tool.

Create the horizontal dimension.
324) Click the **left vertical** line.

325) Click the **right vertical** line.

326) Click a **position** above the profile.

327) Enter **3.100**in, **[78.74]**. View the results.

Create the vertical dimension.

328) Click the **Origin** ⊾ .

329) Click the **top horizontal** line.

330) Click a **position** to the right of the profile.

331) Enter **1.600**in, **[40.64]**.

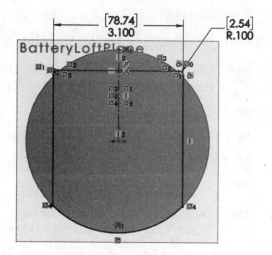

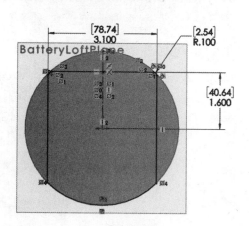

Modify the fillet dimension.

332) Double-click the **.100** fillet dimension.

333) Enter **.500**in, **[12.7]**.

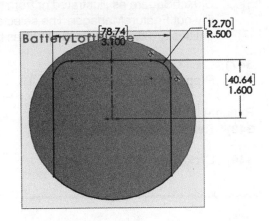

Remove all sharp edges.

334) Click the **Sketch Fillet** Sketch tool. The Sketch Fillet PropertyManager is displayed.

335) Enter **.500**in, **[12.7]** for Radius.

336) Click the **lower left corner point**.

337) Click the **lower right corner point**.

338) Click **OK** ✔ from the Sketch Fillet PropertyManager.

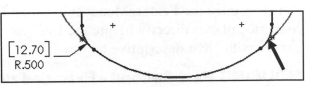

339) Click **OK** ✔ from the PropertyManager.

Close and rename the sketch.

340) Right-click **Exit Sketch**.

341) Rename the **Sketch3** to **SketchSquare**.

Save the model.

342) Click **Save** .

The Loft feature is composed of the SketchSquare and the SketchCircle. Select two individual profiles to create the Loft. The Isometric view provides clarity when selecting Loft profiles.

Display an Isometric view.

343) Click **Isometric view** .

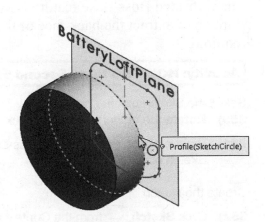

Insert a Lofted Boss feature.

344) Click the **Lofted Boss/Base** feature tool. The Loft PropertyManager is displayed.

345) Click the **upper right side** of the SketchCircle as illustrated.

346) Click the **upper right side** of the SketchSquare as illustrated or from the Fly-out FeatureManager. The selected entities are displayed in the Profiles box.

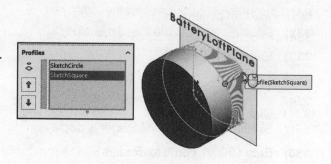

347) Click **OK** ✔ from the Loft PropertyManager.

Rename Loft1. Save the model.

348) Rename **Loft1** to **Boss-Loft1**.

349) Click **Save** 💾 .

💡 Organize the FeatureManager to locate Loft profiles and planes. Insert the Loft reference planes directly before the Loft feature. Rename the planes, profiles and guide curves with clear descriptive names.

HOUSING Part-Second - Extruded Boss/Base Feature

Create the second Extruded Boss/Base feature from the square face of the Loft. How do you estimate the depth of the second Extruded Boss/Base feature? Answer: The Boss-Extrude1 feature of the BATTERY is 4.100in, [104.14mm].

Ribs are required to support the BATTERY. Design for Rib construction. Ribs add strength to the HOUSING and support the BATTERY. Use a 4.400in, [111.76mm] depth as the first estimate. Adjust the estimated depth dimension later if required in the FLASHLIGHT assembly.

The Extruded Boss/Base feature is symmetric about the Right Plane. Utilize Convert Entities to extract the back face of the Loft Base feature. No sketch dimensions are required.

Activity: HOUSING Part - Second Extruded Boss/Base Feature

Select the Sketch plane.

350) **Rotate** the model to view the back.

351) Right-click the **back face** of Boss-Loft1. This is your Sketch plane.

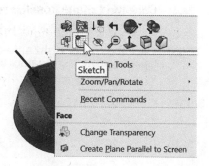

Create the sketch.

352) Click **Sketch** 🖉 from the Context toolbar. The Sketch toolbar is displayed.

353) Click the **Convert Entities** 🗖 Sketch tool.

Insert the second Extruded Boss/Base feature.

354) Click the **Extruded Boss/Base** feature tool. The Boss-Extrude PropertyManager is displayed.

355) Enter **4.400**in, [**111.76**] for Depth in Direction 1.

356) Click the **Draft On/Off** box.

357) Enter **1**deg for Draft Angle.

358) Click **OK** from the Boss-Extrude PropertyManager.

Display a Right view.
359) Click **Right view**.

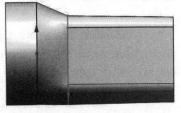

Rename Boss-Extrude2 and save the model.
360) Rename **Boss-Extrude2** to **Boss-Battery**.

361) Click **Save**.

HOUSING Part - Shell Feature

The Shell feature removes material. Use the Shell feature to remove the front face of the HOUSING. In the injection-molded process, the body wall thickness remains constant.

A dialog box is displayed if the thickness value is greater than the Minimum radius of Curvature.

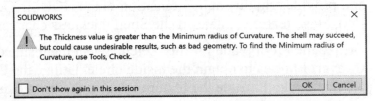

SOLIDWORKS

The Thickness value is greater than the Minimum radius of Curvature. The shell may succeed, but could cause undesirable results, such as bad geometry. To find the Minimum radius of Curvature, use Tools, Check.

☐ Don't show again in this session　　　　OK　Cancel

Activity: HOUSING Part - Shell Feature

Insert the Shell feature.

362) Click **Isometric view**. If needed **Show** the BatteryLoftPlane from the FeatureManager.

363) Click **View**, **Hide/Show**, **Planes** from the Main menu.

364) Click the **Shell** feature tool. The Shell1 PropertyManager is displayed.

365) Click the **front face** of the Base Extrude feature as illustrated.

366) Enter **.100**in, [**2.54**] for Thickness.

367) Click **OK** ✔ from the Shell1 PropertyManager. Shell1 is displayed in the FeatureManager.

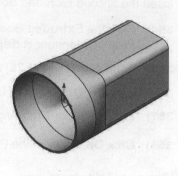

💡 The Shell feature position in the FeatureManager determines the geometry of additional features. Features created before the Shell contained the wall thickness specified in the Thickness option. Position features of different thickness such as the Rib feature and Thread Swept feature after the Shell. Features inserted after the Shell remain solid.

HOUSING Part - Third Extruded Boss/Base Feature

The third Extruded Boss/Base feature creates a solid circular ring on the back circular face of the Boss-Extrude1 feature. The solid ring is a cosmetic stop for the LENSCAP and provides rigidity at the transition of the HOUSING. Design for change. The Extruded Boss/Base feature updates if the Shell thickness changes.

Utilize the Front plane for the sketch. Select the inside circular edge of the Shell. Utilize Convert Entities to obtain the inside circle. Utilize the Circle Sketch tool to create the outside circle. Extrude the feature towards the front face.

Activity: HOUSING Part - Third Extruded Boss Feature

Select the Sketch plane.
368) Right-click **Front Plane** from the FeatureManager. This is your Sketch plane.

Create the sketch.

369) Click **Sketch** 🖉 from the Context toolbar. The Sketch toolbar is displayed.

370) **Zoom-in** and click the **front inside circular edge** of Shell1 as illustrated.

371) Click the **Convert Entities** 🗍 Sketch tool.

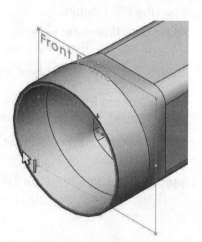

Create the outside circle.
372) Click **Front view** 🗊.

373) Click the **Circle** ⊙ Sketch tool. The Circle PropertyManager is displayed.

374) Sketch a **circle** centered at the Origin.

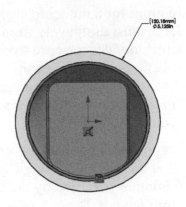

Add a dimension.

375) Click the **Smart Dimension** ⤢ Sketch tool.

376) Click the **circumference** of the circle.

377) Enter **5.125**in, [**130.18**].

Insert the third Extruded Boss/Base feature.

378) Click the **Extruded Boss/Base** 🔘 feature tool. The Boss-Extrude PropertyManager is displayed.

379) Enter **.100**in, [**2.54**] for Depth in Direction 1. The Extrude arrow points to the front.

380) Click **OK** ✔ from the Boss-Extrude PropertyManager.

Display an Isometric view.

381) Click **Isometric view** 🧊.

Rename Boss-Extrude3. Hide the BatteryLoftPlane. Save the model.

382) Rename **Boss-Extrude3** to **Boss-Stop**.

383) If needed, **hide** the BatteryLoftPlane as illustrated.

384) Click **Save** 💾.

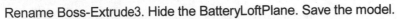

HOUSING Part - Draft Feature

The Draft feature tapers selected model faces by a specified angle by utilizing a Neutral Plane or Parting Line. The Neutral Plane option utilizes a plane or face to determine the pull direction when creating a mold.

The Parting Line option drafts surfaces around a parting line of a mold. Utilize the Parting Line option for non-planar surfaces. Apply the Draft feature to solid and surface models.

A 5deg draft is required to ensure proper thread mating between the LENSCAP and the HOUSING. The LENSCAP Extruded Boss/Base (Boss-Extrude1) feature has a 5deg draft angle.

The outside face of the Extruded Boss/Base (Boss-Extrude1) feature HOUSING requires a 5° draft angle. The inside HOUSING wall does not require a draft angle. The Extruded Boss/Base (Boss-Extrude1) feature has a 5° draft angle. Use the Draft feature to create the draft angle. The front circular face is the Neutral Plane. The outside cylindrical surface is the face to draft.

You created the Extruded Boss/Base and Extruded Cut features with the Draft Angle option. The Draft feature differs from the Extruded feature, Draft Angle option. The Draft feature allows you to select multiple faces to taper.

In order for a model to eject from a mold, all faces must draft away from the parting line which divides the core from the cavity. Cavity side faces display a positive draft, and core side faces display a negative draft. Design specifications include a minimum draft angle, usually less than 5deg.

For the model to eject successfully, all faces must display a draft angle greater than the minimum specified by the Draft Angle. The Draft feature, Draft Analysis Tools and DraftXpert utilize the draft angle to determine what faces require additional draft base on the direction of pull.

☀ You can apply a draft angle as a part of an Extruded Boss/Base or Extruded Cut feature.

The Draft PropertyManager provides the ability to select either the *Manual* or *DraftXpert* tab. Each tab has a separate menu and option selections. The Draft PropertyManager displays the appropriate selections based on the type of draft you create.

☀ The DraftXpert PropertyManager provides the ability to manage the creation and modification of all Neutral Plane drafts. Select the draft angle and the references to the draft. The DraftXpert manages the rest.

Activity: HOUSING Part - Draft Feature

Insert the Draft Neutral Plane feature.

385) Click the **Draft** 🎲 feature tool. The Draft PropertyManager is displayed.

386) Click the **Manual** tab.

387) Click **Neutral** for Type of draft.

388) Click inside the **Neutral Plane** box.

389) Zoom-in and click the thin **front circular face** of Base Extrude.

390) Click inside the **Faces to draft** box.

391) Click the **outside cylindrical face** as illustrated.

392) Enter **5**deg for Draft Angle.

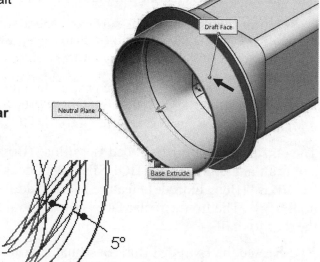

393) Click **OK** ✔ from the Draft PropertyManager. Draft1 is displayed in the FeatureManager.

Display the draft angle and the straight interior.

394) Click **Right view** ⬛.

395) Click **Hidden Lines Visible** ⬛ from the Heads-up View toolbar.

Save the model.

396) Click **Save** 💾.

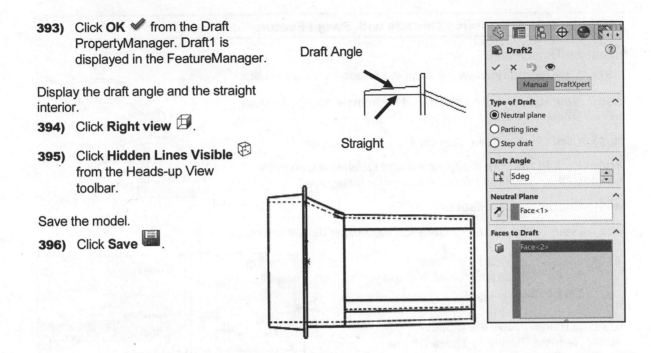

Draft Angle

Straight

💡 The order of feature creation is important. Apply threads after the Draft feature for plastic parts to maintain a constant thread thickness.

HOUSING Part - Threads with Swept Feature

The HOUSING requires a thread. Create the threads for the HOUSING on the outside of the Draft feature. Create the thread with the Swept feature. The thread requires two sketches: ThreadPath and ThreadSection. The LENSCAP and HOUSING Thread utilize the same technique. Create a ThreadPlane. Utilize Convert Entities to create a circular sketch referencing the HOUSING Extruded Boss/Base (Boss-Extrude1) feature. Insert a Helix/Spiral curve to create the path.

Reuse geometry between parts. The ThreadSection is copied from the LENSCAP and is inserted into the HOUSING Top Plane.

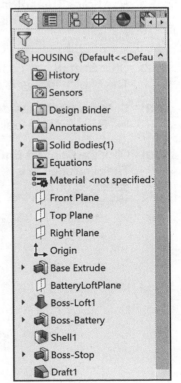

Activity: HOUSING Part - Threads with Swept Feature

Insert the ThreadPlane.

397) Click **Isometric view** ⬛ from the Heads-up View toolbar.

398) Click **Hidden Lines Removed** ⬜ from the Heads-up View toolbar.

399) Click the **thin front circular face**, Base Extrude.

400) Click **Plane** from the Consolidated Reference Geometry toolbar. The Plane PropertyManager is displayed.

401) Check the **Flip offset** box.

402) Enter **.125**in, **[3.18]** for Distance. Accept the default settings.

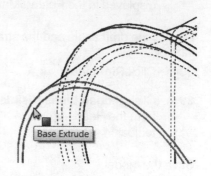

403) Click **OK** ✔ from the Plane PropertyManager. Plane2 is displayed in the FeatureManager.

Rename Plane2. Save the model.

404) Rename **Plane2** to **ThreadPlane**.

405) Click **Save** 🖫.

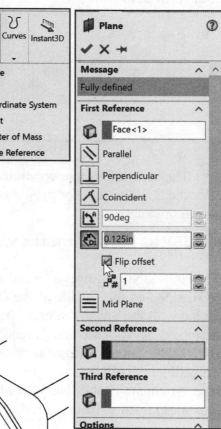

Insert the ThreadPath.

406) Right-click **ThreadPlane** from the FeatureManager. This is your Sketch plane.

407) Click **Sketch** 📝 from the Context toolbar. The Sketch toolbar is displayed.

408) Click the **front outside circular edge** of the Base Extrude feature as illustrated.

409) Click the **Convert Entities** 🗗 Sketch tool. The circular edge is displayed on the ThreadPlane.

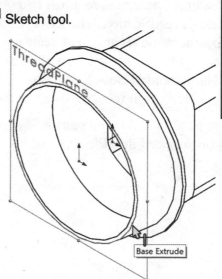

Insert the Helix/Spiral Constant pitch curve.

410) Click the **Helix and Spiral** tool from the Consolidated Curves toolbar as illustrated. The Helix/Spiral PropertyManager is displayed.

411) Enter **.250**in, **[6.35]** for Pitch.

412) Click the **Reverse Direction** box.

413) Enter **2.5** for Revolution.

414) Enter **180** in the Start angle box. The Helix start point and end point are Coincident with the Top Plane.

415) Click the **Taper Helix** box.

416) Enter **5**deg for Angle.

417) Check the **Taper outward** box.

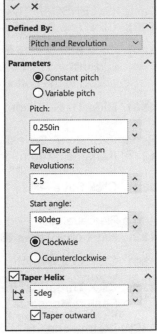

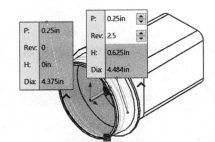

418) Click **OK** ✔ from the Helix/Spiral PropertyManager. Helix/Spiral1 is displayed in the FeatureManager.

419) Rename **Helix/Spiral1** to **ThreadPath**.

Display an Isometric view.

420) Click **Isometric view** 🔲.

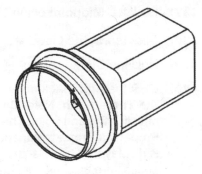

Save the model.

421) Click **Save** 💾.

Copy the LENSCAP ThreadSection.

422) **Open** the LENSCAP part. The LENSCAP FeatureManager is displayed.

423) **Expand** the Thread feature from the FeatureManager.

424) Click the **ThreadSection** sketch. ThreadSection is highlighted.

425) Click **Edit**, **Copy** from the Menu bar.

426) **Close** the LENSCAP.

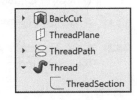

Open the HOUSING.

427) **Return** to the Housing.

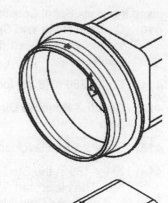

Paste the LENSCAP ThreadSection.

428) Click **Top Plane** from the HOUSING FeatureManager.

429) Click **Edit**, **Paste** from the Menu bar. The ThreadSection is displayed on the Top Plane. The new Sketch7 name is added to the bottom of the FeatureManager.

430) **Hide** ThreadPlane.

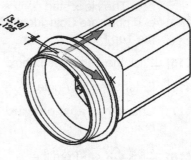

Rename Sketch7. Save the model.

431) Rename **Sketch7** to **ThreadSection**.

432) Click **Save** 💾.

Add a Pierce relation.

433) Right-click **ThreadSection** from the FeatureManager.

434) Click **Edit Sketch**.

435) Click **ThreadSection** from the HOUSING FeatureManager.

436) **Zoom in** on the Midpoint of the ThreadSection.

437) Click the **Midpoint** of the ThreadSection.

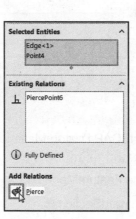

Midpoint of the ThreadSection

438) Click **Isometric view** 🔲.

439) Hold the **Ctrl** key down.

440) Click the **right back edge of the ThreadPath**. Note: Do not click the end point. The Properties PropertyManager is displayed. The selected entities are displayed in the Selected Entities box.

441) Release the **Ctrl** key.

442) Click **Pierce** from the Add Relations box.

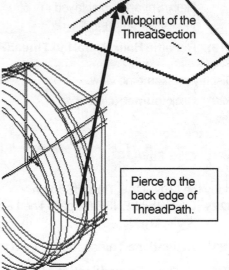

Pierce to the back edge of ThreadPath.

443) Click **OK** ✔ from the Properties PropertyManager.

Caution: Do not click the front edge of the Thread path. The Thread is then created out of the HOUSING.

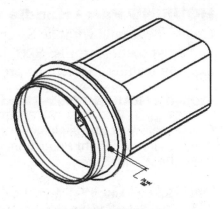

Close the sketch.
444) Right-click **Exit Sketch**.

Insert the Swept feature.

445) Click the **Swept Boss/Base** 🪱 feature tool. The Swept PropertyManager is displayed.

446) Select **Sketch Profile**.

447) **Expand** HOUSING from the fly-out FeatureManager.

448) Click inside the **Profile** box.

449) Click **ThreadSection** from the fly-out FeatureManager.

450) Click **ThreadPath** from the fly-out FeatureManager.

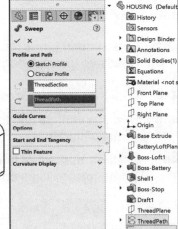

451) Click **OK** ✔ from the Sweep PropertyManager. Sweep1 is displayed in the FeatureManager.

Rename Sweep1. Save the model.
452) Rename **Sweep1** to **Thread**.

453) Click **Save** 💾.

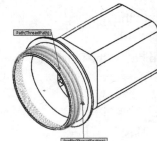

💡 Creating a ThreadPlane provides flexibility to the design. The ThreadPlane allows for a smoother lead. Utilize the ThreadPlane offset dimension to adjust the start of the thread.

HOUSING Part - Handle with Swept Feature

Create the handle with the Swept feature. The Swept feature consists of a sketched path and cross section profile. Sketch the path on the Right Plane. The sketch uses edges from existing features. Sketch the profile on the back circular face of the Boss-Stop feature.

Activity: HOUSING Part - Handle with Swept Feature

Create the Swept path sketch.

454) Right-click **Right Plane** from the FeatureManager.

455) Select **Sketch** from the Context toolbar. The Sketch toolbar is displayed.

456) Click **Right view**.

457) Click **Hidden Lines Removed** from the Heads-up View toolbar.

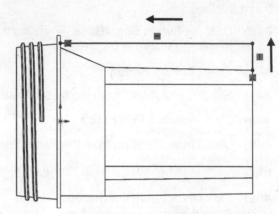

458) Click the **Line** Sketch tool.

459) Sketch a **vertical line** from the right top corner of the Housing upward.

460) Sketch a **horizontal line** below the top of the Boss Stop as illustrated.

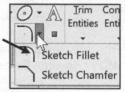

Insert a 2D Fillet.

461) Click the **Sketch Fillet** Sketch tool.

462) Click the **right top corner** of the sketch lines as illustrated.

463) Enter **.500**in, **[12.7]** for Radius.

464) Click **OK** from the Sketch Fillet PropertyManager.

465) Click **OK** from the PropertyManager.

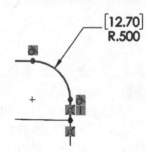

If needed, add a Coincident relation.

466) Click the **left end point** of the horizontal line.

467) Hold the **Ctrl** key down.

468) Click the **right vertical edge** of the Boss Stop.

469) Release the **Ctrl** key.

470) Click **Coincident** from the Add Relations box.

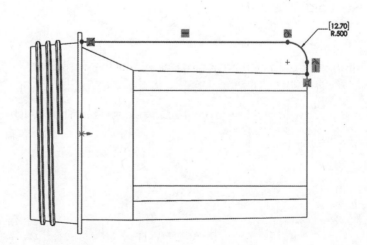

471) Click **OK** ✔ from the Properties PropertyManager.

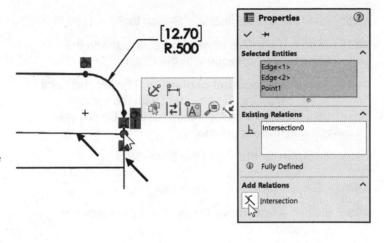

Add an Intersection relation. If needed, use the Add Relations tool.

472) Click the **bottom end point** of the vertical line.

473) Hold the **Ctrl** key down.

474) Click the **right vertical edge** of the Housing.

475) Click the **horizontal edge** of the Housing.

476) Release the **Ctrl** key.

477) Click **Intersection** ✕ from the Add Relations box.

478) Click **OK** ✔ from the Properties PropertyManager.

Add a dimension.

479) Click the **Smart Dimension** ✎ Sketch tool.

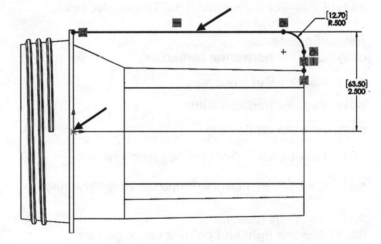

480) Click the **Origin**.

481) Click the **horizontal line**.

482) Click a **position** to the right.

483) Enter **2.500**in, [**63.5**].

Close and rename the sketch.

484) Right-click **Exit Sketch**.

485) Rename **Sketch8** to **HandlePath**.

Save the model.

486) Click **Save** 💾.

Create the Swept Profile.

487) Click **Back view** 🔲.

488) Right-click the **back circular face** of the Boss-Stop feature as illustrated.

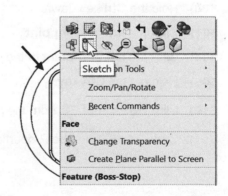

489) Click **Sketch** 🗒 from the Context toolbar. The Sketch toolbar is displayed.

490) Click the **Centerline** ✎ Sketch tool.

491) Sketch a **vertical centerline** collinear with the Right Plane, coincident to the Origin.

492) Sketch a **horizontal centerline**. The left end point of the centerline is coincident with the vertical centerline on the Boss-Stop feature. Do not select existing feature geometry.

493) **Zoom in** on the top of the Boss-Stop.

494) Click the **Line** ✏ Sketch tool.

495) Sketch a **line** above the horizontal centerline as illustrated.

496) Click the **Tangent Arc** ⌐ Sketch tool.

497) Sketch a **90° arc**.

498) Right-click **Select** to exit the Tangent Arc tool.

Add an Equal relation.
499) Click the **horizontal centerline**.

500) Hold the **Ctrl** key down.

501) Click the **horizontal line**.

502) Release the **Ctrl** key.

503) Click **Equal** = from the Add Relations box.

504) Click **OK** ✔ from the Properties PropertyManager.

Add a Horizontal relation.
505) Click the **right end point** of the tangent arc.

506) Hold the **Ctrl** key down.

507) Click the **arc center point**.

508) Click the **left end point** of the centerline.

509) Release the **Ctrl** key.

510) Click **Horizontal** — from the Add Relations box.

511) Click **OK** ✔ from the Properties PropertyManager.

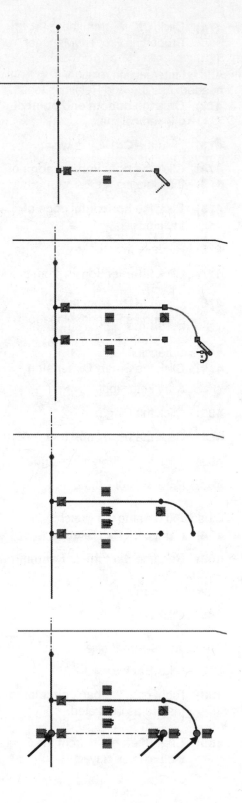

Mirror about the horizontal centerline.

512) Click the **Mirror Entities** ⊞ Sketch tool. The Mirror PropertyManager is displayed.

513) Click the **horizontal** line.

514) Click the **90° arc**. The selected entities are displayed in the Entities to mirror box.

515) Click inside the **Mirror about** box.

516) Click the **horizontal centerline**.

517) Click **OK** ✔ from the Mirror PropertyManager.

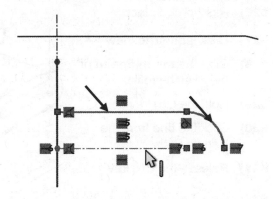

Mirror about the vertical centerline.

518) Click the **Mirror Entities** ⊞ Sketch tool. The Mirror PropertyManager is displayed.

519) Window-Select the **two horizontal lines**, the **horizontal centerline** and the **90° arc** for Entities to mirror. The selected entities are displayed in the Entities to mirror box.

520) Click inside the **Mirror about** box.

521) Click the **vertical centerline**.

522) Click **OK** ✔ from the Mirror PropertyManager.

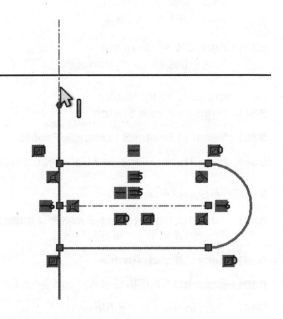

Add dimensions.

523) Click the **Smart Dimension** ✎ Sketch tool.

524) Enter **1.000**in, **[25.4]** between the arc center points.

525) Enter **.100**in, **[2.54]** for Radius.

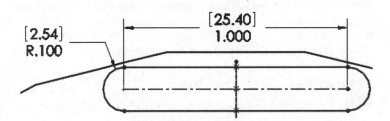

Add a Pierce relation.

526) Right-click **Select**.

527) Click **Isometric view** 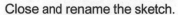.

528) Click the **top midpoint** of the Sketch profile.

529) Hold the **Ctrl** key down.

530) Click the **line** from the Handle Path.

531) Release the **Ctrl** key.

532) Click **Pierce** from the Add Relations box. The sketch is fully defined.

533) Click **OK** ✔ from the Properties PropertyManager.

Close and rename the sketch.

534) Right-click **Exit Sketch**.

535) Rename **Sketch8** to **HandleProfile**.

536) **Hide** ThreadPlane and BatteryLoftPlane if needed.

Insert the Swept feature.

537) Click the **Swept Boss/Base** feature tool. The Sweep PropertyManager is displayed.

538) Select **Sketch Profile**.

539) **Expand** HOUSING from the fly-out FeatureManager.

540) Click inside the **Profile** box.

541) Click **HandleProfile** from the fly-out FeatureManager.

542) Click the **HandlePath** from the fly-out FeatureManager.

543) Click **OK** ✔ from the Sweep PropertyManager. Sweep2 is displayed in the FeatureManager.

Fit the profile to the Graphics window.

544) Press the **f** key.

Display Shaded With Edges.

545) Click **Shaded With Edges** from the Heads-up View toolbar.

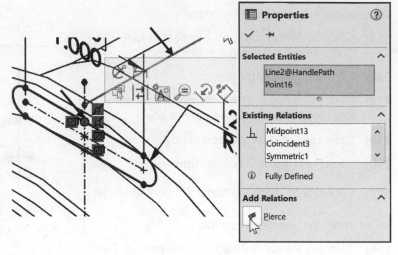

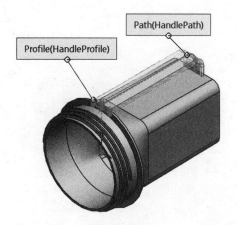

Rename Sweep2. Save the model.
546) Rename **Sweep2** to **Handle**.

547) Click **Save** .

How does the Handle Swept feature interact with other parts in the FLASHLIGHT assembly? Answer: The Handle requires an Extruded Cut to insert the SWITCH.

HOUSING Part - Extruded Cut Feature with Up To Surface

Create an Extruded Cut in the Handle for the SWITCH. Utilize the top face of the Handle for the Sketch plane. Create a circular sketch centered on the Handle.

Utilize the Up To Surface End Condition in Direction 1. Select the inside surface of the HOUSING for the reference surface.

Activity: HOUSING Part - Extruded Cut Feature with Up To Surface

Select the Sketch plane.
548) Right-click the **top face** of the Handle. Handle is highlighted in the FeatureManager. This is your Sketch plane.

Create the sketch.
549) Click **Sketch** from the Context toolbar. The Sketch toolbar is displayed.

550) Click **Top view** .

551) Click **Circle** from the Sketch toolbar. The Circle PropertyManager is displayed.

552) Sketch a **circle** on the Handle near the front as illustrated.

Deselect the circle sketch tool.
553) Right-click **Select**.

Add a Vertical relation.
554) Click the **Origin**.

555) Hold the **Ctrl** key down.

556) Click the **center point** of the circle. The Properties PropertyManager is displayed. The selected entities are displayed in the Selected Entities box.

557) Release the **Ctrl** key.

558) Click **Vertical** | .

559) Click **OK** from the Properties PropertyManager.

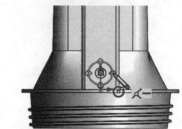

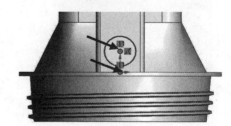

Add dimensions.

560) Click the **Smart Dimension** ✎ Sketch tool.

561) Enter **.510**in, [**12.95**] for diameter.

562) Enter **.450**in, [**11.43**] for the distance from the Origin.

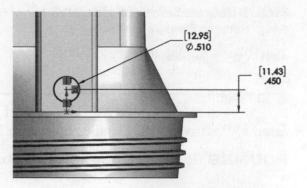

Insert an Extruded Cut feature.

563) **Rotate** the model to view the inside Shell1.

564) Click the **Extruded Cut** ⬚ feature tool. The Cut-Extrude PropertyManager is displayed.

565) Select the **Up To Surface** End Condition in Direction 1.

566) Click the **top inside face** of the Shell1 as illustrated.

567) Click **OK** ✔ from the Cut-Extrude PropertyManager. The Cut-Extrude1 feature is displayed in the FeatureManager.

Rename Cut-Extrude1. Display an Isometric view.

568) Rename the **Cut-Extrude1** feature to **SwitchHole**.

569) Click **Isometric view** 🧊 from the Heads-up View toolbar.

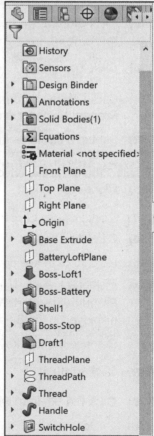

Save the model.

570) Click **Save** 💾.

HOUSING Part - First Rib and Linear Pattern Feature

The Rib ✎ feature adds material between contours of existing geometry. Use Ribs to add structural integrity to a part.

A Rib requires:

- A sketch

- Thickness

- Extrusion direction

The first Rib profile is sketched on the Top Plane. A 1° draft angle is required for manufacturing. Determine the Rib thickness by the manufacturing process and the material.

🔆 Rule of thumb states that the Rib thickness is ½ the part wall thickness. The Rib thickness dimension is .100in [2.54mm] for illustration purposes.

The HOUSING requires multiple Ribs to support the BATTERY. A Linear Pattern feature creates multiple instances of a feature along a straight line. Create the Linear Pattern feature in two directions along the same vertical edge of the HOUSING.

🔆 The Instance to Vary option in the Linear Pattern PropertyManager allows you to vary the dimensions and locations of instances in a feature pattern *after it is created*. You can vary the dimensions of a series of instances, so that each instance is larger or smaller than the previous one. You can also change the dimensions of a single instance in a pattern and change the position of that instance relative to the seed feature of the pattern. For linear patterns, you can change the spacing between the columns and rows in the pattern.

Activity: HOUSING Part - First Rib and Linear Pattern Feature

Display all hidden lines.

571) Click **Hidden Lines Visible** ⬡ from the Heads-up View toolbar.

Create the sketch.

572) Right-click **Top Plane** from the FeatureManager. This is your Sketch plane.

573) Click **Sketch** ⌐ from the Context toolbar.

574) Click **Top view** ⬚.

575) Click the **Line** ╱ Sketch tool.

576) Sketch a **horizontal line** as illustrated. The endpoints are located on either side of the Handle.

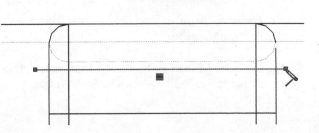

Add a dimension.

577) Click the **Smart Dimension** Sketch tool.

578) Click the **inner back edge**.

579) Click the **horizontal line**.

580) Click a **position** to the right off the profile.

581) Enter **.175**in, **[4.45]**.

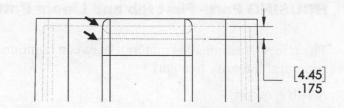

Insert the Rib feature.

582) Click the **Rib** feature tool. The Rib PropertyManager is displayed.

583) Click the **Both Sides** button.

584) Enter **.100**in, **[2.54]** for Rib Thickness.

585) Click the **Parallel to Sketch** button. The Rib direction arrow points to the back. Flip the material side if required. Select the Flip material side check box if the direction arrow does not point towards the back.

586) Click the **Draft On/Off** box.

587) Enter **1**deg for Draft Angle.

588) Click **Front view**.

589) Click the **Rib** sketch inside the HOUSING.

590) Click **OK** from the Rib PropertyManager. Rib1 is displayed in the FeatureManager.

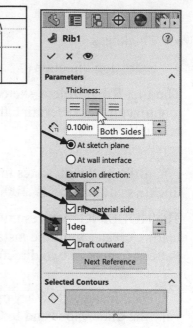

Display an Isometric view. Save the model.

591) Click **Isometric view**.

592) Click **Save**.

Existing geometry defines the Rib boundaries. The Rib does not penetrate through the wall.

Insert the Linear Pattern feature.

593) **Zoom to Area** on Rib1. Click **Rib1** from the FeatureManager.

594) Click the **Linear Pattern** feature tool. Rib1 is displayed in the Features to Pattern box.

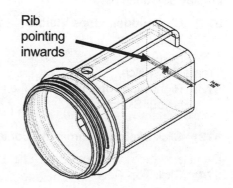

Rib pointing inwards

595) Click inside the **Direction 1 Pattern Direction** box.

596) Click the **hidden upper back vertical edge** of Shell1 in the Graphics window. The direction arrow points upward. Click the Reverse direction button if required.

597) Enter **.500**in, **[12.7]** for Spacing.

598) Enter **3** for Number of Instances.

599) Click inside the **Direction 2 Pattern Direction** box.

600) Click the hidden **lower back vertical edge** of Shell1 in the Graphics window. The direction arrow points downward.

601) Click the Reverse direction button if required. Enter **.500**in, **[12.7]** for Spacing.

602) Enter **3** for Number of Instances.

603) Click the **Pattern seed only** box.

604) Drag the Linear Pattern **Scroll bar** downward to display the Options box.

605) Check the **Geometry pattern** box. Accept the default values. Click **OK** ✔ from the Linear Pattern PropertyManager. LPattern1 is displayed in the FeatureManager.

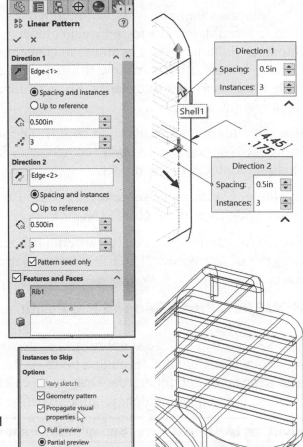

Display an Isometric view. Save the model.

606) Click **Isometric view** ⬛.

607) Click **Save** 💾.

💡 Utilize the Geometry pattern option to efficiently create and rebuild patterns. Know when to check the Geometry pattern.

Check Geometry pattern. You require an exact copy of the seed feature. Each instance is an exact copy of the faces and edges of the original feature. End conditions are not calculated. This option saves rebuild time.

Uncheck Geometry pattern. You require the end condition to vary. Each instance will have a different end condition. Each instance is offset from the selected surface by the same amount.

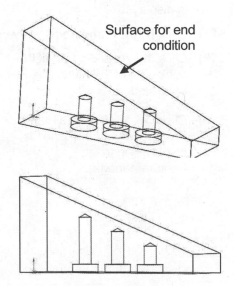

Surface for end condition

💡 Suppress Patterns when not required. Patterns contain repetitive geometry that takes time to rebuild. Pattern features also clutter the part during the model creation process. Suppress patterns as you continue to create more complex features in the part. Unsuppress a feature to restore the display and load into memory for future calculations. Hide features to improve clarity. Show feature to display hidden features.

Rib sketches are not required to be fully defined. The Linear Rib option blends sketched geometry into existing contours of the model.

Example: Create an offset reference plane from the inside back face of the HOUSING.

Sketch two under defined arcs. Insert a Rib feature with the Linear option. The Rib extends to the Shell walls.

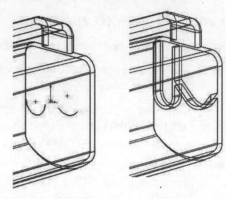

HOUSING Part - Second Rib Feature

The Second Rib feature supports and centers the BATTERY. The Rib is sketched on a reference plane created through a point on the Handle and parallel with the Right Plane. The Rib sketch references the Origin and existing geometry in the HOUSING. Utilize an Intersection and Coincident relation to define the sketch.

Activity: HOUSING Part - Second Rib Feature

Insert a Reference plane for the second Rib feature. Create a Parallel Plane at Point.

608) Click **Wireframe** 🔲 from the Heads-up View toolbar.

609) **Zoom to Area** on the back right side of the Handle.

610) Click **Plane** from the Features toolbar. The Plane PropertyManager is displayed.

611) Click **Right Plane** from the fly-out FeatureManager.

612) Click the **vertex** (point) at the back right of the handle as illustrated.

613) Click **OK** ✔ from the Plane PropertyManager. Plane1 is displayed in the FeatureManager.

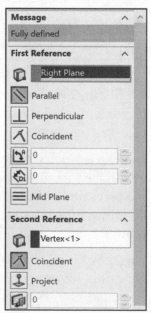

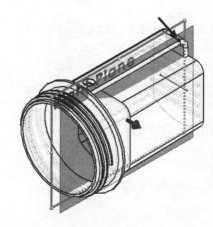

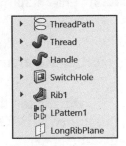

Rename Plane1.
614) Rename **Plane1** to **LongRibPlane**.

Fit to the Graphics window. Save the model.
615) Press the **f** key.

616) Click **Save** 💾.

Create the second Rib.
617) Right-click **LongRibPlane** for the FeatureManager.

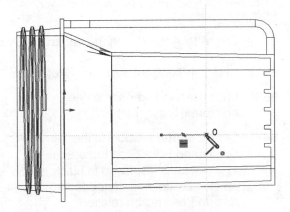

Create the sketch.
618) Click **Sketch** from the Context toolbar.

619) Click **Right view** from the Heads-up View toolbar.

620) Click the **Line** Sketch tool.

621) Sketch a **horizontal line**. Do not select the edges of the Shell1 feature.

Deselect the Line Sketch tool.
622) Right-click **Select**.

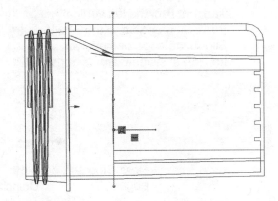

Add a Coincident relation.
623) Click the **left end point** of the horizontal sketch line.

624) Hold the **Ctrl** key down.

625) Click **BatteryLoftPlane** from the FeatureManager.

626) Release the **Ctrl** key.

627) Click **Coincident** ⟨.

628) Click **OK** ✔ from the Properties PropertyManager.

Add a dimension.
629) Click the **Smart Dimension** Sketch tool.

630) Click the **horizontal line**.

631) Click the **Origin**.

632) Click a **position** for the vertical linear dimension text.

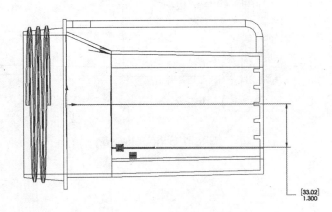

633) Enter **1.300**in, [**33.02**].

💡 When the sketch and reference geometry become complex, create dimensions by selecting Reference planes and the Origin in the FeatureManager.

💡 Dimension the Rib from the Origin, not from an edge or surface for design flexibility. The Origin remains constant. Modify edges and surfaces with the Fillet feature.

Sketch an arc.

634) **Zoom to Area** 🔍 on the horizontal Sketch line.

635) Click the **Tangent Arc** ⟩ Sketch tool.

636) Click the **left end** point of the horizontal line.

637) Click the **intersection** of the Shell1 and Boss Stop features. The sketch is displayed in black and is fully defined. If needed add an Intersection relation between the endpoint of the Tangent Arc, the left vertical Boss-Stop edge and the Shell1 Silhouette edge of the lower horizontal inside wall.

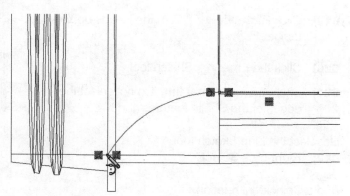

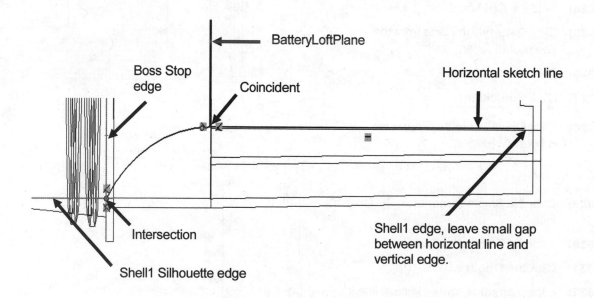

BatteryLoftPlane

Boss Stop edge

Coincident

Horizontal sketch line

Intersection

Shell1 Silhouette edge

Shell1 edge, leave small gap between horizontal line and vertical edge.

Insert the Rib feature.

638) Click the **Rib** feature tool. The Rib PropertyManager is displayed.

639) Click the **Both Sides** box.

640) Enter **.075**in, **[1.91]** for Rib Thickness.

641) Click the **Draft On/Off** box.

642) Enter **1**deg for Angle.

643) Click the **Draft outward** box.

644) Click the **Flip material side** box if required. The direction arrow points towards the bottom.

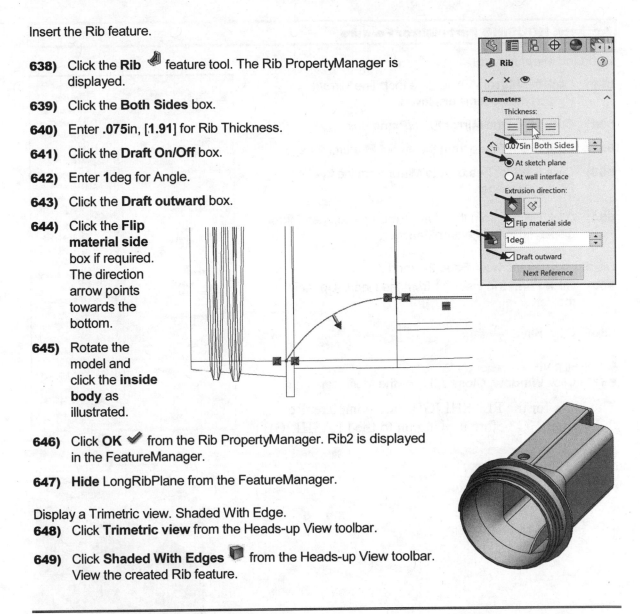

645) Rotate the model and click the **inside body** as illustrated.

646) Click **OK** from the Rib PropertyManager. Rib2 is displayed in the FeatureManager.

647) **Hide** LongRibPlane from the FeatureManager.

Display a Trimetric view. Shaded With Edge.
648) Click **Trimetric view** from the Heads-up View toolbar.

649) Click **Shaded With Edges** from the Heads-up View toolbar. View the created Rib feature.

HOUSING Part - Mirror Feature

An additional Rib is required to support the BATTERY. Reuse features with the Mirror feature to create a Rib symmetric about the Right Plane.

The Mirror feature requires:

- Mirror Face or Plane reference.

- Features or Faces to Mirror.

Utilize the Mirror feature. Select the Right Plane for the Mirror Plane. Select the second Rib for the Features to Mirror.

Activity: HOUSING Part - Mirror Feature

Insert the Mirror feature.

650) Click the **Mirror** ⊞⊟ feature tool. The Mirror PropertyManager is displayed.

651) Click inside the **Mirror Face/Plane** box.

652) Click **Right Plane** from the fly-out FeatureManager.

653) Click **Rib2** for Features to Mirror from the fly-out FeatureManager.

654) Click **OK** ✔ from the Mirror PropertyManager. Mirror1 is displayed in the FeatureManager.

Display a Trimetric view. Save the model.

655) Click **Trimetric view** 🔲 from the Heads-Up View toolbar.

656) Click **Save** 💾.

Close all parts.

657) Click **Window**, **Close All** from the Menu bar.

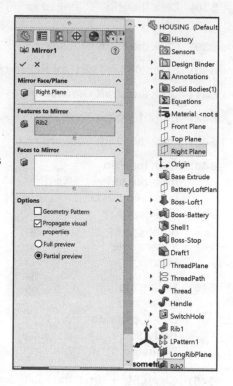

The parts for the FLASHLIGHT are complete. Review the HOUSING before moving on to the FLASHLIGHT assembly.

🔍 Additional information on Extrude Boss/Base, Extrude Cut, Swept, Loft, Helix/Spiral, Rib, Mirror and Reference Planes are found in SOLIDWORKS Help Topics.

Review of the HOUSING Part

The HOUSING utilized the Extruded Boss/Base feature with the Draft Angle option. The Lofted Boss feature was created to blend the circular face of the LENS with the rectangular face of the BATTERY. The Shell feature removed material with a constant wall thickness. The Draft feature utilized the front face as the Neutral plane.

You created a Thread similar to the LENSCAP Thread. The Thread profile was copied from the LENSCAP and inserted into the Top Plane of the HOUSING. The Extruded Cut feature was utilized to create a hole for the Switch. The Rib features were utilized in a Linear Pattern and Mirror feature.

Each feature has additional options that are applied to create different geometry. The Offset From Surface option creates an Extruded Cut on the curved surface of the HOUSING and LENSCAP. The Reverse offset and Translate surface options produce a cut depth constant throughout the curved surface.

Utilize Tools, Sketch Entities, Text to create the text profile on an Offset Plane.

Click the direction arrows in the FeatureManager to expand or collapse the FeatureManager design tree. Your tabs will vary depending on your SOLIDWORKS applications and Add-ins.

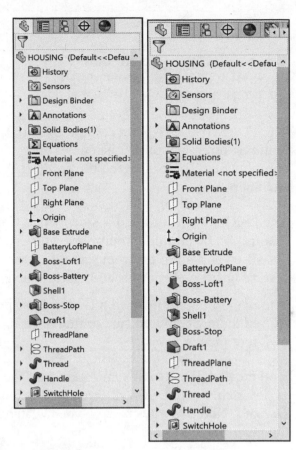

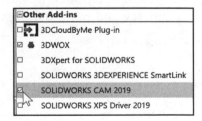

Chapter Summary

You created four parts for the FLASHLIGHT assembly: O-RING, SWITCH, LENSCAP and HOUSING.

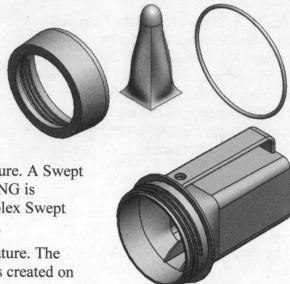

The FLASHLIGHT parts contain over 100 features, reference planes, sketches and components. You organized the features in each part.

The O-RING part utilized a Swept Base feature. A Swept feature can be simple or complex. The O-RING is classified as a simple Swept feature. A complex Swept feature utilizes 3D curves and Guide Curves.

The SWITCH part utilized a Lofted Base feature. The Lofted feature required two or more sketches created on different planes.

You created 2D sketches and addressed the three key states of a sketch: *Fully Defined*, *Over Defined* and *Under Defined*. Note: Always review your FeatureManager for the proper sketch state.

You addressed additional tools that utilized existing geometry: Add Relations, Copy, Save As, Edit Feature and more.

The LENSCAP and HOUSING part utilized a variety of features. You applied design intent to reuse geometry through Geometric relationships, Symmetry and patterns. Review the chapter exercises before moving on to the next chapter.

For many features (Extruded Boss/Base, Extruded Cut, Simple Hole, Revolved Boss/Base, Revolved Cut, Fillet, Chamfer, Scale, Shell, Rib, Circular Pattern, Linear Pattern, Curve Driven Pattern, Revolved Surface, Extruded Surface, Fillet Surface, Edge Flange and Base Flange), you can enter and modify equations directly in the PropertyManager fields that allow numerical inputs.

Create equations with Global Variables, functions and file properties without accessing the Equations, Global Variables and Dimensions dialog box.

For example, in the Extruded Base PropertyManager you can enter equations in:

- Depth fields for Direction 1 and Direction 2

- Draft fields for Direction 1 and Direction 2

- Thickness fields for a Thin Feature with two direction types

- Offset Distance field

To create an equation in a numeric input field, start by entering an = (equal sign). A drop-down list displays options for Global Variables, functions, and file properties. Numeric input fields that contain equations can display either the equation itself or its evaluated value.

Click the What's new ⑦ icon in the PropertyManager to learn what's new about a feature or option.

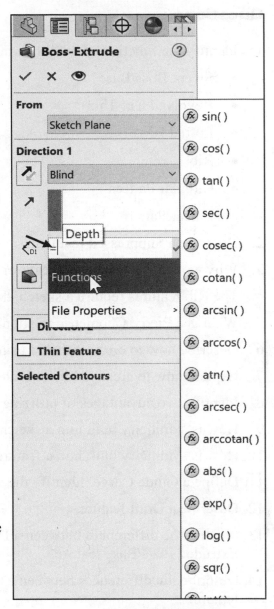

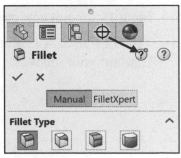

Questions

1. Identify the function of the following features:
   - Swept Boss/Base
   - Revolved Cut Thin
   - Lofted Boss/Base
   - Rib
   - Circular Pattern
   - Linear Pattern

2. Describe a Suppressed feature.

3. Why would you suppress a feature?

4. The Rib features require a sketch, thickness and a _____ direction.

5. What is a Pierce Geometric relation?

6. Describe how to create a thread using the Swept feature. Provide an example.

7. Explain how to create a Linear Pattern feature. Provide an example.

8. Identify two advantages of utilizing Convert Entities in a sketch to obtain the profile.

9. How is symmetry built into a sketch? Provide an example.

10. How is symmetry built into a feature? Provide an example.

11. Define a Guide Curve. Identify the features that utilize Guide Curves.

12. Describe a Draft feature.

13. Identify the differences between a Draft feature and the Draft Angle option in the Extruded Boss/Base feature.

14. Describe the differences between a Circular Pattern feature and a Linear Pattern feature.

15. Identify the advantages of the Convert Entities tool.

16. True or False. A Lofted feature can only be inserted as the first feature in a part. Explain your answer.

Exercises

Exercise 7.1: QUATTRO-SEAL-O-RING Part

Create the QUATTRO-SEAL-O-RING part as a single Swept feature.

- Create a 100mm diameter circle on the Front plane for the path, Sketch1.

- Create the symmetric cross section on the Top Plane for the profile, Sketch2.

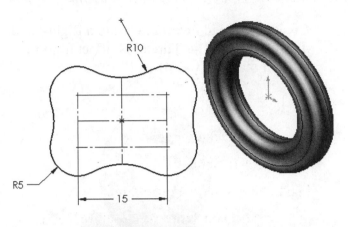

Exercise 7.2: HOOK Part

Create the HOOK part. Utilize Swept Boss/Base for the first feature. View the illustrated FeatureManager.

Not all dimensions are provided. Your HOOK part will vary.

The Swept Boss/Base feature adds material by moving a profile along a path. The Swept Boss/Base feature in this exercise requires a path sketch and a circular profile diameter.

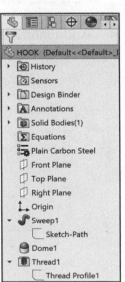

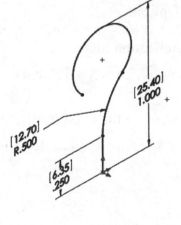

- Utilize the Dome feature (050in, [1.27]) to create a spherical feature on the illustrated circular face.

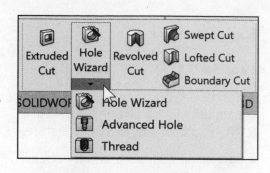

- Create a Thread Feature. Create a Right-hand, #8-36 Thread. The Thread is offset from the bottom face of Sweep1 as illustrated.

- Select the select edge as illustrated for Thread (Edge of Cylinder) loaction.

- Select Right-hand thread and Trim with Start face Thread Options.

- Add material: Plain Carbon Steel.

Exercise 7.3: WEIGHT Part

Create the WEIGHT part. Utilize the Loft Base Feature. Add Material: Plain Carbon Steel.

- The Top Plane and Plane1 are 0.5in, [12.7mm] apart.

- Sketch a rectangle 1.000in, [25.4mm] x .750in, [19.05] on the Top Plane.

- Sketch a square .500in, [12.7mm] on Plane1.

- Create a Lofted feature.

- Add a centered ⌀.150in, [3.81mm] Thru Hole.

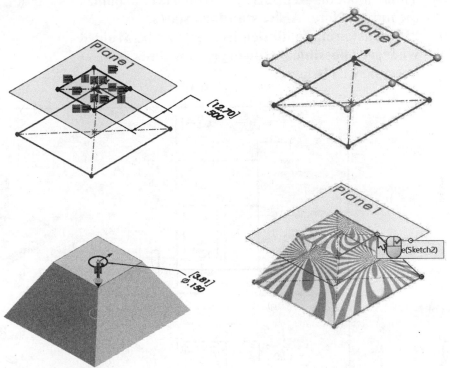

Exercise 7.4: SWEPT-CUT CASE

Create the CASE part.

- Utilize a Swept Cut feature to remove material from the CASE. The profile for the Swept Cut is a semi-circle.

- Dimensions are not provided. Design your case to hold pencils.

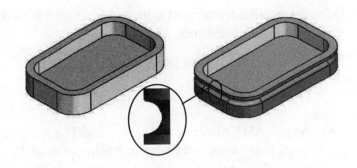

Exercise 7.5: Hole Wizard, Rib and Linear Pattern Features

Create the part from the illustrated A-ANSI, Third Angle drawing: Front, Top, Right and Isometric views.

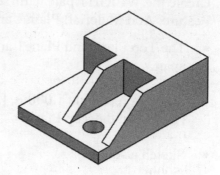

- Apply 6061 Alloy material.

- Calculate the volume of the part and locate the Center of mass.

- Think about the steps that you would take to build the model. **Note: ANSI standard states, "Dimensioning to hidden lines should be avoided wherever possible." However, sometimes it is necessary as below**.

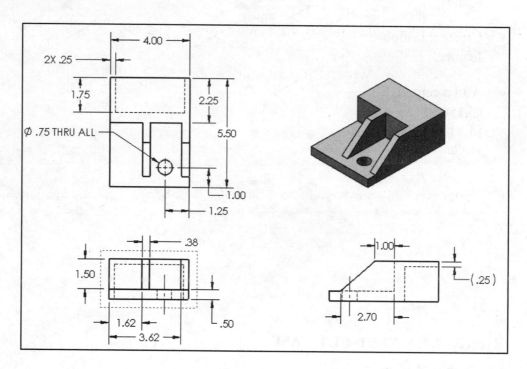

Exercise 7.6: Shell feature

Create the illustrated part with the Extruded Boss/Base, Fillet and Shell features.

- Dimensions are not provided.

- Design your case to hold a bar of soap.

- Apply ABS material to the model. Think about the steps that you would take to build the model.

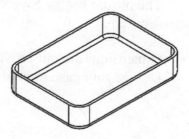

Exercise 7.7: Revolved Base, Hole Wizard, and Circular Pattern features

Create the illustrated ANSI part with the Revolved Base, Hole Wizard, and Circular Pattern features.

- Dimensions are not provided.

- Apply PBT General Purpose Plastic material to the model.

Think about the steps that you would take to build the model.

Exercise 7.8: Extruded Boss/Base and Revolved Boss feature

Create the illustrated ANSI part with the Extruded Boss/Base and Revolved Boss feature. Note: The location of the Origin in the Right and Isometric view.

- Dimensions are not provided.

- Apply 6061 Alloy material to the model.

Think about the steps that you would take to build the model.

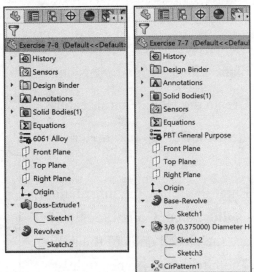

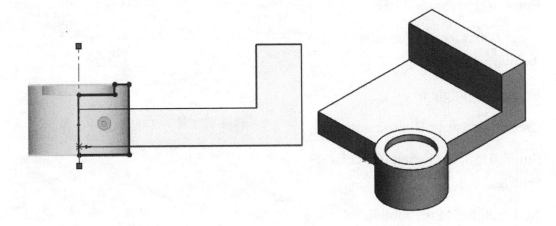

Exercise 7.9: Gem® Paper clip

Create a simple paper clip.

Create an ANSI - IPS model.

Apply material to the model.

Precision = 2.

The paper clip uses (lines and arcs) as the path and a circular diameter profile (0.010in).

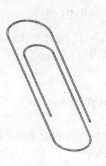

Exercise 7.10: Variable Pitch Spring

Create a Variable pitch spring (ANSI - IPS) with two active coils as illustrated.

Sketch a circle, Coincident to the Origin on the Top plane with a .235in dimension.

Create the Helix/Spiral feature.

Create a Region parameters table.

Enter the following information as illustrated. Coils 1, 2, 5, 6 & 7 are the closed ends of the spring. The pitch needs to be slightly larger than the wire.

Enter .021in for the Pitch.

Enter .080in for the free state of the two active coils.

Enter Start angle of 0deg.

Create the Swept Boss feature.

Enter .015in for Depth (Circular Profile).

Add material to the model.

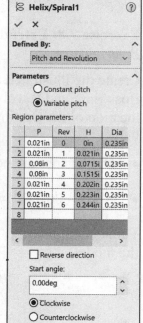

Region parameters:

	P	Rev	H	Dia
1	0.021in	0	0in	0.235in
2	0.021in	1	0.021in	0.235in
3	0.08in	2	0.0715in	0.235in
4	0.08in	3	0.1515i	0.235in
5	0.021in	4	0.202in	0.235in
6	0.021in	5	0.223in	0.235in
7	0.021in	6	0.244in	0.235in
8				

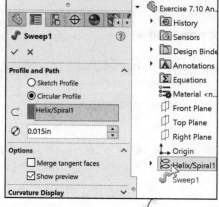

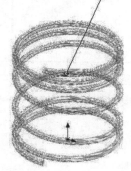

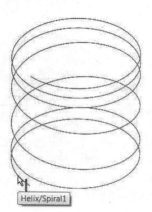

Helix/Spiral1

Exercise 7.11: Water Bottle

Create the container as illustrated.

Create an ANSI - IPS model.

Apply material to the model.

View the sample FeatureManager. Your FeatureManager
can (should) be different. This is just one way to create
this part. Estimate any needed dimension.

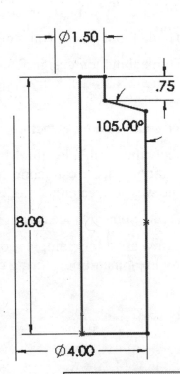

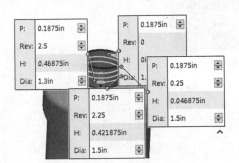

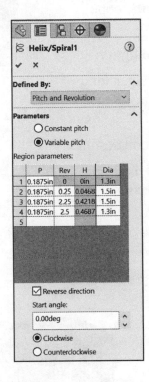

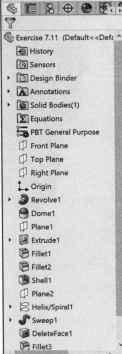

Exercise 7.12: Traditional Ice cream cone

Create a traditional or non-traditional Ice Cream Cone.

Create an ANSI - IPS model.

Think about where you would start.

Think about the design features that create this model. Why does the cone use ribs?

Ribs are used for structural integrity.

View the sample FeatureManager for a traditional ice cream cone. This is just one way to create this model.

Create your own ice cream cone design.

Below are a few sample models from my Freshman Engineering class.

Below are sample models from my Freshman Engineering (Cont:). Create your own ice cream cone design.

Notes:

Chapter 8

Assembly Modeling - Bottom up method

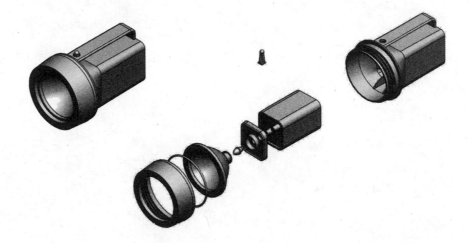

Below are the desired outcomes and usage competencies based on the completion of Chapter 8.

Desired Outcomes:	Usage Competencies:

• Create four assemblies in this project:
 	
   o LENSANDBULB assembly
 	
   o CAPANDLENS assembly
 	
   o BATTERYANDPLATE assembly
 	
   o FLASHLIGHT assembly
	• Develop an understanding of Assembly modeling techniques.
	
	• Combine the LENSANDBULB assembly, CAPANDLENS assembly, BATTERYANDPLATE assembly, HOUSING part and SWITCH part to create the FLASHLIGHT assembly.
	
	• Ability to use the following tools: Insert Component, Hide/Show, Suppress/UnSuppress, Mate, Move Component, Rotate Component, Exploded View, and Interference Detection.

• Create an Inch and Metric Assembly Template.
 	
   o ASM-IN-ANSI
 	
   o ASM-MM-ISO
	• Ability to apply Document Properties and to create Custom Assembly Templates.

Notes:

Chapter 8 - Assembly Modeling – Bottom up method

Chapter Overview

Create four assemblies in this chapter:

1. LENSANDBULB assembly.

2. CAPANDLENS assembly.

3. BATTERYANDPLATE assembly.

4. FLASHLIGHT assembly.

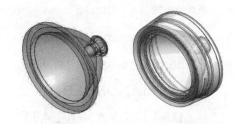

Create an inch and metric Assembly Template.

- ASM-IN-ANSI.

- ASM-MM-ISO.

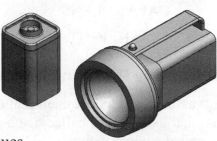

Develop an understanding of assembly modeling techniques. Combine the LENSANDBULB assembly, CAPANDLENS assembly, BATTERYANDPLATE assembly, HOUSING part, and SWITCH part to create the FLASHLIGHT assembly.

Review Standard mate types. Create the following Standard mates:

- Coincident, Concentric, and Distance

Utilize the following tools: Insert Component 🗗 , Hide/Show 🗗 , Suppress 🗗 , Mate 🗗 , Move Component 🗗 , Rotate Component 🗗 , Exploded View 🗗 and Interference Detection 🗗 .

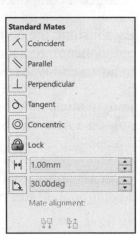

After completing the activities in this chapter, you will be able to:

- Create two Assembly Templates: ASM-IN-ANSI and ASM-MM-ISO.

- Apply the following Standard mates: Coincident, Concentric and Distance.

- Understand the Quick mate procedure.

- Utilize the following tools: Insert Component, Hide/Show Component, Mate, Move Component, Rotate Component, Interference Detection and Suppress/UnSuppress.

- Export a .STL file of the HOUSING part.

- Develop an eDrawing for the FLASHLIGHT assembly.

- Create an Exploded view of the FLASHLIGHT assembly.

- Animate a Collapse view and an Exploded view.

- Organize assemblies into sub-assemblies.

- Create four assemblies:

  o LENSANDBULB.

  o BATTERYANDPLATE.

  o CAPANDLENS.

  o FLASHLIGHT.

Assembly Modeling Overview

An assembly is a document that contains two or more parts. An assembly inserted into another assembly is called a sub-assembly. A part or assembly inserted into an assembly is called a component.

Establishing the correct component relationship in an assembly requires forethought on component interaction. Mates are Geometric relationships that align and fit components in an assembly. Mates remove degrees of freedom from a component.

In dynamics, motion of an object is described in linear and rotational terms. Components possess linear motion along the x, y and z-axes and rotational motion around the x, y and z-axes.

In an assembly, each component has 6 degrees of freedom: 3 translational (linear) and 3 rotational. Mates remove degrees of freedom. All components are rigid bodies.

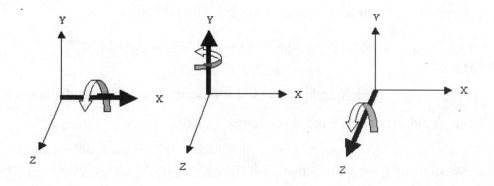

The components do not flex or deform. Components are assembled in this chapter with Standard mate types. There are three mate types displayed in the Mate PropertyManager: Standard, Advanced and Mechanical.

The Standard mate types are Coincident, Parallel, Perpendicular, Tangent, Concentric, Lock, Distance and Angle.

The Advanced mate types are Profile Center, Symmetric, Width, Path Mate, Linear/Linear Coupler, Distance (limit) and Angle.

The Mechanical mate types are Cam, Slot, Gear, Rack Pinion, Screw and Universal Joint.

Mates require geometry from two different components. Selected geometry includes Planar Faces, Cylindrical faces, Linear edges, Circular/Arc edges, Vertices, Axes, Temporary axes, Planes, Points and Origins.

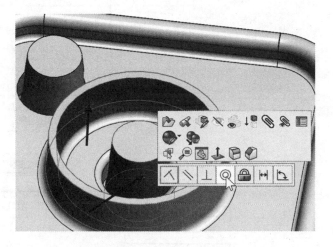

Mates reflect the physical behavior of a component in an assembly. Example: Utilize a Concentric mate between the BATTERY Extruded Boss (Terminal) cylindrical face and the BATTERYPLATE Extruded Boss (Holder) face.

The FLASHLIGHT assembly consists of the following components:

FLASHLIGHT Components:	
BATTERY	BATTERYPLATE
LENS	BULB
O-RING	SWITCH
LENSCAP	HOUSING

How do you organize these components into the FLASHLIGHT assembly? Answer: Create an assembly component layout diagram to determine which components to group into a sub-assembly.

FLASHLIGHT Assembly

Plan the sub-assembly component layout diagram.

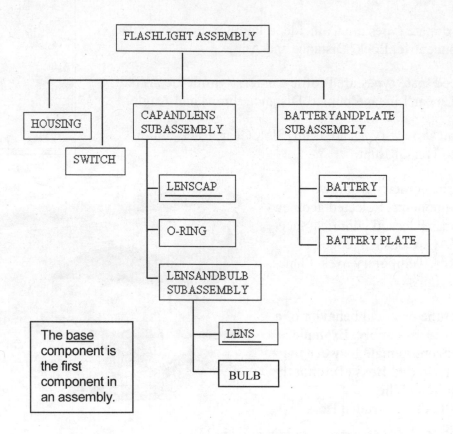

The base component is the first component in an assembly.

Assembly Layout Structure

The FLASHLIGHT assembly steps are as follows:

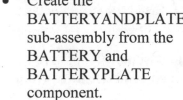

- Create the LENSANDBULB sub-assembly from the LENS and BULB component. The LENS is the Base component.

- Create the BATTERYANDPLATE sub-assembly from the BATTERY and BATTERYPLATE component.

- Create the CAPANDLENS sub-assembly from the LENSCAP, O-RING and LENSANDBULB sub-assembly. The LENSCAP is the Base component.

- Create the FLASHLIGHT assembly. The HOUSING is the Base component. Insert the SWITCH, CAPANDLENS and BATTERYANDPLATE component.

- Modify the dimensions to complete the FLASHLIGHT assembly.

Assembly Techniques

Assembly modeling requires time and practice. Below are helpful hints and techniques to address the Bottom-up design modeling approach.

- Create an assembly layout structure. The layout structure will organize the sub-assemblies and components and save time.

- Insert sub-assemblies and components as lightweight components. Lightweight components save on file size, rebuild time and overall complexity.

- Set Lightweight components in the Options, Performance section.

- Use the Zoom and Rotate commands to select the correct geometry in the mate process. Zoom in to select the correct face, edge, plane, point, etc.

- Improve display. Apply various colors to features and components.

- Mate with Reference planes when addressing complex geometry. Example: An O-RING does not contain a flat surface or edge.

- Activate the Temporary Axes and the required Planes from the Menu bar toolbar.

- Select Reference planes from the fly-out FeatureManager. Expand the component in the FeatureManager to view the planes and features.

Example: Select the Right Plane of the LENS and the Right Plane of the BULB to be collinear. Do not select the Right Plane of the HOUSING if you want to create a reference between the LENS and the BULB.

- Remove display complexity. Hide components and features. Suppress components and features when not required.

- Apply the Move Component and Rotate Component tools if needed before mating. Position the component in the correct orientation.

- Remove unwanted entries. Use Right-click Clear Selections or Right-click Delete from the Assembly Mate Selections text box.

- Verify the position of the mated components. Use Top, Front, Right, and Section views.

- Use caution when you view the color blue in an assembly. Blue indicates that a part is being edited in the context of the assembly.

- Avoid unwanted references. Verify your geometry selections with the PropertyManager.

Assembly Template

An Assembly Document Template is the foundation of the assembly. The FLASHLIGHT assembly and its sub-assemblies require the Assembly Document Template. Utilize the default Assembly Template. Modify the Dimensioning Standard and Units. Create an Assembly Document Template using inch units, ASM-IN-ANSI. Create an Assembly Document Template using millimeter units, ASM-MM-ISO. Save the Templates in the MY-TEMPLATES folder.

Activity: Create the Assembly Templates - ASM-IN-ANSI

Create an Assembly Template.

1) Click **New** 🗋 from the Menu bar.

2) Double-click **Assembly** from the Templates tab. The Begin Assembly PropertyManager and the Window Open dialog box is displayed.

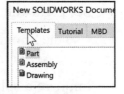

3) Click **Cancel** ✖ from the Window Open dialog box and the Begin Assembly PropertyManager.

Set the Assembly Document Template options.

4) Click **Options** ⚙ from the Main menu.

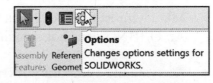

5) Click the **Document Properties** tab from the dialog box.

Set drafting standard, units, and precision.

6) Select **ANSI** for Overall drafting standard.

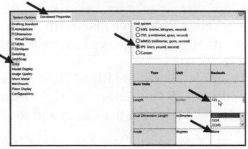

7) Click **Units**. Select **IPS, (inch, pound, second)** for Unit system.

8) Select **.123** in the Length units Decimals drop-down box.

9) Select **None** for Angular units in the Decimals drop-down box.

10) Click **OK** from the Document Properties - Units dialog box.

Save the assembly template. Enter name.
11) Click **Save As** from the drop-down Menu bar.

12) Select the **Assembly Template (*asmdot)** from the Save As type box.

13) Select the **SOLIDWORKS-MODELS 2019\MY-TEMPLATES** folder. Enter **ASM-IN-ANSI** in the File name box.

14) Click **Save**.

Activity: Create the Assembly Templates - ASM-MM-ISO

Create the ASM-MM-ISO assembly template.
15) Click **New** ⬚ from the Menu bar.

16) Double-click **Assembly** from the Templates tab. The Begin Assembly PropertyManager and the Window Open dialog box is displayed.

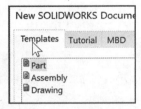

17) Click **Cancel** ✖ from the Window Open dialog box and the Begin Assembly PropertyManager.

Set the Assembly Document Template options.
18) Click **Options** ⚙ from the Main menu.

19) Click the **Document Properties** tab from the dialog box.

20) Select **ISO** for Overall drafting standard.

21) Click **Units**.

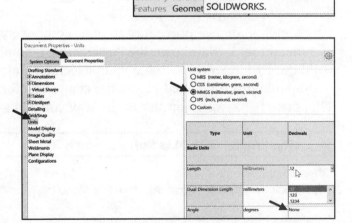

22) Select **MMGS, (millimeter, gram, second)** for Unit system.

23) Select **.12** in the Length units Decimals box.

24) Select **None** in the Angular units Decimals box.

25) Click **OK** from the Document Properties - Units dialog box.

Save the assembly template. Enter name.
26) Click **Save As** from the drop-down Menu bar.

27) Select the **Assembly Template (*asmdot)** from the Save As type box.

28) Select the **SOLIDWORKS-MODELS 2019\MY-TEMPLATES** folder.

29) Enter **ASM-MM-ISO** in the File name box.

30) Click **Save**.

LENSANDBULB Sub-assembly

Create the LENSANDBULB sub-assembly. The LENS is the Base component. LENSANDBULB sub-assembly mates the BULB component to the LENS component. The Right Plane of the LENS and the Right Plane of the BULB are Coincident.

The Top Plane of the LENS and the Top Plane of the BULB are Coincident. The inside Counterbore face of the LENS and the back face of the BULB utilize a Distance mate. The LENS name is added to the LENSANDBULB assembly FeatureManager with the symbol (f). The symbol (f) represents a fixed component. A fixed component cannot move and is locked to the assembly Origin.

Suppress the Lens Shield feature to view all surfaces during the mate process. Utilize Open Part to open the LENS from inside the LENSANDBULB assembly.

Utilize the Suppress ↓🔲 tool from the FeatureManager to Suppress a component. Utilize the UnSuppress ↑🔲 tool to restore the component. Note the Mates of a suppressed component are also suppressed.

☀ Quick mate is a procedure to mate components together in SOLIDWORKS. No command (click Mate from the Assembly CommandManager) needs to be executed. Hold the Ctrl key down, make your selections, release the Ctrl key and a Quick Mate pop-up menu is displayed below the context toolbar. Select the mate and you are finished. This mate behavior is similar to the way you add sketch relations.

Activity: LENSANDBULB Sub-assembly

Close all documents.
31) Click **Windows**, **Close All** from the Menu bar.

Create the LENSANDBULB sub-assembly.
32) Click **New** 🗋 from the Menu bar.

33) Click the **MY-TEMPLATES** tab. Note: Additional templates are displayed.

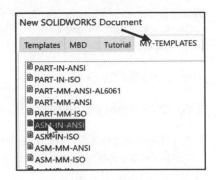

34) Double-click **ASM-IN-ANSI, [ASM-MM-ISO]**.

Insert the LENS. Save the LENS. Enter name. Enter description.
35) Double-click **LENS** from its file location.

36) Click **OK** ✔ from the Begin Assembly PropertyManager. The LENS is fixed to the Origin.

37) Click **Save**.

38) Enter **LENSANDBULB** for File name in the PROJECTS folder.

39) Enter **LENS AND BULB ASSEMBLY** for Description.

40) Click **Save**.

Insert the BULB.

41) Click **Insert Components** 🗁 from the Assembly toolbar.

42) Double-click **BULB** from the PROJECTS folder.

43) Click a **position** in front of the LENS as illustrated.

Fit the model to the Graphics window.
44) Press the **f** key.

Move the BULB.
45) Click and drag the **BULB** in the Graphics window.

Save the LENSANDBULB.
46) Click **Save**.

47) **View** the Assembly FeatureManager.

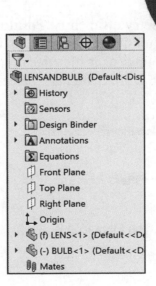

Insert a Coincident mate.

48) Click **Right Plane** of the LENS from the FeatureManager.

49) Hold the **Ctrl** key down.

50) Click **Right Plane** of the BULB.

51) Release the **Ctrl** key. The Mate pop-up menu is displayed.

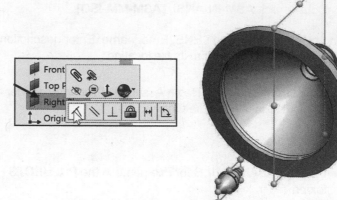

52) Click **Coincident** from the Mate pop-up menu. Coincident1 is created.

Insert the second Coincident mate.

53) Click **Top Plane** of the LENS from the FeatureManager.

54) Hold the **Ctrl** key down.

55) Click **Top Plane** of the BULB from the FeatureManager.

56) Release the **Ctrl** key. The Mate pop-up menu is displayed.

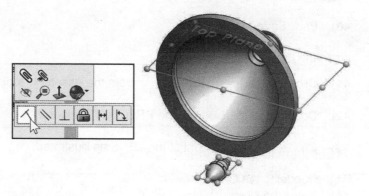

57) Click **Coincident** from the Mate pop-up menu.

Select face geometry efficiently. Position the mouse pointer in the middle of the face. Do not position the mouse pointer near the edge of the face. Zoom in on geometry. Utilize the Face Selection Filter for narrow faces.

Activate the Face Selection Filter.

58) **Right-click** in the Graphics window.

59) Click the **Selection Filters** icon as illustrated.

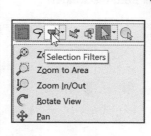

60) Click **Filter Faces**. The Filter icon is displayed on the mouse pointer.

Only faces are selected until the Face Selection Filter is deactivated. Select Clear All Filters 🖫 from the Selection Filter toolbar to deactivate all filters.

Insert the third Coincident mate.

61) **Zoom to Area** 🔍 and **Rotate** ↻ on the CBORE.

62) Click the **BulbHole face** of the LENS in the Graphics window as illustrated.

63) **Rotate** to view the Bottom back flat face, Revolve1 of the BULB as illustrated.

64) Hold the **Ctrl** key down.

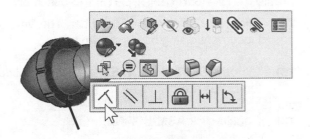

65) Click the **bottom back flat face, Revolve1** of the BULB.

66) Release the **Ctrl** key. The Mate pop-up menu is displayed.

67) Click **Coincident** from the Mate pop-up menu. Coincident3 is created.

Clear the Face filter.
68) **Right-click** in the Graphics window. Click the **Selection Filters** icon.

69) Click **Clear All Filters** 🖫 from the Selection Filter toolbar.

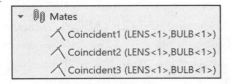

Display the Mate types.
70) **Expand** the Mates folder in the FeatureManager. View the inserted mates.

Display a Right view - Wireframe.
71) Click **Right view** ⬚ from the Heads-up View toolbar. Click **Wireframe** ⬚ from the Heads-up View toolbar. View the results.

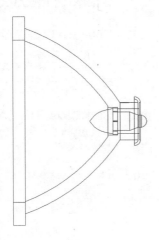

Display an Isometric view - Shaded With Edges.
72) Click **Isometric view** ⬛. Click **Shaded With Edges** ⬛ from the Heads-up View toolbar.

Save the LENSANDBULB.

73) Click **Save** 💾. View the results in the Graphics window.

BATTERYANDPLATE Sub-assembly

Create the BATTERYANDPLATE sub-assembly. Utilize two Coincident Mates and one Concentric Mate to assemble the BATTERYPLATE component to the BATTERY component.

Note: Utilize the Selection Filter required. Select planes from the FeatureManager when the Selection Filters are activated.

Activity: BATTERYANDPLATE Sub-assembly

Create the BATTERYANDPLATE sub-assembly.

74) Click **New** ⬜ from the Menu bar. The New SOLIDWORKS Document dialog box is displayed.

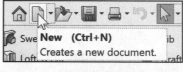

75) Click the **MY-TEMPLATES** tab. Additional document templates are displayed.

76) Double-click **ASM-IN-ANSI**.

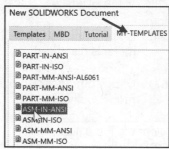

Insert the BATTERY part.
77) Double-click **BATTERY** from the PROJECTS folder.

Fix the BATTERY to the Assembly Origin.

78) Click **OK** ✔ from the Begin Assembly PropertyManager. The BATTERY is fixed to the Origin.

Save the BATTERYANDPLATE sub-assembly. Enter name. Enter description.

79) Click **Save** 💾 .

80) Select the **PROJECTS** folder.

81) Enter **BATTERYANDPLATE** for File name.

82) Enter **BATTERY AND PLATE FOR 6-VOLT FLASHLIGHT** for Description.

83) Click **Save**. The BATTERYANDPLATE FeatureManager is displayed.

Insert the BATTERYPLATE part.

84) Click **Insert Components** 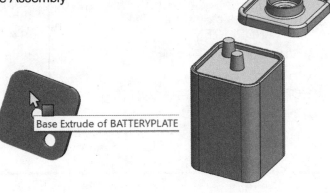 from the Assembly toolbar.

85) **Browse** to the PROJECTS folder.

86) Double-click **BATTERYPLATE**.

87) Click a **position** above the BATTERY as illustrated.

Insert a Coincident mate.
88) Click the **outside bottom face** of the BATTERYPLATE.

89) **Rotate** the model to view the top narrow flat face of the BATTERY Base Extrude feature. Hold the **Ctrl** key down.

90) Click the **top narrow flat face** of the BATTERY Base Extrude feature as illustrated.

91) Release the **Ctrl** key. The Mate Pop-up menu is displayed.

92) Click **Coincident** from the Mate Pop-up menu. Coincident1 is created.

Insert a Coincident mate.
93) Click **Right Plane** of the BATTERY from FeatureManager.

94) Hold the **Ctrl** key down. Click **Right Plane** of the BATTERYPLATE from the FeatureManager.

95) Click **Coincident** from the Mate Pop-up menu. Coincident2 is created.

Insert a Concentric mate.
96) Click the center Terminal feature **cylindrical face** of the BATTERY as illustrated.

97) Hold the **Ctrl** key down. Click the Holder feature **cylindrical face** of the BATTERYPLATE. Release the **Ctrl** key. The Mate Pop-up menu is displayed.

98) Click **Concentric** from the Mate Pop-up menu.

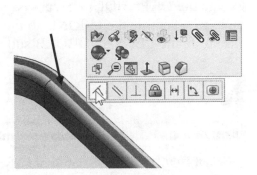

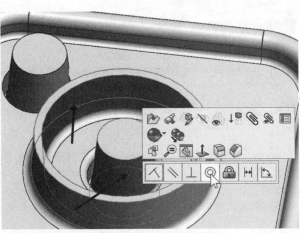

99) **Expand** the Mates folder. View the created mates.

Display an Isometric view. Save the BATTERYANDPLATE.

100) Click **Isometric view**.

101) Click **Save**.

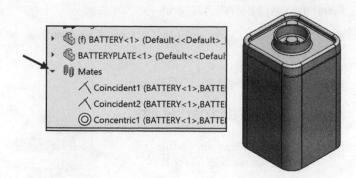

CAPANDLENS Sub-assembly

Create the CAPANDLENS sub-assembly. Utilize two Coincident mates and one Distance mate to assemble the O-RING to the LENSCAP. Utilize three Coincident mates to assemble the LENSANDBULB sub-assembly to the LENSCAP component.

Caution: Select the correct reference. Expand the LENSCAP and O-RING. Click the Right Plane within the LENSCAP. Click the Right Plane within the O-RING.

Activity: CAPANDLENS Sub-assembly

Create the CAPANDLENS sub-assembly.

102) Click **New** from the Menu bar.

103) Click the **MY-TEMPLATES** tab. Additional document templates are displayed.

104) Double-click **ASM-IN-ANSI**.

Insert the LENSCAP sub-assembly.

105) Double-click **LENSCAP** from the PROJECTS folder.

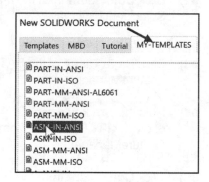

106) Click **OK** from the Begin Assembly PropertyManager. The LENSCAP is fixed to the Origin.

Save the CAPANDLENS assembly. Enter name. Enter description.

107) Click **Save**.

108) Select the **PROJECTS** folder.

109) Enter **CAPANDLENS** for File name.

110) Enter **LENSCAP AND LENS** for Description.

111) Click **Save**. The CAPANDLENS FeatureManager is displayed.

Insert the O-RING part.

112) Click **Insert Components** from the Assembly toolbar.

113) Double-click **O-RING** from the PROJECTS folder.

114) Click a **position** behind the LENSCAP as illustrated.

Insert the LENSANDBULB assembly.

115) Click **Insert Components** from the Assembly toolbar.

116) Double-click **LENSANDBULB** from the PROJECTS folder.

117) Click a **position** behind the O-RING as illustrated.

Display an Isometric view.

118) Click **Isometric view** .

Move and hide components.

119) Click and drag the **O-RING** and **LENSANDBULB** as illustrated in the Graphics window.

120) Right-click **LENSANDBULB** in the FeatureManager.

121) Click **Hide components** from the Context toolbar.

Insert three mates between the LENSCAP and O-RING.

122) Click **Right Plane** of the LENSCAP from the FeatureManager.

123) Hold the **Ctrl** key down.

124) Click **Right Plane** of the O-RING from the FeatureManager.

125) Release **the** Ctrl key.

126) Click **Coincident** from the Mate pop-up menu. Coincident1 is created.

Insert a second Coincident mate.

127) Click **Top Plane** of the LENSCAP from the FeatureManager.

128) Hold the **Ctrl** key down.

129) Click **Top Plane** of the O-RING from the FeatureManager.

130) Release the **Ctrl** key.

131) Click **Coincident**. Coincident2 is created.

Pin the Mate PropertyManager. Insert a Distance mate.

132) Click the **Mate** ✎ Assembly tool. The Mate
PropertyManager is displayed.

133) Click the **Keep Visible** ⟊ icon.

134) Click the Shell1 **back inside face** of the LENSCAP as
illustrated.

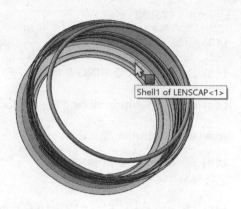

135) Click **Front Plane** of the O-RING in the fly-out
FeatureManager.

136) Click **Distance**.

137) Enter **.125/2**in, [**3.175/2mm**].

138) Click **OK** ✔ from the Distance Mate PropertyManager.

139) Click **OK** ✔ from the Mate PropertyManager.

Display an Isometric view. Expand the Mates folder.

140) Click **Isometric view** 🔷.

141) **Expand** the Mates folder. View the created mates.

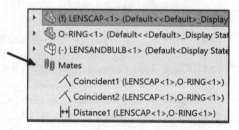

Save the model.

142) Click **Save** 💾.

How is the Distance mate, .0625in, [1.588],
calculated? Answer:

O-RING Radius (.1250in/2) = .0625in.

O-RING Radius [3.175mm/2] = [1.588mm].

Utilize a Section view 🔲 tool from the Heads-up
View toolbar to locate internal geometry for mating
and verify position of components.

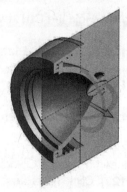

☼ Build flexibility into the mate. A Distance mate
offers additional flexibility over a Coincident mate.
You can modify the value of a Distance mate.

Show the LENSANDBULB.
143) Right-click **LENSANDBULB** in the FeatureManager.

144) Click **Show components** from the Contexts toolbar.

Fit the model to the Graphics window.
145) Press the **f** key.

Insert a Coincident mate.
146) Click **Right Plane** of the LENSCAP from the FeatureManager.

147) Hold the **Ctrl** key down.

148) Click **Right Plane** of the LENSANDBULB from the FeatureManager. Release the **Ctrl** key. The Mate pop-up menu is displayed.

149) Click **Coincident** from the Mate pop-up menu.

Insert a Coincident Mate.
150) Click **Top Plane** of the LENSCAP from the FeatureManager.

151) Hold the **Ctrl** key down.

152) Click **Top Plane** of the LENSANDBULB from the FeatureManager.

153) Release the **Ctrl** key. The Mate pop-up menu is displayed.

154) Click **Coincident** from the Mate pop-up menu.

Insert a Coincident Mate.
155) Click the flat inside **narrow back face** of the LENSCAP.

156) **Rotate** the model to view the front flat face of the LENSANDBULB.

157) Hold the **Ctrl** key down.

158) Click the **front flat face** of the LENSANDBULB.

159) Release the **Ctrl** key. The Mate pop-up menu is displayed.

160) Click **Coincident** from the Mate pop-up menu.

161) **Expand** the Mates folder. View the created mates.

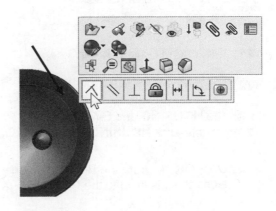

Confirm the location of the O-RING.

162) Click **Right Plane** of the CAPANDLENS from the FeatureManager.

163) Click **Section view** 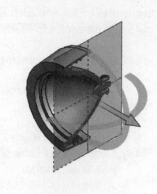 from the Heads-up View toolbar.

164) Click **Isometric view** from the Heads-up View toolbar.

165) **Expand** the Section 2 box. Click **inside** the Reference Section Plane box.

166) Click **Top Plane** of the CAPANDLENS from the fly-out FeatureManager.

167) Click **OK** ✔ from the Section View PropertyManager.

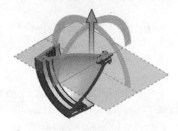

Save the CAPANDLENS sub-assembly. Return to a full view.

168) Click **Section view** from the Heads-up View toolbar.

169) Click **Save** .

The LENSANDBULB, BATTERYANDPLATE, and CAPANDLENS sub-assemblies are complete. The components in each assembly are fully defined. No minus (-) sign or red error flags exist in the FeatureManager. Insert the sub-assemblies into the final FLASHLIGHT assembly.

FLASHLIGHT Assembly

Create the FLASHLIGHT assembly. The HOUSING is the Base component. The FLASHLIGHT assembly mates the HOUSING to the SWITCH component. The FLASHLIGHT assembly mates the CAPANDLENS and BATTERYANDPLATE.

Activity: FLASHLIGHT Assembly

Create the FLASHLIGHT assembly.

170) Click **New** from the Menu bar.

171) Click the **MY-TEMPLATES** tab.

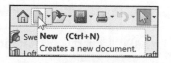

172) Double-click **ASM-IN-ANSI**.

Insert the HOUSING and SWITCH.

173) Double-click **HOUSING** from the PROJECTS folder.

174) Click **OK** ✔ from the Begin Assembly PropertyManager. The HOUSING is fixed to the Origin.

175) Click **Insert Components** from the Assembly toolbar.

176) Double-click **SWITCH** from the PROJECTS folder.

177) Click a **position** in front of the HOUSING as illustrated.

Save the FLASHLIGHT assembly. Enter name. Enter description.

178) Click **Save**. Select the **PROJECTS** folder.

179) Enter **FLASHLIGHT** for File name.

180) Enter **FLASHLIGHT ASSEMBLY** for Description.

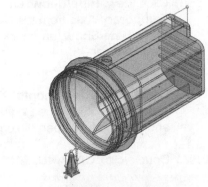

181) Click **Save**. The FLASHLIGHT FeatureManager is displayed.

Pin the Mate PropertyManager. Insert a Coincident mate.

182) Click the **Mate** Assembly tool.

183) If needed, click the **Keep Visible** icon.

184) Click **Right Plane** of the HOUSING from the fly-out FeatureManager.

185) Click **Right Plane** of the SWITCH from the fly-out FeatureManager. Coincident is selected by default.

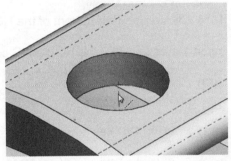

186) Click **OK** from the Mate PropertyManager.

Insert a Coincident mate.

187) Click **View, Hide/Show,** check **Temporary Axes** from the Menu bar.

188) Click the **Temporary axis** inside the Switch Hole of the HOUSING.

189) Click **Front Plane** of the SWITCH from the fly-out FeatureManager. Coincident is selected by default. Click **OK** from the Mate PropertyManager.

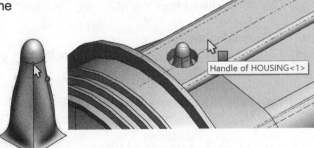

Insert a Distance mate.

190) Click the **top face** of the Handle.

191) Click the **Vertex** on the Loft top face of the SWITCH. Click **Distance**.

192) Enter **.100**in, **[2.54]**. Check the **Flip Direction** box if needed.

193) Click **OK** ✔ from the Mate
PropertyManager.

194) Un-pin the Mate PropertyManager. Click the
Keep Visible 📌 icon.

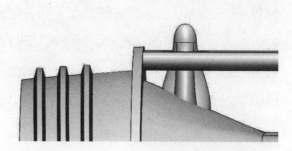

195) Click **OK** ✔ from the Mate
PropertyManager.

Insert the CAPANDLENS assembly.
196) Click **View**, **Hide/Show**, un-check
Temporary Axis from the Menu bar. Click
View, **Hide/Show**, un-check **Origins** from
the Menu bar.

197) Click **Insert Components** 🗁 from the
Assembly toolbar. The Begin Assembly PropertyManager
and the Window Open dialog box is displayed.

198) Double-click **CAPANDLENS** from the PROJECTS folder.

Place the sub-assembly.
199) Click a **position** in front of the HOUSING as illustrated.

Insert Mates between the HOUSING component and the
CAPANDLENS sub-assembly.
200) Click **Right Plane** of the HOUSING from the
FeatureManager.

201) Hold the **Ctrl** key down.

202) Click **Right Plane** of the CAPANDLENS from the fly-out
FeatureManager.

203) Release the **Ctrl** key. The Mate pop-up menu is displayed.

204) Click **Coincident** from the Mate pop-up menu.

Insert a Coincident mate.
205) Click **Top Plane** of the HOUSING from the
FeatureManager.

206) Hold the **Ctrl** key down. Click **Top Plane** of the
CAPANDLENS from the FeatureManager.

207) Release the **Ctrl** key. The Mate pop-up menu is displayed.

208) Click **Coincident** from the Mate pop-up menu.

Insert a Coincident mate.

209) Click the **front face** of the Boss-Stop on the HOUSING.

210) **Rotate** the model to view the back face of the CAPANDLENS as illustrated.

211) Hold the **Ctrl** key down. Click the **back face** of the CAPANDLENS.

212) Release the **Ctrl** key.

213) Click **Coincident** from the Mate pop-up menu.

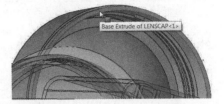

Display an Isometric view. Save the FLASHLIGHT assembly.

214) Click **Isometric view** 🧊.

215) Click **Save** 💾.

Insert the BATTERYANDPLATE sub-assembly.

216) Click **Insert Components** 📂 from the Assembly toolbar. The Begin Assembly PropertyManager and the Window Open dialog box is displayed.

217) Double-click **BATTERYANDPLATE** from the PROJECTS folder.

218) Click a **position** to the left of the HOUSING as illustrated.

Rotate the part.

219) Click **BATTERYANDPLATE** in the FeatureManager.

220) Click **Rotate Component** 🔄 from the Assembly toolbar. Rotate the **BATTERYANDPLATE** until it is approximately parallel with the HOUSING.

221) Click **OK** ✔ from the Rotate Component PropertyManager.

Insert a Coincident mate.

222) Click **Right Plane** of the HOUSING from the FeatureManager.

223) Hold the **Ctrl** key down.

224) Click **Front Plane** of the BATTERYANDPLATE from the FeatureManager.

225) Release the **Ctrl** key. The Mate pop-up menu is displayed. Click **Coincident** from the Mate pop-up menu.

226) Move the **BATTERYANDYPLATE** in front of the
HOUSING.

Insert a Coincident mate.
227) Click **Top Plane** of the HOUSING from the
FeatureManager.

228) Hold the **Ctrl** key down.

229) Click **Right Plane** of the BATTERYANDPLATE
from the FeatureManager.

230) Release the **Ctrl** key. The Mate pop-up menu is
displayed.

231) Click **Coincident** from the Mate pop-up menu.

Display the Section view.
232) Click **Right Plane** in the FLASHLIGHT Assembly
FeatureManager.

233) Click **Section view** 📖 from the Heads-up View
toolbar.

234) Click **OK** ✔ from the Section View
PropertyManager.

Move the BATTERYANDPLATE in front
of the HOUSING.
235) Click and drag the
BATTERYANDPLATE in front of
the HOUSING as illustrated.

Insert a Coincident mate.
236) Click the **back center Rib1 face** of the HOUSING
as illustrated.

237) **Rotate** the model to view the bottom face of the
BATTERYANDPLATE.

238) Hold the **Ctrl** key down. Click the **bottom face** of the
BATTERYANDPLATE.

239) Release the **Ctrl** key. The Mate pop-up menu is
displayed.

240) Click **Coincident** from the Mate pop-up menu.

Display an Isometric view.
241) Click **Isometric view** 🧊 .

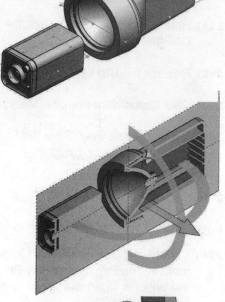

Base Extrude of BATTERY<1>

Rib1 of HOUSING<1>

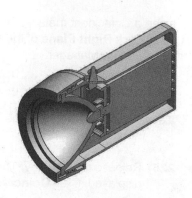

Display the Full view.

242) Click **Section view**  from the Heads-up View toolbar.

Save the FLASHLIGHT assembly.

243) Click **Save** .

🔍   Additional information on Assembly, Move Component, Rotate Component, and Mates is available in SOLIDWORKS Help Topics.

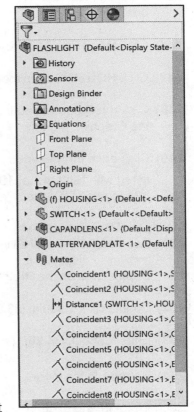

⚙️   Review of the FLASHLIGHT assembly

The FLASHLIGHT assembly consisted of the HOUSING part, SWITCH part, CAPANDLENS sub-assembly and BATTERYANDPLATE sub-assembly.

The CAPANDLENS sub-assembly contained the BULBANDLENS sub-assembly, the O-RING and the LENSCAP part. The BATTERYANDPLATE sub-assembly contained the BATTERY and BATTERYPLATE part.

You inserted eight Coincident mates and a Distance mate. Through the Assembly Layout illustration you simplified the number of components into a series of smaller assemblies. You also enhanced your modeling techniques and skills.

You still have a few more areas to address. One of the biggest design issues in assembly modeling is interference. Let's

investigate the FLASHLIGHT assembly.

💡   Clearance Verification checks the minimum distance between components and reports any value that fails to meet your input value of the minimum clearance. View SOLIDWORKS Help for additional information.

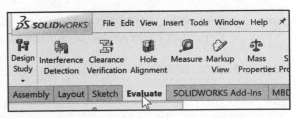

Addressing Interference Issues

There is an interference issue between the FLASHLIGHT components. Address the design issue. Adjust Rib2 on the HOUSING. Test with the Interference Check command. The FLASHLIGHT assembly is illustrated in inches.

Activity: Addressing Interference Issues

Check for interference.

244) Click the **Interference Detection** 🗐 tool from the Evaluate tab in the CommandManager.

245) Delete **FLASHLIGHT.SLDASM** from the Selected Components box.

246) Click **BATTERYANDPLATE** from the fly-out FeatureManager.

247) Click **HOUSING** from the fly-out FeatureManager.

248) Click the **Calculate** button. The interference is displayed in red in the Graphics window.

249) Click each **Interference** in the Results box to view the interference in red with Rib2 of the HOUSING. Click **OK** ✔ from the Interference Detection PropertyManager.

Modify the Rib2 dimension to address the interference issue.

250) Expand the HOUSING in the FeatureManager.

251) Double-click on the **Rib2** feature.

252) Double click **1.300**in, [33.02].

253) Enter **1.350**in, [34.29].

254) Rebuild 🔵 the model.

Recheck for Interference.

255) Click **Interference Detection** 🗐 from the Evaluate tab. The Interference dialog box is displayed.

256) Delete **FLASHLIGHT.SLDASM** from the Selected Components box. Click **BATTERYANDPLATE** from the FeatureManager.

257) Click **HOUSING** from the FeatureManager.

258) Click the **Calculate** button. No Interference is displayed in the Results box. The FLASHLIGHT design is complete.

259) Click **OK** ✔ from the Interference Detection PropertyManager.

Save the FLASHLIGHT assembly.

260) Click **Save** 💾.

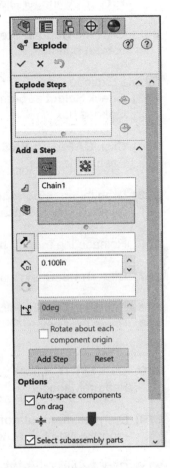

Exploded View

The Exploded View illustrates how to assemble the components in an assembly. Create an Exploded View of the FLASHLIGHT assembly. Click and drag components in the Graphics window. The Manipulator icon indicates the direction to explode.

Select an alternate component edge for the Explode direction. Drag the component in the Graphics window or enter an exact value in the Explode distance box. In this activity, manipulate the top-level components in the assembly.

In the chapter exercises, create exploded views for each sub-assembly and utilize the Regular step ☜ option in the top level assembly.

Access the Explode view option as follows:

- Right-click the configuration name in the ConfigurationManager.

- Select the Exploded View tool in the Assembly toolbar.

- Select Insert, Exploded View from the Menu bar.

The Exploded View feature uses the Explode PropertyManager as illustrated. You can go through an exploded view step-by-step. You can roll back an exploded view to see the results of each step.

🔅 You can create exploded view steps that rotate a component with or without linear translation. Use the rotation and translation handles of the triad. You can include rotation and translation in the same explode step. You can also edit the explode step translation distance and rotation angle values in the PropertyManager.

Activity: FLASHLIGHT Assembly-Exploded View

Insert an Exploded view.

261) Click the **Exploded View** 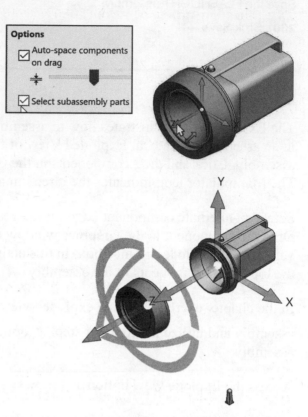 tool from the Assembly toolbar. The Explode PropertyManager is displayed.

262) Click the **Regular step (translate and rotate)** button.

Create Explode Chain1.

263) Check **Select sub-assembly parts** in the Options box.

264) Click **CAPANDLENS** from the fly-out FeatureManager.

265) Drag the **manipulator handle** to the front (approximate 9.5 inches).

266) Release the **mouse** button.

267) Click **Done**. Chain1 is created.

Fit the model to the Graphics window.

268) Press the **f** key.

Create Chain2.

269) Click **SWITCH** from the fly-out FeatureManager.

270) Drag the **manipulator handle** straight up (approximate 5.0 inches).

271) Click **Done**. The SWITCH moves upward. Chain2 is created.

Create Chain3.

272) Click **LENS** from the fly-out FeatureManager.

273) Enter **4.6in** in the Explode Distance box.

274) Click the **Reverse Direction** button.

275) Click **Add Step**. Chain3 is created. The LENS moves backwards.

Create Chain4.
276) Click **O-RING** from the fly-out FeatureManager.

277) Drag the **manipulator handle** to the front of the LENS as illustrated.

278) Click **Done**. Chain4 is created.

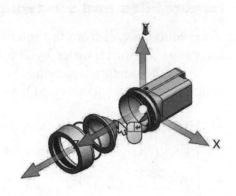

Create Chain5.
279) Click **HOUSING** from the fly-out FeatureManager.

280) Drag the **manipulator handle** backwards to expose the BATTERYANDPLATE.

281) Click **Done**. Chain5 is created.

Create Chain6.
282) Click **BATTERYPLATE** from the fly-out FeatureManager.

283) Drag the **manipulator handle** forward.

284) Click **Done**. Chain6 is created.

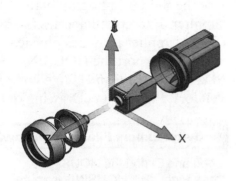

Create Chain7.
285) Click **BULB** from the fly-out FeatureManager.

286) Drag the **manipulator handle** to the back of the LENS as illustrated.

287) Click **Done**. Chain7 is created.

288) Click **OK** ✔ for the Explode PropertyManager.

Fit the model to the Graphics window.
289) Press the **f** key.

Remove the Exploded State.
290) **Right-click** in the Graphics window.

291) Click **Collapse** from the Pop-up menu.

Display an Isometric view. Save the model.
292) Click **Isometric view** 🗍.

293) Click **Save** 💾.

Export Files and eDrawings

You receive a call from the sales department. They inform you that the customer increased the initial order by 200,000 units. However, the customer requires a prototype to verify the design in six days. What do you do? Answer: Contact a Rapid Prototype supplier. You export three SOLIDWORKS files:

- HOUSING.

- LENSCAP.

- BATTERYPLATE.

Use the Stereo Lithography (STL) format. Email the three files to a Rapid Prototype supplier. Example: Paperless Parts Inc. (www.paperlessparts.com). A Stereolithography (SLA) supplier provides physical models from 3D drawings. 2D drawings are not required. Export the HOUSING. SOLIDWORKS eDrawings provides a facility for you to animate, view and create compressed documents to send to colleagues, customers and vendors. Publish an eDrawing of the FLASHLIGHT assembly.

Activity: Export Files and eDrawings

Open and Export the HOUSING.

294) Right-click **HOUSING** from the FeatureManager.

295) Click **Open Part** from the Context toolbar.

296) Click **Save As** from the drop-down Menu bar.

297) Select **STL (*.stl)** from the Save as type drop-down menu. The dialog box is displayed.

298) Click the **Options** button.

299) Click the **Binary** box from the Output as format box.

300) Click the **Course** box for Resolution.

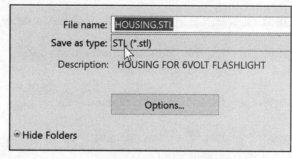

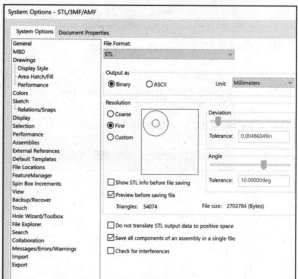

Create the binary STL file.

301) Click **OK** from the Export Options dialog box.

302) Click **Save** from the Save dialog box. A status report is provided.

303) Click **Yes**.

Publish an eDrawing and email the document to a colleague.

Create the eDrawing and animation.

304) Click **File**, **Publish to eDrawings** from the Menu bar.

305) Click the **Play Animate** button. Click **Play**. View the results.

Stop the animation.
306) Click the **Stop** button. Click the **Reset** button.

Save the eDrawing.
307) Click **Save** from the eDrawing Main menu.

308) Select the **PROJECTS** folder. Enter **FLASHLIGHT** for File name.

309) Click **Save**.

Close the eDrawing dialog box. Close all models.
310) **Close** ☒ the eDrawing dialog box.

311) **Close** all models in the session.

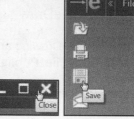

It is time to go home. The telephone rings. The customer is ready to place the order. Tomorrow you will receive the purchase order.

The customer also discusses a new purchase order that requires a major design change to the handle. You work with your industrial designer and discuss the two options. The first option utilizes Guide Curves on a Swept feature.

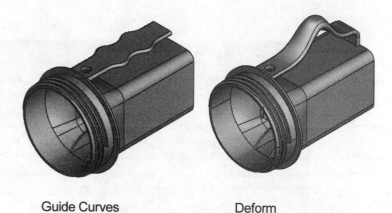

Guide Curves   Deform

The second option utilizes the Deform feature.

You contact your mold maker and send an eDrawing of the LENSCAP.

The mold maker recommends placing the parting line at the edge of the Revolved Cut surface and reversing the Draft Angle direction. The mold maker also recommends a snap fit versus a thread to reduce cost. The Core-Cavity mold tooling is explored in the project exercises.

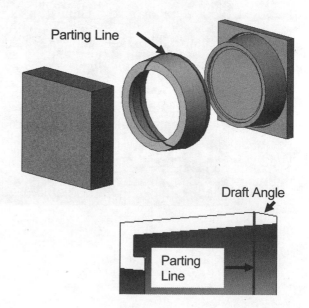

Parting Line

Draft Angle

Parting Line

🔍 Additional information on Interference Detection, eDrawings, STL files (stereolithography), Guide Curves, Deform and Mold Tools are available in SOLIDWORKS Help Topics.

Chapter Summary

The FLASHLIGHT contains over 100 features, Reference planes, sketches and components. You organized the features in each part. You developed an assembly layout structure to organize your components.

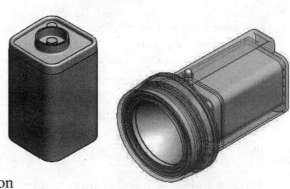

The O-RING utilizes a simple Swept Base feature. The SWITCH utilizes the Lofted Base feature. The LENSCAP and HOUSING utilize the Swept Boss and Lofted Boss feature.

A simple Swept Base feature requires a *path* and *profile*. The Lofted feature requires a minimum of two profiles sketched on different planes.

The LENSCAP and HOUSING utilized a variety of features. You applied design intent to reuse geometry through Geometric relationships, symmetry and patterns.

The assembly required an Assembly Template. You utilized the ASM-IN-ANSI Template to create the LENSANDBULB, CAPANDLENS, BATTERYANDPLATE and FLASHLIGHT assemblies.

You created an STL file of the Housing and an eDrawing of the FLASHLIGHT assembly to communicate with your vendor, mold maker and customer. Review the chapter exercises before moving on to the next chapter.

💡 Quick mate is a procedure to mate components together in SOLIDWORKS. No command (click Mate from the Assembly CommandManager) needs to be executed. Hold the Ctrl key down, make your selections, release the Ctrl key and a Quick Mate pop-up menu is displayed below the context toolbar. Select the mate and you are finished. This mate behavior is similar to the way you add sketch relations.

Questions

1. True or False. A Part Template is the foundation for an assembly document. Explain your answer.

2. Describe the difference between a Distance Mate and a Coincident Mate. Provide an example of each.

3. Describe an assembly or sub-assembly. Are they the same?

4. Describe five proven assembly modeling techniques. Can you add a few more?

5. Explain how to determine an interference between components in an assembly. Provide an example.

6. Describe the Deform feature.

7. Describe Mates. Why are Mates important in assembling components?

8. In an assembly, each component has_____# degrees of freedom. Name them.

9. True or False. A fixed component cannot move.

10. Describe a Section view.

11. Describe a Suppressed feature and component. Provide an example.

12. Identify the type of faces utilized for a Concentric mate.

13. List the Standard mate types. Where would you locate additional information on a Tangent Mate?

14. True or False. Only planes are utilized for Mate References. Explain your answer.

15. True or False. Only faces are utilized for Mate References. Explain your answer.

16. Define the steps required to create an Exploded view.

Exercises

Exercise 8.1: Weight-Hook Assembly

Create an ANSI, IPS Weight-Hook assembly. The Weight-Hook assembly has two components: WEIGHT and HOOK.

- Create a new assembly document. Copy and insert the WEIGHT part from the Chapter 8 Homework folder.

- Fix the WEIGHT to the Origin as illustrated in the Assem1 FeatureManager.

- Insert the HOOK part from the Chapter 8 Homework folder into the assembly.

- Insert a Concentric mate between the inside top cylindrical face of the WEIGHT and the cylindrical face of the thread. Concentric is the default mate.

- Insert the first Coincident mate between the top edge of the circular hole of the WEIGHT and the top circular edge of Sweep1, above the thread.

- Coincident is the default mate. The HOOK can rotate in the WEIGHT.

- Fix the position of the HOOK.

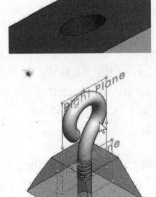

- Insert the second Coincident mate between the Right Plane of the WEIGHT and the Right Plane of the HOOK. Coincident is the default mate.

- Expand the Mates folder and view the created mates.

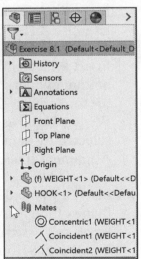

Exercise 8.2: Weight-Link Assembly

Create an ANSI, IPS Weight-Link assembly. The
Weight-Link assembly has two components and a sub-
assembly: Axle component, FLATBAR component and
the Weight-Hook sub-assembly that you created in
Exercise 8.1.

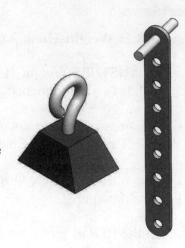

- Create a new assembly document. Copy and insert the
  Axle part from the Chapter 8 Homework folder.

- Fix the Axle component to the Origin.

- Copy and insert the FLATBAR part from the Chapter
  8 Homework folder.

- Insert a Concentric mate between the Axle cylindrical face
  and the FLATBAR inside face of the top circle.

- Insert a Coincident mate between the Front Plane of the
  Axle and the Front Plane of the FLATBAR.

- Insert a Coincident mate between the Right Plane of the
  Axle and the Top Plane of the FLATBAR. Position the
  FLATBAR as illustrated.

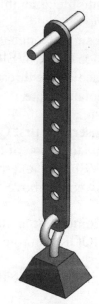

- Insert the Weight-Hook sub-assembly that you created in
  Exercise 8.1.

- Insert a Tangent mate between the inside
  bottom cylindrical face of the FLATBAR and
  the top circular face of the HOOK, in the
  Weight-Hook assembly. Tangent mate is
  selected by default. Click Flip Mate
  Alignment if needed.

- Insert a Coincident mate between the Front
  Plane of the FLATBAR and the Front Plane of the Weight-
  Hook sub-assembly. Coincident mate is selected by default.
  The Weight-Hook sub-assembly is free to move in the
  bottom circular hole of the FLATBAR.

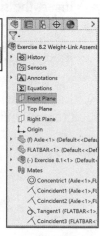

Exercise 8.3: Binder Clip

- Create a Gem® binder clip.

- Create an ANSI - IPS assembly.

- Create two components - Binder and Binder Clip.

- Apply material to each component and address all needed mates.

What is your Base Sketch for each component?

What are the dimensions?

View SOLIDWORKS Help or the chapter on the Swept Base feature to create the Binder Clip component.

Determine the static and dynamic behavior of mates in each sub-assembly before creating the top level assembly.

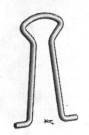

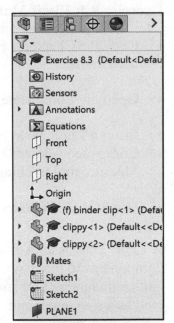

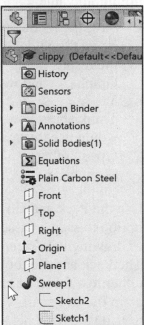

Exercise 8.4: Limit Mate (Advanced Mate Type)

- Copy the Chapter 8 Homework\Limit Mate folder to your hard drive.

- Open the Limit Mate assembly.

- Insert a Limit Mate to restrict the movement of the Slide Component - lower and upper movement. (Use the Measure tool to obtain max and min distances).

- Use SOLIDWORKS Help for additional information.

A Limit Mate is an Advanced Mate type. Limit mates allow components to move within a range of values for distance and angle mates. You specify a starting distance or angle as well as a maximum and minimum value.

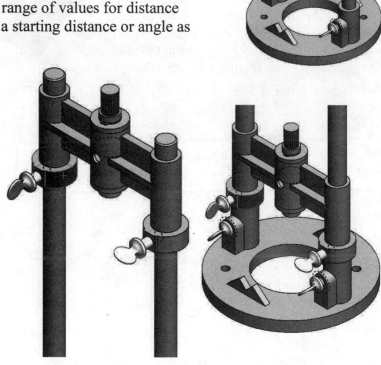

- Save the model and move the Slide to view the results in the Graphics window.

- Think about how you would use this mate type in other assemblies.

Use the Pack and Go option to save an assembly or drawing with references. The Pack and Go tool saves either to a folder or creates a zip file to e-mail. View SOLIDWORKS help for additional information.

Exercise 8.5: Screw Mate (Mechanical Mate Type)

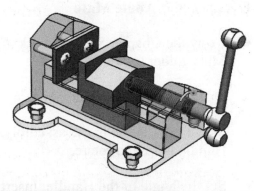

- Copy the Chapter 8 Homework\Screw Mate folder to your hard drive.

- Open the Screw Mate assembly.

- Insert a Screw mate between the inside Face of the Base and the Face of the vice and any other mates that are required. A Screw is a Mechanical Mate type. View the avi file for proper movement.

A Screw mate constrains two components to be concentric and also adds a pitch relationship between the rotation of one component and the translation of the other. Translation of one component along the axis causes rotation of the other component according to the pitch relationship. Likewise, rotation of one component causes translation of the other component. Use SOLIDWORKS Help if needed.

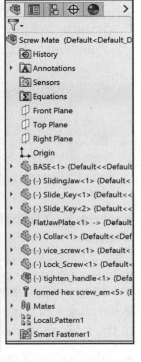

- Rotate the handle and view the results. Think about how you would use this mate type in other assemblies.

💡 Use the Select Other tool (See SOLIDWORKS Help if needed) to select faces and edges that are hidden in an assembly.

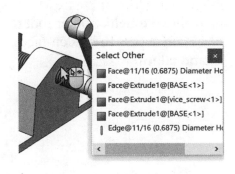

Exercise 8.6: Angle Mate

- Copy the Chapter 8 Homework\Angle Mate folder to your hard drive.

- Open Angle Mate assembly.

- Move the Handle in the assembly. The Handle is free to rotate.

- Set the angle of the Handle. Insert an Angle mate (165 degrees) between the Handle and the Side of the valve using Planes. An Angle mate places the selected items at the specified angle to each other.

- The Handle has a 165 degree Angle mate to restrict flow through the valve. Think about how you would use this mate type in other assemblies.

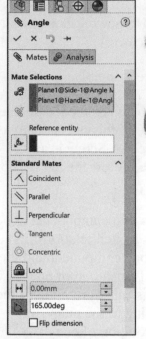

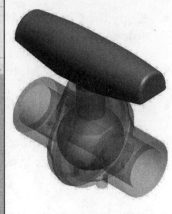

Exercise 8.6A: Angle Mate (Cont:)

Create two end caps (lids) for the valve using the Top-down Assembly method. Note the Reference - In-Content symbols in the FeatureManager.

- Modify the Appearance of the body to observe the change - enhance visualization.

- Apply the Select-other tool to obtain access to hidden faces and edges.

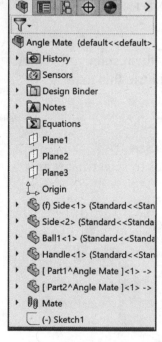

Exercise 8.7: Symmetric Mate (Advanced Mate)

- Copy the Chapter 8 Homework\Symmetric Mate folder to your hard drive.

- Open the Symmetric Mate assembly.

- View the movement.

- Insert a Symmetric Mate for the Guild Rollers.

- Think about how you would use this mate type in other assemblies.

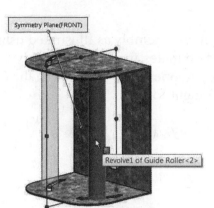

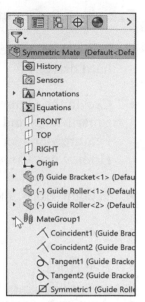

Exercise 8.8: Gear Mate (Mechanical mate)

- View the ppt located in the Chapter 8 Homework\Gears General folder.

- Create the Gear assembly as illustrated.

- Create the needed components and mates.

- Use the SOLIDWORKS Toolbox. The SOLIDWORKS Toolbox is an add-in.

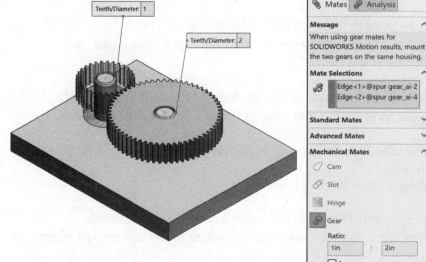

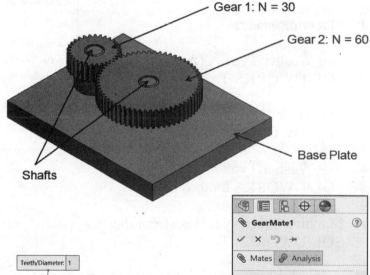

Exercise 8.9: Counter Weight Assembly

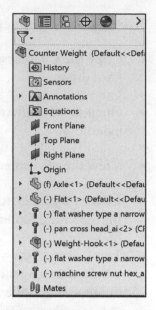

- Copy the Chapter 8 Homework\Counter Weight folder to your hard drive.

- Create the Counter Weight assembly as illustrated using SmartMates and Standard mates from the Assembly FeatureManager. All components are supplied in the Chapter 8 Homework\Counter-Weight folder.

- Weight-Hook sub-assembly.

- Weight.

- Eye Hook.

- Axle component (f). Fixed to the origin.

- Flat component.

- Flat Washer Type A (from the SOLIDWORKS Toolbox).

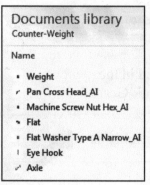

- Pan Cross Head Screw (from the SOLIDWORKS Toolbox).

- Flat Washer Type A (from the SOLIDWORKS toolbox).

- Machine Screw Nut Hex (from the SOLIDWORKS Toolbox).

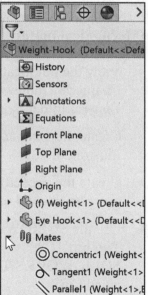

Apply SmartMates with the Flat Washer Type A Narrow_AI, Machine Screw Nut Hex_AI and the Pan Cross Head_AI components.

Use a Distance mate to fit the Axle in the middle of the Flat. Note a Symmetric mate could replace the Distance mate.

Think about the design of the assembly.

Apply all needed Lock mates.

The symbol (f) represents a fixed component. A fixed component cannot move and is locked to the assembly Origin.

Additional Project Exercises

Exercise 8.10: Butterfly Valve Assembly Project

- Copy the Chapter 8 Homework\Butterfly Valve Assembly Project folder to your local hard drive.

- Create an ANSI IPS Butterfly Valve assembly document.

- Create and insert all needed components and mates to assemble the assembly and to simulate proper movement.

- You are the designer. Address all tolerance and dimension modifications if needed. Use Standard, Advanced and Mechanical Mates. Create and insert any additional components if needed.

- Create a C-ANSI Landscape - Third Angle Isometric Exploded Drawing document with Explode lines of the assembly using your knowledge of SOLIDWORKS. Insert a BOM with Balloons. Insert all needed General notes in the Title Block.

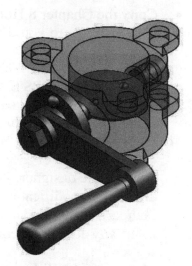

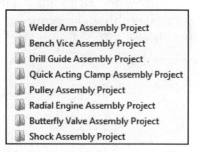

Welder Arm Assembly Project
Bench Vice Assembly Project
Drill Guide Assembly Project
Quick Acting Clamp Assembly Project
Pulley Assembly Project
Radial Engine Assembly Project
Butterfly Valve Assembly Project
Shock Assembly Project

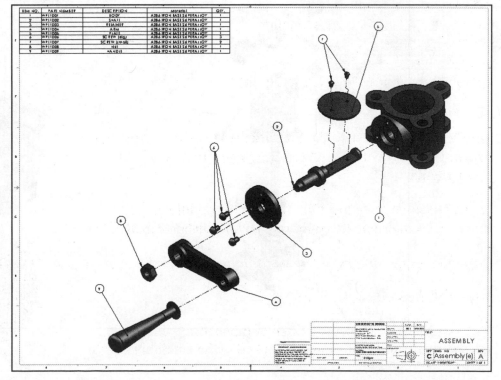

Exercise 8.11: Shock Assembly Project

- Copy the Chapter 8 Homework\Shock Assembly Project folder to your hard drive. View all components.

- Create an ANSI IPS Shock assembly document. Create and insert all needed components and mates to assemble the assembly and to simulate proper movement.

- You are the designer. Address all tolerance and dimension modifications if needed. Be creative. Use Standard, Advanced and Mechanical Mates. Create and insert any additional components if needed.

- Create a C-ANSI Landscape - Third Angle Isometric Exploded Drawing document with Explode lines of the assembly using your knowledge of SOLIDWORKS. Insert a BOM with Balloons. Insert all needed General notes in the Title Block.

Name
Welder Arm Assembly Project
Bench Vice Assembly Project
Drill Guide Assembly Project
Quick Acting Clamp Assembly Project
Pulley Assembly Project
Radial Engine Assembly Project
Butterfly Valve Assembly Project
Shock Assembly Project

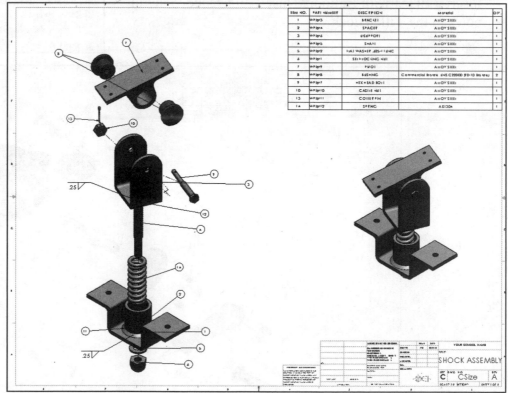

Exercise 8.12: Clamp Assembly Project

- Copy the Chapter 8 Homework\Clamp
  Assembly Project folder to your hard drive.
  View all components.

- Create an ANSI IPS Clamp assembly
  document. Create and insert all needed
  components and mates to assemble the
  assembly and to simulate proper
  movement.

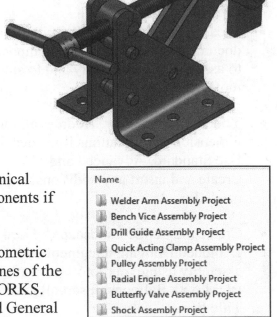

- You are the designer. Address all tolerance
  and dimension modifications if needed. Be
  creative. Use Standard, Advanced and Mechanical
  Mates. Create and insert any additional components if
  needed.

Name
Welder Arm Assembly Project
Bench Vice Assembly Project
Drill Guide Assembly Project
Quick Acting Clamp Assembly Project
Pulley Assembly Project
Radial Engine Assembly Project
Butterfly Valve Assembly Project
Shock Assembly Project

- Create a C-ANSI Landscape - Third Angle Isometric
  Exploded Drawing document with Explode lines of the
  assembly using your knowledge of SOLIDWORKS.
  Insert a BOM with Balloons. Insert all needed General
  notes in the Title Block.

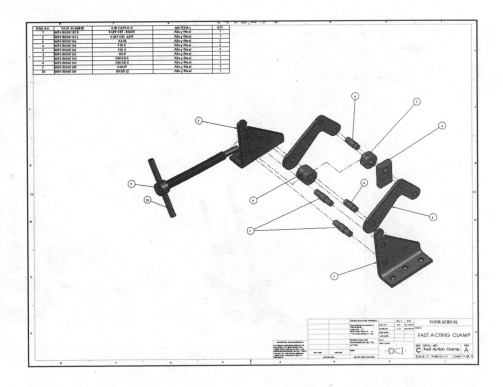

Exercise 8.13: Drill Guide Assembly Project

- Copy the Chapter 8 Homework\Drill Guide Assembly Project folder to your hard drive. View all components.

- Create an ANSI IPS Drill Guides assembly document. Insert all needed components and mates to assemble the assembly and to simulate proper movement.

- You are the designer. Address all tolerance and dimension modifications if needed. Be creative. Use Standard, Advanced and Mechanical Mates. Create and insert any additional components if needed.

- Create a C-ANSI Landscape - Third Angle Isometric Exploded Drawing document with Explode lines of the assembly using your knowledge of SOLIDWORKS. Insert a BOM with Balloons. Insert all needed General notes in the Title Block.

Name
- Welder Arm Assembly Project
- Bench Vice Assembly Project
- Drill Guide Assembly Project
- Quick Acting Clamp Assembly Project
- Pulley Assembly Project
- Radial Engine Assembly Project
- Butterfly Valve Assembly Project
- Shock Assembly Project

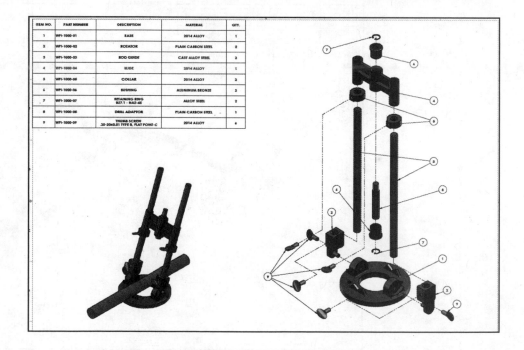

ITEM NO.	PART NUMBER	DESCRIPTION	MATERIAL	QTY.
1	WP1-1000-01	BASE	2014 ALLOY	1
2	WP1-1000-02	ROTATOR	PLAIN CARBON STEEL	2
3	WP1-1000-03	ROD GUIDE	CAST ALLOY STEEL	2
4	WP1-1000-04	SLIDE	2014 ALLOY	1
5	WP1-1000-05	COLLAR	2014 ALLOY	2
6	WP1-1000-06	BUSHING	ALUMINUM BRONZE	2
7	WP1-1000-07	RETAINING RING B27.1 - NA2-65	ALLOY STEEL	2
8	WP1-1000-08	DRILL ADAPTOR	PLAIN CARBON STEEL	1
9	WP1-1000-09	THUMB SCREW .25-20x0.51 TYPE B, FLAT POINT-C	2014 ALLOY	6

Exercise 8.14: Pulley Assembly Project

- Copy the Chapter 8 Homework\Pulley Assembly Project folder to your hard drive. View all components.

- Create an ANSI Pulley assembly document. Insert all needed components and mates to assemble the assembly and to simulate proper movement.

- You are the designer. Address all tolerance and dimension modifications if needed. Be creative. Use Standard, Advanced and Mechanical Mates. Create and insert any additional components if needed.

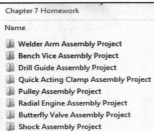

- Create a C-ANSI Landscape - Third Angle Isometric Exploded Drawing document with Explode lines of the assembly using your knowledge of SOLIDWORKS. Insert a BOM with Balloons. Insert all needed General notes in the Title Block.

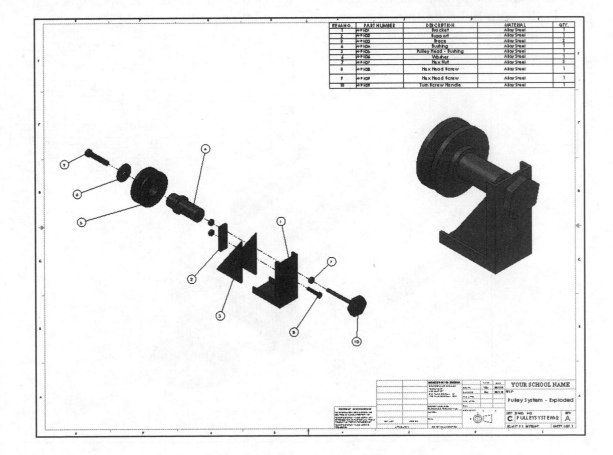

ITEM NO.	PART NUMBER	DESCRIPTION	MATERIAL	QTY.
1	WP101	Bracket	Alloy Steel	1
2	WP102	Support	Alloy Steel	1
3	WP103	Brace	Alloy Steel	2
4	WP104	Bushing	Alloy Steel	1
5	WP105	Pulley Head - Bushing	Alloy Steel	1
6	WP106	Washer	Alloy Steel	1
7	WP107	Hex Nut	Alloy Steel	2
8	WP108	Hex Head Screw	Alloy Steel	1
9	WP109	Hex Head Screw	Alloy Steel	1
10	WP109	Turn Screw Handle	Alloy Steel	1

Exercise 8.15: Welder Arm Assembly Project

- Copy the Chapter 8 Homework\Welder Arm Assembly Project folder to your hard drive. View all components.

- Create an ANSI Welder Arm assembly document. Insert all needed components and mates to assemble the assembly and to simulate proper movement.

- You are the designer. Address all tolerance and dimension modifications if needed. Be creative. Use Standard, Advanced and Mechanical Mates. Create and insert any additional components if needed.

- Create a C-ANSI Landscape - Third Angle Isometric Exploded Drawing document with Explode lines of the assembly using your knowledge of SOLIDWORKS. Insert a BOM with Balloons. Insert all needed General notes in the Title Block.

Name
- Welder Arm Assembly Project
- Bench Vice Assembly Project
- Drill Guide Assembly Project
- Quick Acting Clamp Assembly Project
- Pulley Assembly Project
- Radial Engine Assembly Project
- Butterfly Valve Assembly Project
- Shock Assembly Project

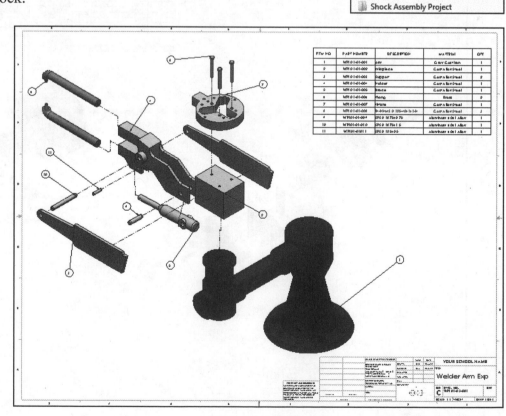

Exercise 8.16: **Radial Engine Assembly Project**

- Copy the Chapter 8 Homework\Radial Engine Assembly Project folder to your hard drive. View all components.

- Create an ANSI Radial Engine assembly document. Insert all needed components and mates to assemble the assembly and to simulate proper movement.

- You are the designer. Address all tolerance and dimension modifications if needed. Be creative.

- Use Standard, Advanced and Mechanical Mates.

- Create and insert any additional components if needed.

- Create a C-ANSI Landscape - Third Angle Isometric Exploded Drawing document with Explode lines of the assembly using your knowledge of SOLIDWORKS.

- Insert a BOM with Balloons. Insert all needed General notes in the Title Block.

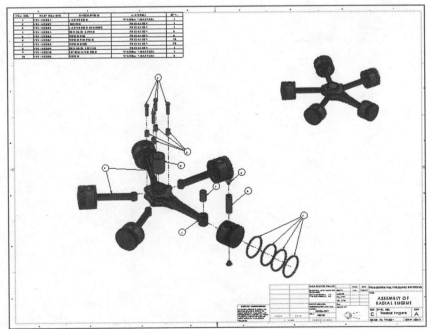

Notes:

Chapter 9

Fundamentals of Drawing

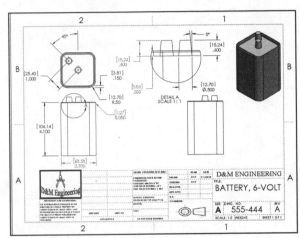
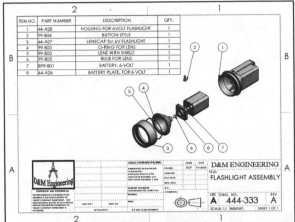

Below are the desired outcomes and usage competencies based on the completion of Chapter 9.

Desired Outcomes:	Usage Competencies:
• Custom Drawing and Sheet Template.	• Define Document Properties: Drafting standard, units, and precision. • Insert Custom Properties, Title block information, Company logo, tolerance and units.

• Two Assembly drawings and a part drawing:
 	
  • BATTERY
 	
  • FLASHLIGHT
 	
  • O-RING-DESIGN-TABLE
	• Create the following drawing views: Front, Top, Right, Isometric, Detail, Section, and Exploded.
	
	• Proficiency to insert and modify dimensions, BOM, Balloon text, Annotations, Centerlines, Center Marks, and Center of Mass point.

• Design Table and various Configurations.	• Capability to create three configurations in a design table: Small, Medium, and Large.

Notes:

Chapter 9 - Fundamentals of Drawing

Chapter Overview

Create three drawings in this chapter:

- BATTERY.

- FLASHLIGHT.

- O-RING-DESIGN-TABLE.

The BATTERY part drawing contains the Front, Top, Right, Detail and Isometric views. Orient the views to fit the drawing sheet. Incorporate the BATTERY part dimensions into the drawing.

The FLASHLIGHT assembly drawing contains an Exploded view, a Bill of Materials and balloon text. The Balloon items correspond to the Item Number in the BOM. The numeric part number is a user-defined property in each part.

Insert a Design Table for the O-RING part. A Design Table is an Excel spreadsheet that contains parameters. Define the Sketch-path and Sketch-profile diameters of the Swept feature.

Create three configurations of the O-RING part:

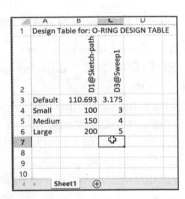

- Small.

- Medium.

- Large.

The O-RING-DESIGN-TABLE drawing contains three configurations of the O-RING. Utilize the View Palette from the Task Pane. Use the drawing view properties to control each part configuration. The three drawings utilize a custom Drawing Template and Sheet Format. The Drawing Template defines the dimensioning standard, units and precision. The Sheet Format contains the Title block information and a Company logo.

There are two major design modes used to develop a drawing: Edit Sheet Format and Edit Sheet. Work between the two drawing modes in this project. The Edit Sheet Format mode provides the ability to:

- Change the Title block size and text headings.

- Incorporate a Company logo.

- Add drawing, design or company text.

The Edit Sheet mode provides the ability to:

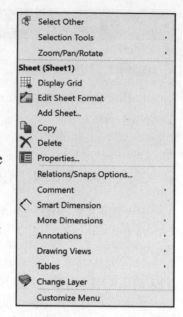

- Add or modify views.

- Add or modify dimensions.

- Add or modify text.

After completing the activities in this chapter, you will be able to:

- Utilize the View Layout tab and the Annotation toolbar for the following tools: Model View, Projected View, Detail View, Note, Model Items, Balloons and Magnetic line.

- Create two Drawing Templates: A-IN-ANSI and A-MM-ISO.

- Insert, move and edit part and drawing dimensions.

- Develop a Design Table for the O-RING part and insert the correct configuration into a drawing.

- Insert an Exploded view with a Bill of Materials and Magnetic lines.

- Apply the Edit Sheet and Edit Sheet Format modes.

- Insert a Center of Mass point.

New Drawing and the Drawing Template

The foundation of a new SOLIDWORKS drawing is the Drawing Template. Drawing size, drawing standards, company information, manufacturing, and/or assembly requirements, units and other properties are defined in the Drawing Template.

The Sheet Format is incorporated into the Drawing Template. The Sheet Format contains the border, title block information, revision block information, company name, and/or logo information, Custom Properties and SOLIDWORKS Properties.

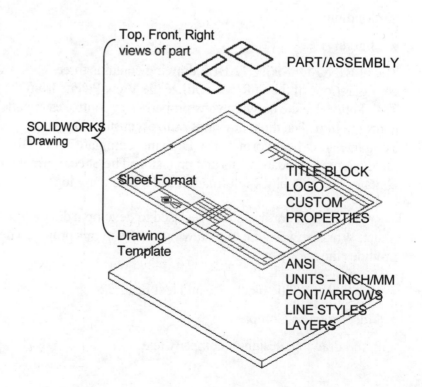

Custom Properties and SOLIDWORKS Properties are shared values between documents.

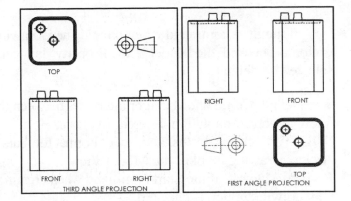

Views from the part or assembly are inserted into the SOLIDWORKS Drawing. Views are inserted in Third or First Angle projection. Notes and dimensions for millimeter drawings are provided in brackets [x] for this chapter.

Utilize an A size Drawing Template with Sheet Format for the BATTERY drawing and FLASHLIGHT assembly drawing. A copy of the default Drawing Template illustrated in this activity is contained in the SOLIDWORKS-MODELS 2019\MY-TEMPLATES folder. The default Drawing Templates contain predefined Title block Notes linked to Custom Properties and SOLIDWORKS Properties.

💡 For printers supporting millimeter paper sizes, utilize the Printer, Properties and Scale to Fit option.

Activity: New Drawing and the Drawing Template

Close all parts and drawings.

1)   Click **Windows**, **Close All** from the Menu bar.

Create a new drawing.

2)   Click **New** 🗋 from the Menu bar.

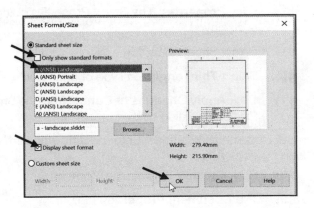

3)   Double-click **Drawing** from the Templates tab. Un-check the Only show standard formats box if needed.

4)   Select **A (ANSI) Landscape**.

5)   Click **OK** from the Sheet Format/Size dialog box.

6)   Click **Cancel** ✕ from the Model View PropertyManager. The Draw1 FeatureManager is displayed. The Draw1 FeatureManager is displayed to the left of the Graphics window.

The A (ANSI) Landscape paper is displayed in the Graphics window. The sheet border defines the drawing size, 11″ × 8.5″ or (279.4mm × 215.9mm).

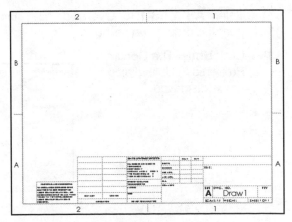

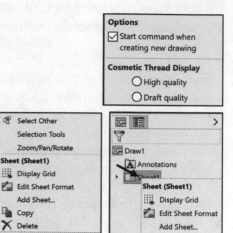

💡 If the Start command when creating new drawing option is checked, the Model View PropertyManager is selected by default.

The Default CommandManager alternates between the View Layout, Annotation, Sketch, Evaluate, SOLIDWORKS Add-Ins, and Sheet Format toolbars. A New Drawing invokes the Model View PropertyManager if the Start command when creating new drawing option is checked.

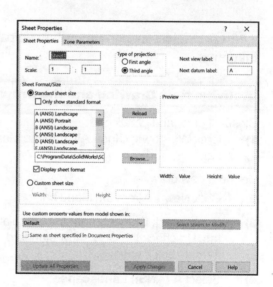

Set Sheet Properties and Document Properties for the Drawing Template. Sheet Properties control the Sheet Size, Sheet Scale, Type of Projection and more.

Document Properties control the display of dimensions, annotations, symbols and more in the drawing.

Set Sheet Properties.

7) Right-click on **Sheet1** in the FeatureManager

8) Click **Properties**. The Sheet Properties dialog box is displayed.

9) Select Sheet Scale **1:1**.

10) Select **Third angle** for Type of projection.

11) Click **Apply Changes** or **Cancel** from the Sheet Properties dialog box.

Set Document Properties. Enter drafting standard, units, and precision.

12) Click **Options** ⚙ from the Menu bar.

13) The **Document Properties** tab.

14) Select **ANSI**, **[ISO]** for Overall drafting standard from the drop-down menu.

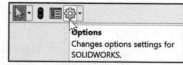

15) Click **Units**. The Document Properties - Units dialog box is displayed.

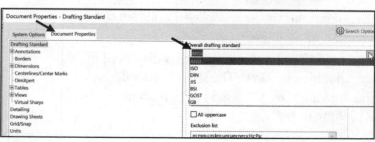

16) Select **IPS**, **[MMGS]** for Unit system.

17) Select **.123**, **[.12]** for Length units Decimal places from the drop-down menu.

18) Select **None** for Angular units Decimal places.

19) Click **OK** from the Document Properties - Units dialog box.

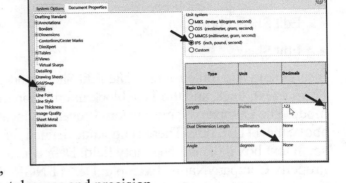

Detailing options provide the ability to address dimensioning standards, text style, center marks, extension lines, arrow styles, tolerance and precision.

There are numerous text styles and sizes available in SOLIDWORKS.

Save the Drawing.

20) Click **Save** 💾. The Draw1 FeatureManager is displayed.

Title Block

The Title block contains text fields linked to System Properties and Custom Properties. System Properties are determined from the SOLIDWORKS documents. Custom Property values are assigned to named variables. Save time. Utilize System Properties and define Custom Properties in your Sheet Formats.

System Properties and Custom Properties for Title Block:			
System Properties Linked to fields in default Sheet Formats:	**Custom Properties of drawings linked to fields in default Sheet Formats:**		**Custom Properties of parts and assemblies linked to fields in default Sheet Formats:**
SW-File Name (in DWG. NO. field):	CompanyName:	EngineeringApproval:	Description (in TITLE field):
SW-Sheet Scale:	CheckedBy:	EngAppDate:	Weight:
SW-Current Sheet:	CheckedDate:	ManufacturingApproval:	Material:
SW-Total Sheets:	DrawnBy:	MfgAppDate:	Finish:
	DrawnDate:	QAApproval:	Revision:
	EngineeringApproval:	QAAppDate:	

The drawing document contains two modes:

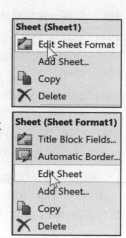

- Edit Sheet Format.

- Edit Sheet.

Insert views and dimensions in the Edit Sheet mode. Modify the Sheet Format text, lines, and the Title block information in the Edit Sheet Format mode. The CompanyName Custom Property is located in the Title block above the TITLE box. There is no value defined for CompanyName. A small text box indicates an empty field. Define a value for the Custom Property CompanyName. Example: D&M ENGINEERING. The Tolerance block is located in the Title block.

The Tolerance block provides information to the manufacturer on the minimum and maximum variation for each dimension on the drawing.

If a specific tolerance or note is provided on the drawing, the specific tolerance or note will override the information in the Tolerance block. General tolerance values are based on the design requirements and the manufacturing process. Modify the Tolerance block in the Sheet Format for ASME Y14.5 machined parts. Delete unnecessary text. The FRACTIONAL text refers to inches. The BEND text refers to sheet metal parts.

Activity: Title Block

Invoke the Edit Sheet Format Mode.

21)   Right-click **Edit Sheet Format** from the Pop-up menu in the Graphics window. The Title block lines are displayed in blue.

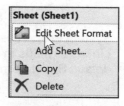

22)   **Zoom in** on the Sheet Format Title block.

Define COMPANYNAME Custom Property.

23)   Click a **center position** above the TITLE box as illustrated. The Note text $PRP:"COMPANYNAME" is displayed. The Note PropertyManager is displayed.

24)   Click the **Link to Property** icon as illustrated from the Text Format box. The Link to Property dialog box is displayed.

25) Click the **File Properties** button. The Summary Information dialog box is displayed.

26) Click the **Custom** tab from the Summary Information dialog box.

27) Select **CompanyName** from the Property Name box. Click inside the **Value/Text Expression** box.

28) Enter **D&M ENGINEERING** or your **company or school** name.

29) Click inside the **Evaluated Value** box. The CompanyName is displayed.

30) Click **OK** from the Summary Information dialog box.

31) Click **Cancel** from the Link to Property dialog box. D&M ENGINEERING is displayed.

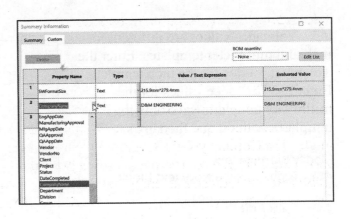

Modify the font size.

32) Uncheck the **Use document's font** box in the Note PropertyManager.

33) Click the **Font** button. The Choose Font dialog box is displayed.

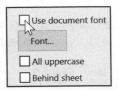

Select a new font size.

34) Click **Bold**.

35) Click **Points**.

36) Select **18**.

37) Click **OK** from the Choose Font dialog box. The text is displayed in the Title block.

38) Click **OK** ✔ from the Note PropertyManager.

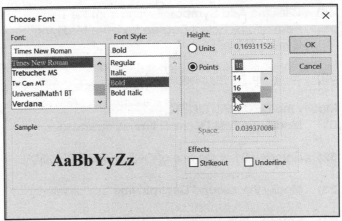

Modify the Tolerance Note in the text box.

39) **Zoom in** on the Tolerance Note in the text box.

40) Double-click the text **INTERPRET GEOMETRIC TOLERANCING PER:** The Note PropertyManager is displayed.

41) Enter **ASME Y14.5** as illustrated.

42) Click **OK** ✔ from the Note PropertyManager.

💡 Click outside the Note text box to end the Note, or Click OK ✔ from the Note PropertyManager.

Tolerance values are different for inch and millimeter templates. Enter the Tolerance values for the inch template. Enter the Tolerance values for the millimeter template. Save drawing templates with unique filenames.

Modify the Note text.
43) Right-click the **Note text block**.

44) Click **Edit Text**.

45) Delete the line **FRACTIONAL +-**.

46) Delete the text **BEND +-**.

Enter ANGULAR tolerance.
47) Click a **position** at the end of the line.

48) Enter **0**.

49) Click the **Add Symbol** icon from the Text Format box. Select **Degree** from the Modifying Symbols library.

50) Enter **30′** for minutes of a degree.

Modify the DECIMAL LINES.
51) Modify the **first Decimal line** as illustrated.

52) Enter **+- .01** at the end of ONE PLACE DECIMAL.

53) Modify the **second Decimal line** as illustrated.

54) Enter **+- .005** at the end of TWO PLACE DECIMAL.

55) Click **OK** from the Note PropertyManager.

Fit the sheet to the Graphics window.
56) Press the **f** key.

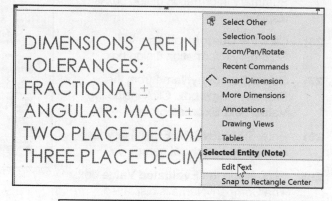

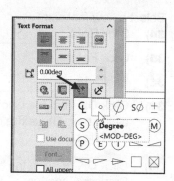

The Add Symbol $\overset{+}{\mathcal{L}}$ icon is accessible through the Note PropertyManager. The ± symbol is located in the Modify Symbols list. The ± symbol is displayed as <MOD-PM>. The degree symbol ° is displayed as <MOD-DEG>. Select icon symbols or enter values from the keyboard.

Interpretation of tolerances is as follows for dimensions:

- The angular dimension 110° is machined between 109.5° and 110.5°.

- The dimension 2.04 is machined between 2.03 and 2.05.

Additional Custom Properties and Notes are added later in this project.

Company Logo and Save Sheet Format

A Company logo is normally located in the Title block of the drawing. You can create your own Company logo or copy and paste an existing picture.

If you have your own Logo, skip the process of copying and applying the LOGO folder and file shown below.

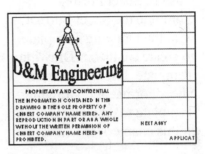

Activity: Insert the Logo and Save the Sheet Format

Insert a Company or School Logo.

57) Copy the **LOGO** folder to your hard drive. Note: If you have your own Logo, use it for the drawing.

58) Click **Insert**, **Picture** from the Menu bar. The Open dialog box is displayed.

59) Double-click the **Logo.jpg** file. The Sketch Picture PropertyManager is displayed.

60) Un-check the **Enable scale tool** box.

61) Un-check the **Lock aspect ratio** box.

62) Drag the picture handles to size the **picture** to the left side of the Title block.

63) Click **OK** ✓ from the Sketch Picture PropertyManager.

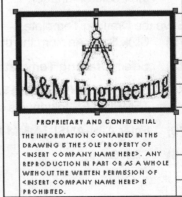

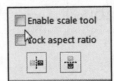

Return to the Edit Sheet mode.
64) Right-click in the **Graphics window**.

65) Click **Edit Sheet**. The Title block is displayed in black.

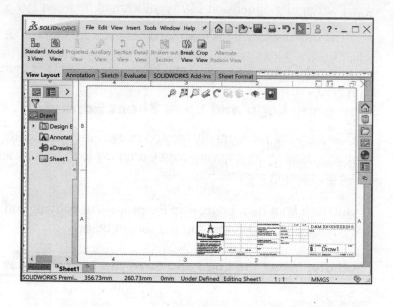

Fit the Sheet Format to the Graphics window.
66) Press the **f** key.

Save the Sheet Format as a Custom Sheet Format. Combine the Sheet Format with the Drawing Template to create a Custom Drawing Template. Use the Custom Sheet Format and Drawing Template to create drawings in this chapter.

Save the Sheet Format.
67) Click **File**, **Save Sheet Format** from the Menu bar.

68) **Select** your working folder.

69) Enter **MY-A-FORMAT** for File name. Note: .slddrt file extension.

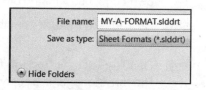

70) Click **Save**.

71) **Rebuild** the drawing. The Sheet Format1 icon is displayed in the FeatureManager.

Save the Drawing Template.
72) Click **Save As** from the drop-down Menu bar.

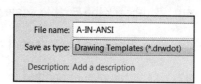

73) Select **Drawing Templates (*.drwdot)** for Save as type.

74) Select the **MY-TEMPLATES** folder. Note: The MY-TEMPLATES folder was created in Chapter 5.

75) Enter **A-IN-ANSI**, **[A-MM-ISO]** for File name.

76) Click **Save**.

Close all documents.

77) Click **Window**, **Close All** from the Menu bar.

Verify the template.

78) Click **New** 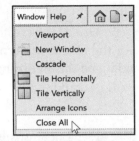 from the Menu bar.

79) Click the **MY-TEMPLATES** tab. Additional document templates are displayed.

80) Double-click the **A-IN-ANSI**, **[A-MM-ISO]** Drawing Template. The Draw2 FeatureManager is displayed.

The Draw2-Sheet1 drawing is displayed in the Graphics window. The Draw# - Sheet1 is a sequential number created in the current SOLIDWORKS session.

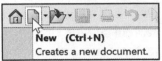

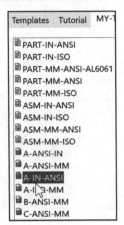

Your default system document templates may be different if you are a new user vs. an existing user who has upgraded from a previous version.

Utilize the A-IN-ANSI, [A-MM-ISO] Drawing Template for the BATTERY drawing.

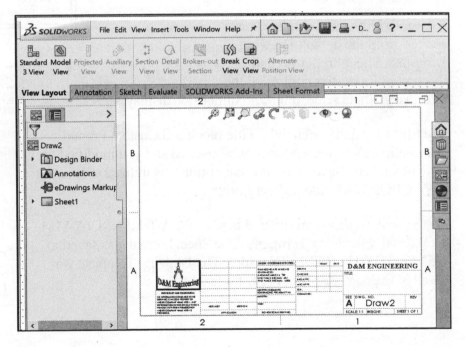

Utilize descriptive filenames for the Drawing Template that contains the size, dimension standard and units.

File Locations is a System Option. The option is active for the current session of SOLIDWORKS in some network environments.

Combine custom Drawing Templates and Sheet Formats to match your company's drawing standards. Save the empty Drawing Template and Sheet Format separately to reuse information.

Review Drawing Templates

A custom Drawing Template was created from the default Drawing Template. Sheet Properties and Document Properties controlled the Sheet size, Sheet scale, units and dimension display.

The Sheet Format contained a Title block and Custom Property information. A Company Logo was inserted and you modified the Title block. The Save Sheet Format option was utilized to save the MY-A-FORMAT.slddrt Sheet Format.

The Save As option was utilized to save the A-IN-ANSI, [A-MM-ISO].drwdot Drawing Template. The Sheet Format was saved in the MY-SHEETFORMATS folder. The Drawing Template was saved in the MY-TEMPLATES folder.

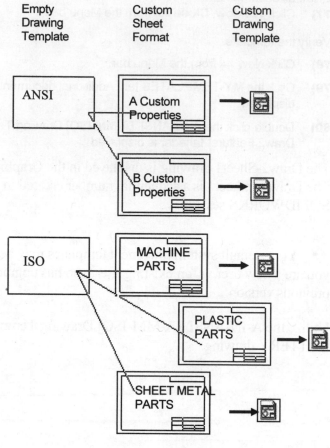

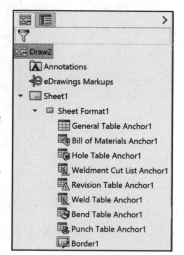

BATTERY Drawing

A drawing contains part views, geometric dimensioning and tolerances, notes, and other related design information. When a part is modified, the drawing automatically updates. When a driving dimension in the drawing is modified, the part is automatically updated.

Create the BATTERY drawing from the BATTERY part. Utilize the Model View feature in the Drawings toolbar. The Front view is the first view inserted into the drawing. The Top view and Right view are projected views from the Front view. Insert dimensions into the drawing with the Insert Model Items feature.

Activity: Insert Three Standard Views - BATTERY Drawing

Insert three standard views using the Model View tool. Note: You can also insert views using the View Palette.

81)   Click **Model View** 🖼 from the View Layout tab.

82)   **Browse** to the PROJECTS folder.

83)   Double-click the **BATTERY** part.

Insert the Front, Top and Right view.

84)   Check the **Create multiple views** box.

85)   Click **Front**, **Top** and **Right** view from the Orientation box. Front, Top and Right are selected. De-activate the Isometric view if needed.

86)   Display **WireFrame**.

87)   Click **OK** ✔ from the Model View PropertyManager. Three views are displayed on Sheet1.

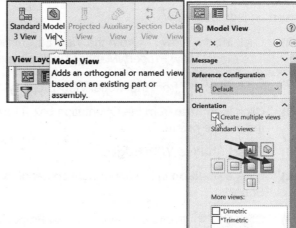

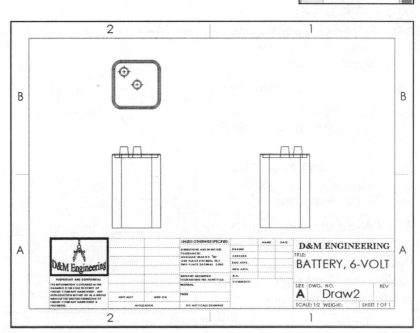

A part can't be inserted into a drawing when the Edit Sheet Format mode is selected.

Move parent and child views independently by dragging their view boundary. Hold the Shift key down and select multiple views to move as a group.

Click the View Palette icon in the Task Pane. Click the drop-down arrow to view an open part or click the Browse button to locate a part. Click and drag the desired view/views into the active drawing sheet.

Activity: BATTERY Drawing-Insert an Isometric View

Insert an Isometric view.

88) Click **Model View** 🌐 from the View Layout tab in the CommandManager. The Model View PropertyManager is displayed.

89) Click **Next** ➡ from the Model View PropertyManager.

90) Click **Isometric** from the Orientation box. The Isometric view is placed on the mouse pointer.

91) Select **Shaded With Edges**.

92) Click a **position** in the upper right corner of the Graphics window on Sheet1.

93) Click **OK** ✔ from the Drawing View4 PropertyManager.

View the Sheet Scale.

94) Right-click a **position** inside the Graphics window, Sheet1 boundary.

95) Click the **drop-down** arrow.

96) Click **Properties** from the drop-down menu. The Sheet Scale is 1:2.

97) Click **Cancel**. Note: Click Apply if you changed the Sheet Properties.

The SW-Sheet scale Property 1:2 is linked to the Title block through the Sheet Format. Later, change the Sheet scale to fit the BATTERY dimensions.

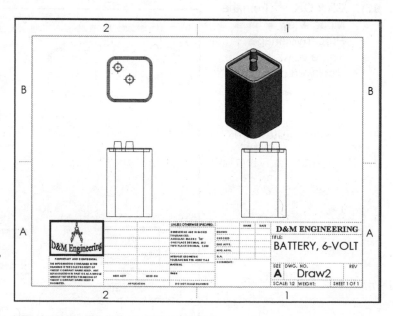

Save the Drawing.

98)   Click **Save As** 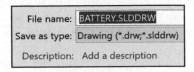 from the drop-down Menu bar.

99)   Select the **PROJECTS** folder. The default name is BATTERY.

100) Click **Save** from the Save As dialog box.

Each drawing has a unique file name. Drawing file names end with a .slddrw suffix in SOLIDWORKS. Part file names end with a .sldprt suffix.

Text in the Title block is linked to the Filename and description created in the part. The DWG. NO. text box utilizes the Property, $PRP: "SW-File Name" passed from the BATTERY part to the BATTERY drawing. The TITLE text box utilizes the Property, $PRPSHEET: "Description."

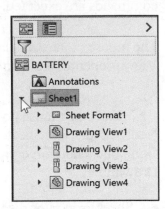

The filename BATTERY is displayed in the DWG. NO. box. The Description BATTERY, 6-VOLT is displayed in the Title box.

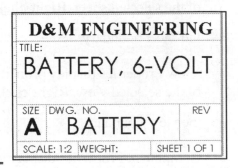

The BATTERY drawing contains three Principle views (Standard views): Front, Top, Right and an Isometric view. You created the views with the Model View tool. Drawing views can be inserted with the following processes:

- Utilize the Model View tool.

- Click and drag a part into the drawing or select the Standard 3 Orientation view option.

- Predefine views in a custom Drawing Template.

- Drag a hyperlink through Internet Explorer.

- Utilize the View Palette located in the Task Pane to the right of the Graphics window.

The mouse pointer provides feedback in both the Drawing Sheet and Drawing View modes. The mouse pointer displays the Drawing Sheet icon when the Sheet properties and commands are executed.

The mouse pointer displays the Drawing View icon when the View properties and commands are executed.

View the mouse pointer for feedback to select Sheet, View, Component and Edge properties in the Drawing.

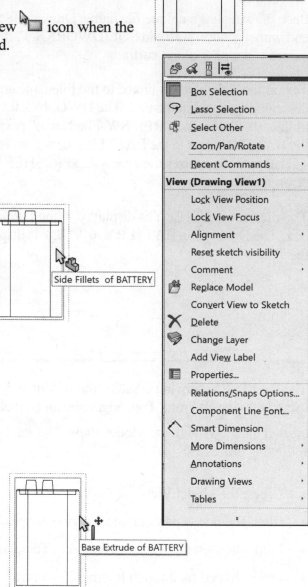

- Sheet Properties display properties of the selected sheet. Right-click in the sheet boundary .

- View Properties display properties of the selected view. Right-click on the view boundary .

- Component Properties display properties of the selected component. Right-click on the face of the component .

- Edge Properties display properties of the geometry. Right-click on an edge .

The Drawing views are complete. Move the views to allow for ample spacing for the dimensions and notes. Zoom in on narrow view boundary boxes if required.

Drawing View1

Side Fillets of BATTERY

Base Extrude of BATTERY

Box Selection
Lasso Selection
Select Other
Zoom/Pan/Rotate
Recent Commands
View (Drawing View1)
Lock View Position
Lock View Focus
Alignment
Reset sketch visibility
Comment
Replace Model
Convert View to Sketch
Delete
Change Layer
Add View Label
Properties...
Relations/Snaps Options...
Component Line Font...
Smart Dimension
More Dimensions
Annotations
Drawing Views
Tables

Detail View

A Detail view enlarges an area of an existing view. You need to specify location, name, shape and scale. Create a Detail view named A with a circle and a 1:1 scale.

A Detail circle specifies the area of an existing view to enlarge. The circle contains the same letter as the view name.

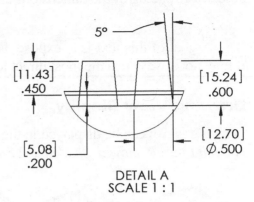

5°

[11.43]
.450

[15.24]
.600

[12.70]
Ø.500

[5.08]
.200

DETAIL A
SCALE 1 : 1

Activity: BATTERY Drawing-Detail View

Add a Detail view to the drawing.

101) Click the **Detail View** 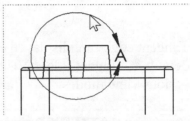 tool from the View Layout toolbar. The circle sketch tool is selected.

Sketch a circle with the center point located between the two terminals in Drawing View1 (Front).
102) **Sketch** the circle as illustrated.

Position the Detail View.
103) Click a **position** to the right of the Drawing View3 (Top).

104) Click the **Use custom scale** box.

105) Enter **1:1** in the Custom Scale box.

106) Click **OK** ✔ from the Drawing View A PropertyManager.

107) **Rebuild** 🛢 the model.

Fit the Drawing to the Graphics window.
Save the drawing.
108) Press the **f** key.

109) Click **Save** 💾.

Center marks are displayed by default. Center marks and centerlines are controlled by the Tools, Options, Document Properties, Auto insert on view creation option.

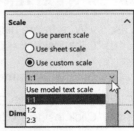

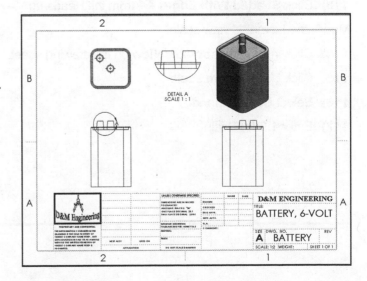

Allow for spacing of the dimensions and notes. Move the views by their view boundary.

A goal of this text is to expose the new user to various tools, techniques and procedures. It may not always use the most direct tool or process.

Drawing View Display

Drawing views can be displayed in the following modes: *Wireframe*, *Hidden Lines Visible*, *Hidden Lines Removed*, *Shaded* and *Shaded With Edges mode*.

Tangent edges are displayed either in Visible, With Font, or Removed mode. Note: System default is Tangent Edges Visible. Display hidden lines, profile lines and tangent edges in various view modes.

Activity: BATTERY Drawing-View Display

Display Wire frame display in the Detail View.
110) Click inside the **Detail view** on the drawing sheet.

111) Click **Wire frame** ⬡ from the Heads-up View toolbar.

Display Shaded With Edges in the Isometric view.
112) Click inside the **Isometric view** on the drawing sheet.

113) Click **Shaded With Edges** ⬛ from the Heads-up View toolbar.

Modify the Isometric view scale.

114) Click inside the **Isometric view** on the drawing sheet.

115) Click **Use custom scale**.

116) Select **User Defined**.

117) Enter **1:3** for Scale.

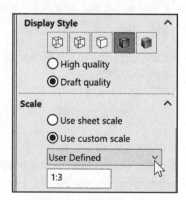

Insert Model Items and Move Dimensions

Dimensions created for each feature in the part are inserted into the drawing. Dimensions are inserted by sheet, view or feature. Select inside the sheet boundary to insert the dimensions into Sheet1. Move the dimensions off the profile lines in each view.

Illustrations for this project are provided in inches and millimeters. The Detailing option and Dual Dimensions Display produce both inches and millimeters for each dimension. The primary units are set to inches. The secondary units are set to millimeters.

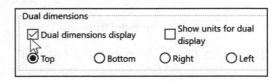

The drawing dimension location is dependent on Feature dimension creation and Selected drawing views.

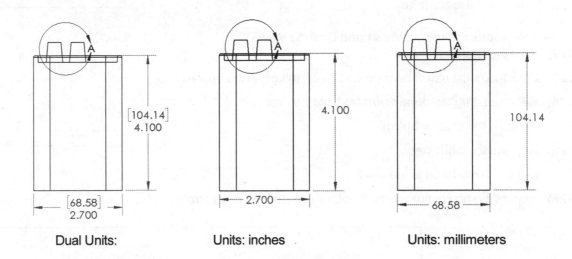

Dual Units: Units: inches   Units: millimeters

Move dimensions within the same view. Use the mouse pointer to drag dimensions and leader lines to new locations.

Move dimensions to a different view. Utilize the Shift key to drag a dimension to another view.

Leader lines reference the size of the profile. A gap must exist between the profile lines and the leader lines. Shorten the leader lines to maintain a drawing standard. Use the Arrow buttons in the PropertyManager to flip dimension arrows.

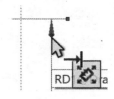

Activity: BATTERY Drawing-Insert Model Items and Move Dimensions

Insert dimensions into the Sheet.

118) Click inside the **sheet boundary**.

119) Click the **Model Items** tool from the Annotation toolbar. The Model Items PropertyManager is displayed.

120) Select **Entire model** from the Source/Destination box.

121) Click **OK** ✔ from the Model Items PropertyManager. If needed, change the Sheet scale to address the dimensions. The dimensions in your Detail view may be different, due to the size of your circle sketch.

Dimensions appear in different view. To move a dimension from a view, utilize the Shift key and only drag the dimension text. Hold the Shift key down. Click and drag the dimension. Release the mouse button and then the Shift key.

Move dimensions in Drawing View1 and Drawing View3.

122) Hold the **Shift** key down.

123) Click the horizontal dimension text **2.700**, **[68.58]** in the Top view.

124) Click and drag the **dimension text** into the Front view.

125) Release the **mouse button**.

126) Release the **Shift** key.

127) Flip the **arrowhead** to the inside.

128) Repeat the above procedure for **other dimensions** as illustrated.

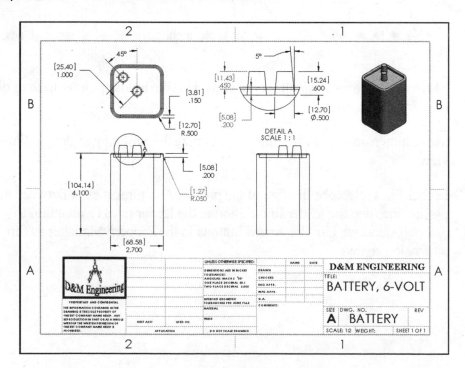

129) Click the vertical dimension text **4.100**, [**104.14**].

130) Drag the **text** to the right of the BATTERY.

131) Move the **leader line endpoints** if required. The end points of the leader line are displayed.

132) Drag each **square control end point** to the left until it is off the profile line.

133) Create a **gap** between the profile line and the leader lines.

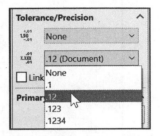

Modify the radius text in Drawing View1 and Drawing View3.

134) Click the **R.050**, [Ø**1.27**] text on Drawing View1. The Dimension PropertyManager is displayed.

135) Select **.12**, 2 places in the Tolerance/Precision box. R.05 is displayed.

136) Click the **R.500**, [Ø**12.70**] text on Drawing View3.

137) Select **.12**, 2 place in the Tolerance/Precision box. R.50 is displayed.

138) Click **OK** ✔ from the Dimension PropertyManager.

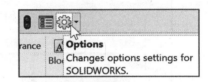

Save the drawing.

139) Click **Save** 💾.

Shorten the Bent leader length.

140) Click **Options** ⚙ from the Menu bar.

141) Click the **Document Properties** tab.

142) Click the **Dimensions** folder.

143) Enter **.25**in, [**6.35**] for Bent leaders length.

144) Click **OK**.

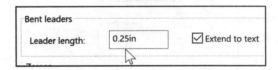

Activity: BATTERY Drawing - Insert a Note

Insert a Note.

145) Click the **Note** 𝐀 tool from the Annotation toolbar. The Note PropertyManager is displayed.

146) Click a **position** in the COMMENTS box.

147) Enter **CHAPTER 9**.

148) Click **OK** ✔ from the Note PropertyManager.

Save the BATTERY drawing.

149) Click **Save** 💾.

	NAME	DATE
DRAWN		
CHECKED		
ENG APPR.		
MFG APPR.		
Q.A.		
COMMENTS:		
CHAPTER 9		

There are hundreds of Document Property options. Where do you go for additional information on these Properties? Answer: Select the Help button in the lower right-hand corner of the Document Property dialog box.

In SOLIDWORKS, inserted dimensions in the drawing are displayed in gray. Imported dimensions from the part are displayed in black.

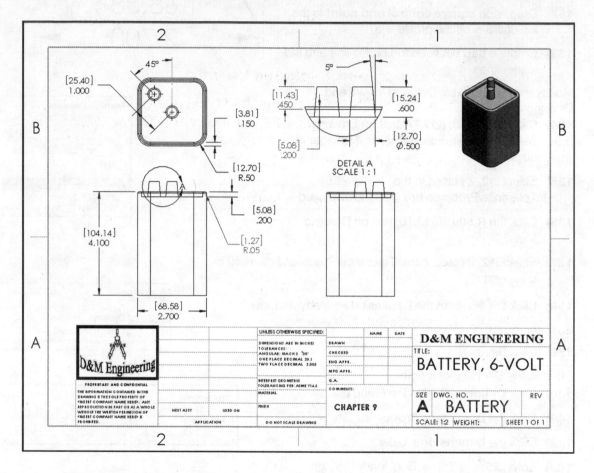

As an exercise, insert additional Custom Properties: Revision, Drawnby, DrawnDate, PartNo (DWG.NO.), etc.

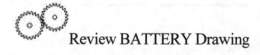   Review BATTERY Drawing

You created a new drawing, BATTERY with the A-IN-ANSI, [A-MM-ISO] Drawing Template. The BATTERY drawing utilized the BATTERY part in the Model View PropertyManager. The Model View PropertyManager provides the ability to insert new views with the View Orientation option.

You selected Front, Top, Right, and Isometric to position the BATTERY views. The Detail View tool inserted a Detail view of the BATTERY. You moved the views by dragging the blue view boundary. You inserted the dimensions and annotations to detail the BATTERY drawing.

You inserted part dimensions and annotations into the drawing with the Insert Model Items tool. Dimensions were moved to new positions. Leader lines and dimension text were repositioned. Annotations were edited to reflect the drawing standard. You modified the dimension text by inserting additional text.

New Assembly Drawing and Exploded View

Create a new drawing named FLASHLIGHT. Insert the FLASHLIGHT assembly Isometric view. Modify the view properties to display the Exploded view. The Bill of Materials reflects the components of the FLASHLIGHT assembly. Create a drawing with a Bill of Materials. Label each component with Balloon text.

Activity: New Assembly Drawing and Exploded View

Close all parts and drawings.

150) Click **Windows**, **Close All** from the Menu bar.

Create a new drawing.

151) Click **New** ⬜ from the Menu bar.

152) Click the **MY-TEMPLATES** tab. The MY-TEMPLATES tab was created in Chapter 5.

153) Double-click the **A-IN-ANSI, [A-MM-ISO]** Drawing Template.

Insert the FLASHLIGHT assembly.

154) **Browse** to the PROJECTS folder.

155) Double-click the **FLASHLIGHT** assembly.

156) Select *Isometric for the Orientation box.

157) Click **Shaded With Edges** from the Display Style box.

158) Click a **position** on the right side of the drawing.

159) Click **OK** ✔ from the Drawing View1 PropertyManager.

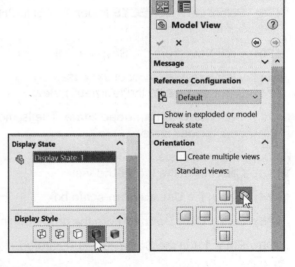

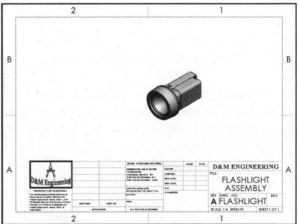

Edit the Title text.

160) Right-click a **position** in the sheet boundary.

161) Click **Edit Sheet Format**.

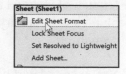

162) Double-click the title text **FLASHLIGHT ASSEMBLY**. The Note PropertyManager and the Formatting dialog box are displayed.

163) Enter **.19**in for Text Height.

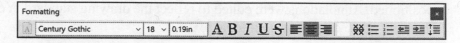

164) Click **OK** ✔ from the Note PropertyManager.

165) Right-click a **position** in the Sheet boundary.

166) Click **Edit Sheet**. The FLASHLIGHT ASSEMBLY text is sized to the Title box.

Save the drawing.

167) Click **Save** 💾 .

168) Click **Save All**. The Save As dialog box is displayed.

169) Select the **PROJECTS** folder. FLASHLIGHT is the default file name.

170) Click **Save** from the Save As dialog box.

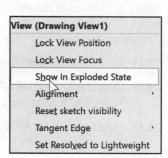

Display the Exploded view of the assembly.

171) Right-click inside the **Isometric view**.

172) Click **Show in Exploded State**. The Isometric view is displayed in an Exploded state.

Modify the view scale.

173) Click inside the **Isometric view**.

174) Check the **Use custom scale** box.

175) Select **User Defined** from the drop-down menu.

176) Enter **1:4** in the Scale box as illustrated.

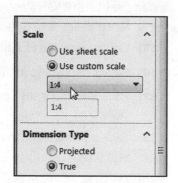

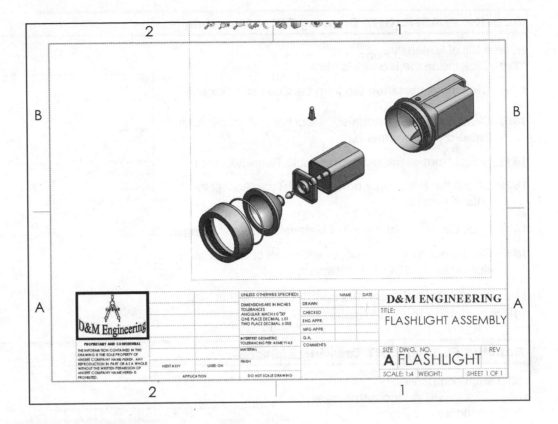

177) Click **OK** ✔ from the Model View1 PropertyManager. View the Exploded view.

Bill of Materials and Balloons

Apply the Bill of Materials 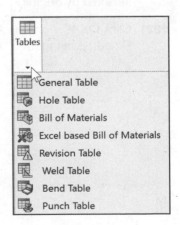 tool from the Consolidated Tables drop-down menu. A Bill of Materials (BOM) is a table inserted into a drawing to keep a record of the parts used in an assembly. The default BOM template contains the Item Number, Quantity, Part No. and Description.

The default Item number is determined by the order in which the component is inserted into the assembly. Quantity is the number of instances of a part or assembly. Part No. is determined by the following: file name, default and the User Defined option, Part Number used by the Bill of Materials. Description is determined by the description entered when the document is saved.

Activity: FLASHLIGHT Drawing-Bill of Materials

Insert a Bill of Materials.

178) Click inside the **Isometric view**.

179) Click the **Annotation** tab from the CommandManager.

180) Click the **Bill of Materials** 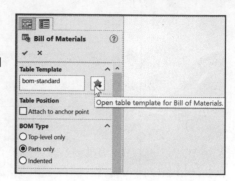 tool from the Consolidated Table drop-down menu.

181) Select **bom-standard** for the Table Template.

182) Check the **Parts only** box for BOM Type. Accept the default settings.

183) Click **OK** ✔ from the Bill of Materials PropertyManager.

184) Click a **position** in the upper left corner of Sheet1 as illustrated. The BOM is displayed.

The Bill of Materials requires additional work that you will complete in the next section.

Activity: FLASHLIGHT Drawing-Balloons

Label each component.

185) Click inside the **Isometric view** boundary.

186) Click the **Auto Balloon** 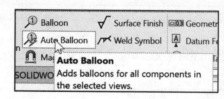 tool from the Annotation toolbar. The Auto Balloon PropertyManager is displayed. Accept the defaults. The Insert magnetic line(s) option is selected by default.

187) Click **OK** ✔ from the Auto Balloon PropertyManager.

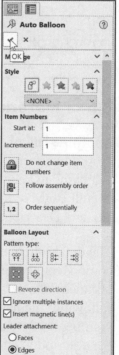

188) Click and drag the **balloons** inside the view boundary.

Insert a balloon for the O-RING.

189) Click the **Balloon** ⌕① tool from the Annotation toolbar.

190) Click the **O-RING** in the Graphics window.

191) **Position** the Balloon in the drawing.

192) Click **OK** ✔ from the Balloon PropertyManager.

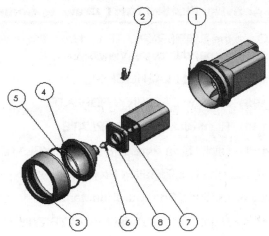

The Balloon note displays an arrowhead on a selected edge and a filled dot on a selected face when the Drawing Standard is set to ANSI. The Balloon note displays a "?" if no edge or face is selected.

☀ Magnetic lines are a convenient way to align balloons along a line at any angle. You attach balloons to magnetic lines, choose to space the balloons equally or not, and move the lines freely, at any angle, in the drawing.

Part Numbers

Use the following prefix codes to categorize created parts and drawings. The part names and part numbers are as follows:

Category:	Prefix:	Part Name:	Part Number:
Molded Parts	44-	BATTERYPLATE	44-A26
		LENSCAP	44-A27
		HOUSING	44-A28
Purchased Parts	B99-	BATTERY	B99-B01
	99-	LENS	99-B02
		O-RING	99-B03
		SWITCH	99-B04
		BULB	99-B05
Assemblies	10-	FLASHLIGHT	10-F123

The Bill of Materials requires editing. The current part file name determines the PART NUMBER parameter values. The Configuration Properties controls the display of the PART NUMBER in the Bill of Materials. Redefine the PART NUMBER for each part.

Activity: FLASHLIGHT Drawing-ConfigurationManager

Open the BATTERYPLATE part from the drawing.

193) **Expand** Drawing View1.

194) **Expand** FLASHLIGHT.

195) **Expand** BATTERYANDPLATE.

196) Right-click **BATTERYPLATE**.

197) Click **Open Part**. The BATTERYPLATE part is displayed.

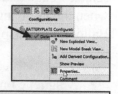

Display the Configuration Properties.

198) Click the **ConfigurationManager** tab.

199) Right-click the **Default [BATTERYPLATE]** configuration.

200) Click **Properties**.

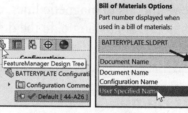

201) Select **User Specified Name** from the drop-down menu.

202) Enter **44-A26**.

203) Click **OK** ✔ from the Configuration Properties PropertyManager. The ConfigurationManager displays the Default [44-A26] configuration.

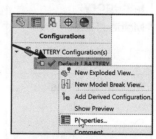

204) Click the **FeatureManager** tab.

Return to the FLASHLIGHT drawing.

205) Click **Window**, **FLASHLIGHT-Sheet1** from the Menu bar.

Activity: FLASHLIGHT Drawing-Update the Bill of Materials

Update the Bill of Materials.

206) **Rebuild** the model.

Enter the BATTERY PART NUMBER.

207) Right-click **BATTERY** part from the FeatureManager.

208) Click **Open Part**. The BATTERY part is displayed.

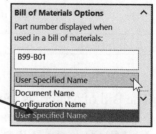

209) Click the BATTERY **ConfigurationManager** tab.

210) Right-click **Default [BATTERY]** from the ConfigurationManager.

211) Click **Properties**.

212) Select **User Specified Name** from the drop-down menu.

213) Enter **B99-B01**.

214) Click **OK** ✔ from the Configuration Properties PropertyManager.

Return to the FeatureManager.

215) Click the **FeatureManager** tab.

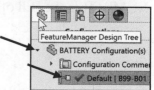

Return to the FLASHLIGHT drawing.
216) Click **Window**, **FLASHLIGHT-Sheet1** from the Menu bar.

Update the Bill of Materials.
217) **Rebuild** the model.

Resize the PART NUMBER column.
218) Drag the **vertical line** between the PART NUMBER and the DESCRIPTION column to the left.

219) Drag the **vertical line** between the DESCRIPTION and the QTY to the left.

As an exercise, complete the PART NUMBER column. The User Specified Name for the remaining PART numbers. Insert the Third Angle Projection icon. Insert additional Custom Properties: Revision, Drawnby, DrawnDate, PartNo. (DWG. NO.), etc.

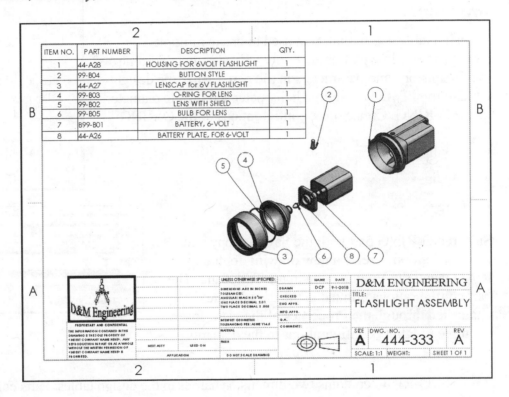

Save the FLASHLIGHT drawing.
220) Click **Save** 💾 .

Close All documents.
221) Click **Windows**, **Close All** from the Menu bar.

Review the FLASHLIGHT Drawing

The FLASHLIGHT drawing contained an Exploded view. The Exploded view was created in the FLASHLIGHT assembly. The Bill of Materials listed the Item Number, Part Number, Description and Quantity of components in the assembly. Balloons were inserted to label the top-level components in the FLASHLIGHT assembly. You developed Properties in the part to modify the Part Number utilized in the Bill of Materials.

Design Tables and O-RING-DESIGN-TABLE Drawing

A design table is a spreadsheet used to create multiple configurations in a part or assembly. The design table controls the dimensions and parameters in the part. Utilize the design table to modify the overall path diameter and profile diameter of the O-RING. Create three configurations of the O-RING:

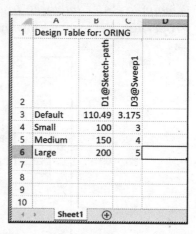

- Small.

- Medium.

- Large.

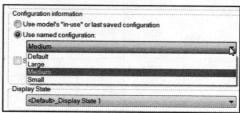

Save the O-RING part with the Save as copy and continue option. Create a new drawing for the O-RING using the View Palette in the Task Pane. The part configurations utilized in the drawing are controlled through the Properties of the view. Insert the three O-RING configurations in the drawing.

The O-RING contains two dimension names in the design tables. Parts contain hundreds of dimensions and values. Rename dimension names for clarity.

Activity: O-RING Part-Design Table

Open the O-RING part.

222) Click **Open** from the Menu bar.

223) Double-click **O-RING** from the PROJECTS folder.

Save a copy of the O-RING. Enter name and description.
224) Click **Save As** from the drop-down Menu bar.

225) Check the **Save as copy and continue** box.

226) Enter **O-RING-DESIGN-TABLE** for File name.

227) Enter **O-RING WITH DESIGN TABLE** for Description.

228) Click **Save**.

Open the O-RING-DESIGN-TABLE part.

229) Click **Open** 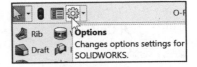 from the Menu bar.

230) Double-click **O-RING-DESIGN-TABLE** from the PROJECTS folder.

Modify the Primary Units.

231) Click **Options** ⚙, **Document Properties** tab.

232) Click **Units**. ANSI is the default Overall Drafting standard.

233) Select **MMGS**. Select **.12** for Length units Decimal.

234) Click **OK** from the Document Properties - Units dialog box.

235) Double-click the **face** of the O-RING. The diameter dimensions are displayed in millimeters.

Insert a Design Table.

236) Click **Insert**, **Tables**, **Design Table**  from the Menu bar. The Auto-create option is selected. Accept the defaults.

237) Click **OK** ✔ from the Design Table PropertyManager.

238) Hold the **Ctrl** key down.

239) Select **D1@Sketch-path** and **D3@Sweep1**.

240) Release the **Ctrl** key.

241) Click **OK** from the Dimensions dialog box.

The dimension variable name will be different if sketches or features were deleted. The input dimension names and default values are automatically entered into the Design Table. The value Default is entered in Cell A3. The values for the O-RING are entered in Cells B3 through C6. The sketch-path diameter is controlled in Column B. The sketch-profile diameter is controlled in Column C.

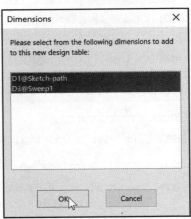

Enter the three configuration names.
242) Click **Cell A4**.

243) Enter **Small**.

244) Click **Cell A5**.

245) Enter **Medium**.

246) Click **Cell A6**.

247) Enter **Large**.

Enter the dimension values for the Small configuration.
248) Click **Cell B4**. Enter **100**.

249) Click **Cell C4**. Enter **3**.

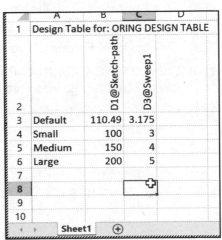

	A	B	C	D
1	Design Table for: ORING DESIGN TABLE			
2		D1@Sketch-path	D3@Sweep1	
3	Default	110.49	3.175	
4	Small	100	3	
5	Medium	150	4	
6	Large	200	5	
7				
8				
9				
10				

Enter the dimension values for the Medium configuration.
250) Click **Cell B5**.

251) Enter **150**.

252) Click **Cell C5**.

253) Enter **4**.

Enter the dimension values for the Large configuration.
254) Click **Cell B6**.

255) Enter **200**.

256) Click **Cell C6**.

257) Enter **10**.

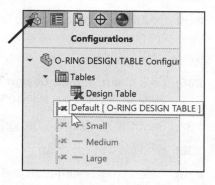

Build the three configurations.
258) Click a **position** outside the EXCEL Design Table in the Graphics window.

259) Click **OK** to generate the configurations.

Display the configurations.
260) Double-click **Small**.

261) Double-click **Medium**.

262) Double-click **Large**.

263) Double-click **Default**.

Return to the FeatureManager.
264) Click the **O-RING-DESIGN-TABLE FeatureManager** 🐭 tab.

Activity: Create the O-RING-DESIGN-TABLE Drawing

Create a new drawing.
265) Click **New** 🗋 from the Menu bar.

266) Click the **MY-TEMPLATES** tab.

267) Double-click the **A-IN-ANSI**, **[A-ISO-MM]** Drawing Template. The Model View PropertyManager is displayed.

268) Click **Cancel** ✖ from the Model View PropertyManager.

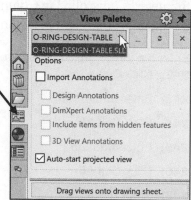

Insert an Isometric view using the View Palette.
269) Click the **View Palette** tab from the Task Pane.

270) Select **O-RING-DESIGN-TABLE** from the drop-down menu. The available views are displayed.

271) Drag and drop the **Isometric view** to the left side of the drawing.

272) Select **1:2** for Scale.

273) Click **OK** ✔ from the Drawing View1 PropertyManager.

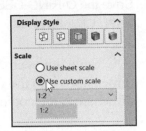

Deactivate the Origin if needed.
274) Click **View**, **Hide/Show**, uncheck **Origins** from the Menu bar.

Save the drawing.
275) Click **Save** 💾 . Accept the defaults.

276) Click **Save**. If needed, adjust the font size in the drawing for the Title box and DWG. No.

Activity: O-RING Drawing-Design Table

Copy the Isometric view.
277) Click inside the **Isometric view** boundary.

278) Click **Edit**, **Copy** from the Menu bar.

Paste the Isometric view.
279) Click a **position** to the right of the Isometric view.

280) Click **Edit**, **Paste** from the Menu bar.

Display the medium configuration.
281) Right-click inside the second **Isometric view** boundary.

282) Click **Properties** from the drop-down menu. Expand the drop-down menu is needed.

283) Select **Medium** from the Use named configuration list.

284) Click **OK** from the Drawing View Properties dialog box.

285) Click **OK** ✔ from the Drawing View2 PropertyManager.

286) **Rebuild** 🔘 the model.

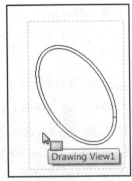

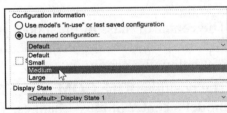

As an exercise, insert material and the Large Configuration as illustrated. Insert the Third Angle Projection icon, with Custom Properties: Revision, Drawnby, DrawnDate, PartNo. (DWG. NO.) etc.

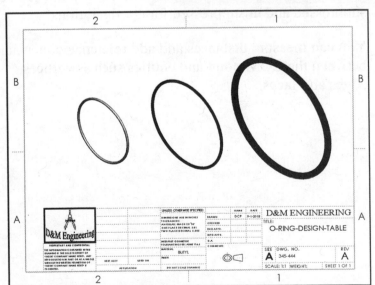

Save the O-RING-DESIGN-TABLE drawing.

287) Click **Save** 💾 .

☀️ Save time with repetitive dimensioning in configurations. Insert dimensions into the first view. Copy the view. The view and the dimensions are copied to the new view.

Add a Center of Mass point

Add a center of mass (COM) point to parts, assemblies and drawings.

COM points added in component documents also appear in the assembly document. In a drawing document of parts or assemblies that contain a COM point, you can show and reference the COM point.

Add a COM to a part or assembly document by clicking **Center of Mass** (Reference Geometry toolbar) or **Insert, Reference Geometry, Center of Mass** ✪ or checking the **Create Center of Mass feature** box in the Mass Properties dialog box.

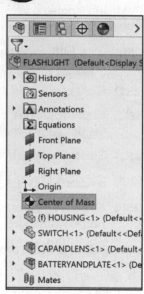

The center of mass of the model is displayed in the Graphics window and in the FeatureManager design tree just below the origin.

The position of the **COM** point ✪ updates when the model's center of mass changes. The COM point can be suppressed and unsuppressed for configurations.

You can measure distances and add reference dimensions between the COM point and entities such as vertices, edges and faces.

If you want to display a reference point where the CG was located at some particular point in the FeatureManager, you can insert a Center of Mass Reference Point. See SOLIDWORKS Help for additional information.

Add a center of mass (COM) point to a drawing view. The center of mass is a selectable entity in drawings and you can reference it to create dimensions.

In a drawing document, click **Insert**, **Model Items**. The Model Items PropertyManager is displayed. Under Reference Geometry, click the **Center of Mass** icon. Enter any needed additional information. Click **OK** from the Model Items PropertyManager. View the results in the drawing.

The part or assembly needs to have a **COM before** you can view the COM in the drawing. To view the center of mass in a drawing, click **View**, **Hide/Show**, **Center of mass**.

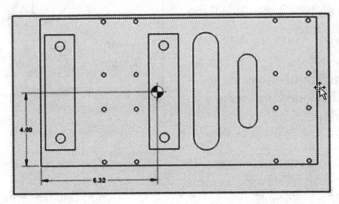

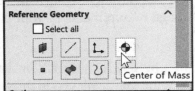

Chapter Summary

You produced three drawings: BATTERY, FLASHLIGHT, and O-RING-DESIGN-TABLE. The drawings contained Standard views, Detail view and Isometric views.

The drawings utilized a custom Sheet Format and custom Drawing Template. The Sheet Format contained the Company logo and Title block information.

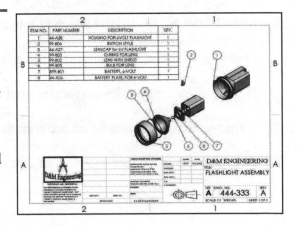

You incorporated the BATTERY part dimensions into the drawing. You obtained an understanding of displaying views with the ability to insert, add and modify dimensions.

You used two major design modes in the drawings: Edit Sheet Format and Edit Sheet.

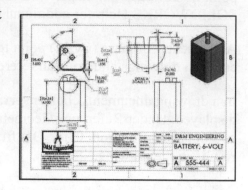

The FLASHLIGHT assembly drawing contained an Exploded view and Bill of Materials. The Properties for the Bill of Materials were developed in each part with a user defined Part Number.

You created three configurations of the O-RING part with a Design Table. A Design Table controlled parameters and dimensions. You utilized the three configurations in the O-RING-DESIGN-TABLE drawing using the View Palette tool.

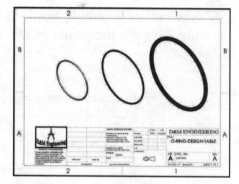

Drawings are an integral part of the design process. Part, assemblies and drawings all work together to fulfill the design requirements of your customer.

In SOLIDWORKS, inserted dimensions in the drawing are displayed in gray. Imported dimensions from the part are displayed in black.

Questions

1. Identify the differences between a Drawing Template and a Sheet Format. Provide an example.

2. Identify the command to save the Sheet Format.

3. Identify the command to save the Drawing Template.

4. Describe a Bill of Materials. Provide an example.

5. Name the two major design modes used to develop a drawing in SOLIDWORKS.

6. Name seven components that are commonly found in a title block.

7. Describe the procedure to insert an Isometric view into a drawing.

8. In SOLIDWORKS, drawing file names end with a _____ suffix.

9. True or False. Most engineering drawings use the following font: Times New Roman - All small letters. Explain your answer.

10. Describe Leader lines. Provide an example.

11. Describe a Note on a drawing. Provide an example.

12. Explain the procedure to display an Exploded Assembly view in a drawing.

13. Explain the procedure on labeling components in an Exploded view on an assembly drawing. Provide an example.

14. Describe the procedure to create a Design Table.

15. True or False. You cannot display different configurations in the same drawing. Explain your answer.

16. True or False. The Part Number is only entered in the Bill of Materials. Explain your answer.

17. There are hundreds of options in the Document Properties, Drawings and Annotations toolbars. How would you locate additional information on these options and tools?

18. Describe the View Palette.

19. Describe the procedure to insert a Center of Mass point into a drawing either for an assembly or part.

Exercises

Exercise 9.1: FLATBAR - 3 HOLE Drawing

Create the A (ANSI) Landscape - IPS - Third Angle 3HOLES drawing as illustrated. Do not display Tangent Edges. Do not dimension to hidden lines.

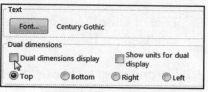

First create the part from the drawing, then create the drawing. Use the default A (ANSI) Landscape Sheet Format/Size.

Insert the Front, Top and Shaded Isometric view as illustrated.

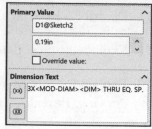

Insert dimensions.

Address gaps between feature lines. Address drawing view display modes.

Add a Smart (Linked) Parametric note for MATERIAL THICKNESS in the drawing as illustrated. Hide the dimension in the Top view. Insert needed Centerlines.

Modify the Hole dimension text to include 3X THRU EQ. SP. and 2X as illustrated.

Insert Custom Properties: Material, Number, Revision, Description, DrawnBy, DrawnDate, CompanyName, etc.

Material is 1060 Alloy.

Insert Company and Third Angle projection icons. The icons are available in the homework folder.

	Property Name	Type	Value / Text Expression	Evaluated Value
1	CompanyName	Text	D&M ENGINEERING	D&M ENGINEERING
2	DrawnBy	Text	DCP	DCP
3	DrawnDate	Text	9-1-2018	9-1-2018
4	Number	Text	556-099	556-099
5	Description	Text	3 HOLES	3 HOLES
6	Revision	Text	A	A
7	SWFormatSize	Text	8.5in*11in	8.5in*11in

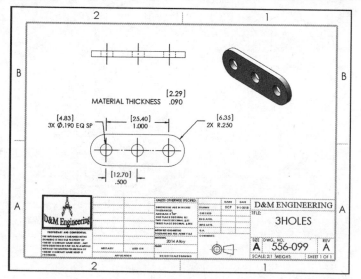

Exercise 9.2: CYLINDER Drawing

Create the A (ANSI) Landscape - IPS - Third Angle CYLINDER drawing as illustrated. Do not display Tangent Edges. Do not dimension to hidden lines.

First create the part from the drawing; then create the drawing. Use the default A (ANSI) Landscape Sheet Format/Size.

Insert views as illustrated.

Insert dimensions.

Address gaps between feature lines.

Address drawing view display modes.

Insert Company and Third Angle projection icons. The icons are available in the homework folder.

Insert Centerlines, Center Marks and Annotations.

Insert Custom Properties: Material, Number, Revision, Description, DrawnBy, DrawnDate, CompanyName, etc.

Material is AISI 1020.

Sheet Format/Size

⦿ Standard sheet size

☐ Only show standard formats

| A (ANSI) Landscape |
| A (ANSI) Portrait |
| B (ANSI) Landscape |
| C (ANSI) Landscape |
| D (ANSI) Landscape |
| E (ANSI) Landscape |
| A0 (ANSI) Landscape |

a - landscape.slddrt　　　Browse...

	Property Name	Type	Value / Text Expression	Evaluated Value
1	Material	Text	"SW-Material@CylinderPart1.SLDPRT"	AISI 1020
2	Description	Text	CYLINDER	CYLINDER

	Property Name	Type	Value / Text Expression	Evaluated Value
1	DrawnDate	Text	11-30-2018	11-30-2018
2	DrawnBy	Text	DCP	DCP
3	CompanyName	Text	D&M ENGINEERING	D&M ENGINEERING
4	Revision	Text	A	A
5	PartNo	Text	667-888	667-888
6	SWFormatSize	Text	8.5in*11in	8.5in*11in

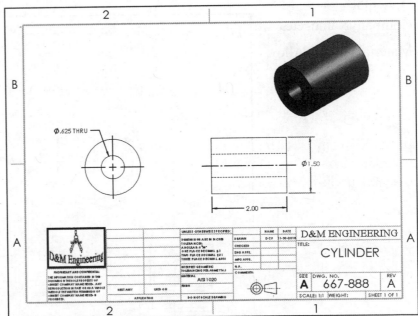

Exercise 9.3: PRESSURE PLATE Drawing

Create the A (ANSI) Landscape - IPS - Third Angle PRESSURE PLATE drawing. Do not display Tangent edges. Do not dimension to hidden lines.

First create the part from the drawing, then create the drawing. Use the default A (ANSI) Landscape Sheet Format/Size.

Insert the Front and Right view as illustrated.

Insert dimensions.

Address gaps between feature lines.

Address drawing view display modes.

Insert Company and Third Angle projection icons. The icons are available in the homework folder.

Insert Centerlines, Center Marks and Annotations.

Insert Custom Properties: Material, Number, Revision, Description, DrawnBy, DrawnDate, CompanyName, etc.

Material is 1060 Alloy.

	Property Name	Type	Value / Text Expression	Evaluated Value
1	Material	Text	"SW-Material@Circle-hole dimensionPart4ee.SLDPRT"	1060 Alloy
2	Description	Text	PRESSURE PLATE	PRESSURE PLATE
3	Number	Text	55-34	55-34

	Property Name	Type	Value / Text Expression	Evaluated Value
1	DrawnBy	Text	DCP	DCP
2	DrawnDate	Text	9-1-2018	9-1-2018
3	CompanyName	Text	D&M ENGINEERING	D&M ENGINEERING
4	Revision	Text	A	A
5	PartNo	Text	55-568	55-568
6	SWFormatSize	Text	8.5in*11in	8.5in*11in

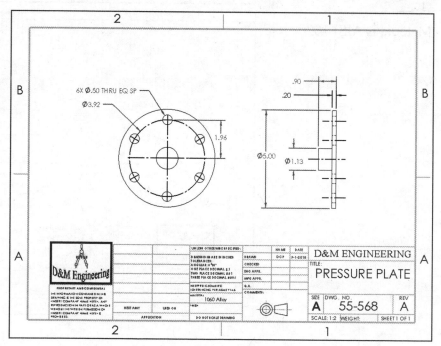

Exercise 9.4: LINKS Assembly Drawing

Create the LINK assembly. Utilize three different FLATBAR configurations and a SHAFT-COLLAR. The parts are located in the chapter folder.

Create the LINK assembly drawing as illustrated. Use the default A (ANSI) Landscape Sheet Format/Size.

Insert Company and Third Angle projection icons. The icons are available in the homework folder.

Remove all Tangent Edges.

Insert Custom Properties: Number, Revision, Description, DrawnBy, DrawnDate, CompanyName, etc.

Insert a Bill of Materials as illustrated with Balloons.

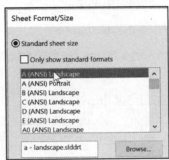

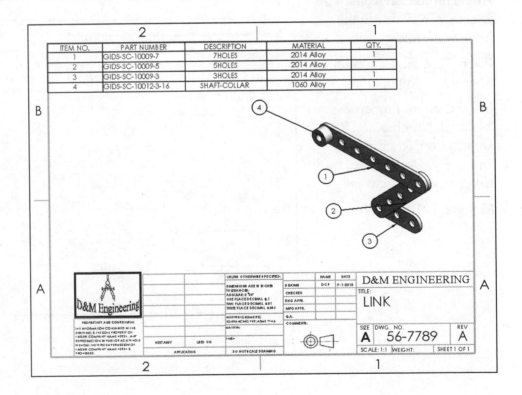

Exercise 9.5: PLATE-1 Drawing

Create the A (ANSI) Landscape - MMGS - Third Angle PLATE-1 drawing as illustrated below. Do not display Tangent edges. Do not dimension to Hidden lines.

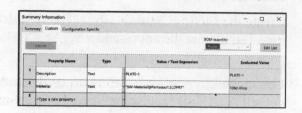

First create the part from the drawing, then create the drawing. Use the default A (ANSI) Landscape Sheet Format/Size.

Insert the Front and Right view as illustrated.

Insert dimensions.

Address gaps between feature lines.

Address drawing view display modes.

Insert Company and Third Angle projection icons. The icons are available in the homework folder.

Insert Centerlines, Center Marks and Annotations.

Insert Custom Properties: Material, Number, Description, Revision, DrawnBy, DrawnDate, CompanyName, etc.

Material is 1060 Alloy.

	Property Name	Type	Value / Text Expression	Evaluated Value
1	Description	Text	PLATE-1	PLATE-1
2	DrawnBy	Text	DCP	DCP
3	DrawnDate	Text	11-30-2018	11-30-2018
4	CompanyName	Text	D&M ENGINEERING	D&M ENGINEERING
5	Revision	Text	A	A
6	PartNo	Text	544-6689	544-6689
7	SWFormatSize	Text	215.9mm*279.4mm	215.9mm*279.4mm

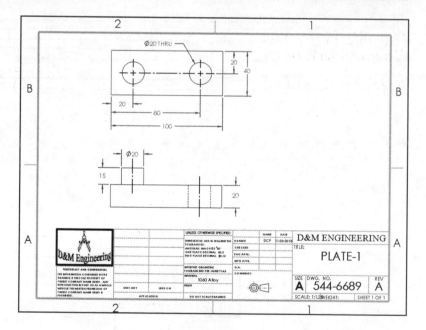

Exercise 9.6: FLAT-PLATE Drawing

Create the A (ANSI) Landscape - IPS - Third Angle FLAT-PLATE drawing. Do not display Tangent edges. Do not dimension to Hidden lines.

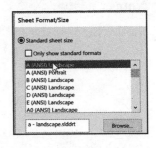

First create the part from the drawing, then create the drawing. Use the default A (ANSI) Landscape Sheet Format/Size.

Insert the Front, Top, Right and Isometric views as illustrated.

Insert dimensions.

Address gaps between feature lines.

Address drawing view display modes.

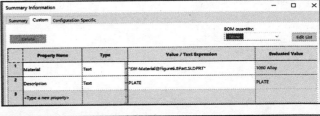

Insert Company and Third Angle projection icons. The icons are available in the homework folder.

Insert Centerlines, Center Marks and Annotations.

Insert Custom Properties: Material, Number, Revision, Description, DrawnBy, DrawnDate, CompanyName, etc.

Material is 1060 Alloy.

	Property Name	Type	Value / Text Expression	Evaluated Value
1	DrawnBy	Text	DCP	DCP
2	DrawnDate	Text	9-1-2018	9-1-2018
3	CompanyName	Text	D&M ENGINEERING	D&M ENGINEERING
4	Revision	Text	A	A
5	PartNo	Text	2445-990	2445-990
6	SWFormatSize	Text	8.5in*11in	8.5in*11in

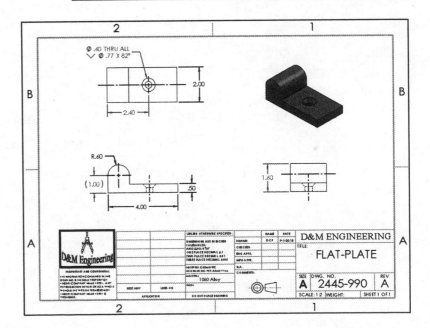

Notes:

Chapter 10

Introduction to the Certified SOLIDWORKS Associate - (CSWA) Exam

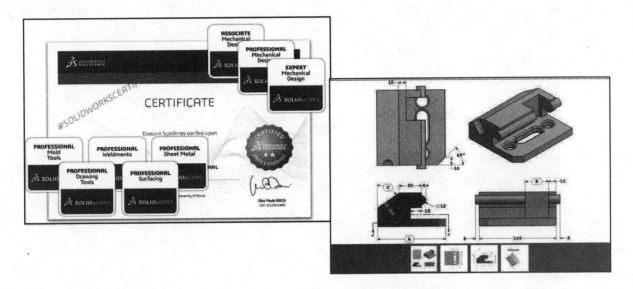

Below are the desired outcomes and usage competencies based on the completion of Chapter 10.

Desired Outcomes:	**Usage Competencies:**
• Procedure and process knowledge • Exam categories: ○ Drafting Competencies, Basic Part Creation and Modification, Intermediate Part Creation and Modification, Advanced Part Creation and Modification, and Assembly Creation and Modification.	• Familiarity of the Certified SOLIDWORKS Associate CSWA exam. • Comprehension of the skill sets, to past the Certified SOLIDWORKS Associate CSWA exam. • Awareness of the question types. • Capability to locate additional Certified SOLIDWORKS Associate CSWA exam information.

Notes:

Chapter 10 - Certified SOLIDWORKS Associate - (CSWA) Exam

Chapter Objective

Provide a basic introduction into the curriculum and categories of the Certified SOLIDWORKS Associate (CSWA) exam. Awareness of the exam procedure, process, and required model knowledge needed to take the CSWA exam. The five exam categories are Drafting Competencies, Basic Part Creation and Modification, Intermediate Part Creation and Modification, Advanced Part Creation and Modification, and Assembly Creation and Modification.

Introduction

DS SOLIDWORKS Corp. offers various types of certification. Each stage represents increasing levels of expertise in 3D CAD: Certified SOLIDWORKS Associate CSWA, Certified SOLIDWORKS Professional CSWP and Certified SOLIDWORKS Expert CSWE along with specialty fields.

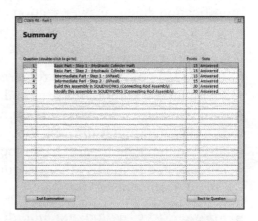

The CSWA Academic exam is provided either in a single 3 hour segment, or 2 – 90 minute segments.

Part 1 of the CSWA Academic exam is 90 minutes, minimum passing score is 80, with 6 questions. There are two questions in the Basic Part Creation and Modification category, two questions in the Intermediate Part Creation and Modification category and two questions in the Assembly Creation and Modification category.

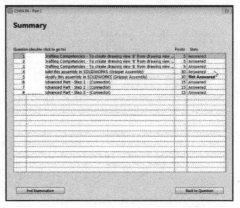

Part 2 of the CSWA Academic exam is 90 minutes; minimum passing score is 80 with 8 questions. There are three questions on the CSWA Academic exam in the Drafting Competencies category, three questions in the Advanced Part Creation and Modification category and two questions in the Assembly Creation and Modification category.

The CSWA exam for industry is only provided in a single 3 hour segment. The exam consists of 14 questions in five categories worth a total of 240 points. All exams cover the same material.

ASSOCIATE
Mechanical Design

ƏS SOLIDWORKS

Part 1:

Basic Part Creation and Modification, Intermediate Part Creation and Modification

There are **two questions** on the CSWA Academic exam in the *Basic Part Creation and Modification* category and **two questions** in the *Intermediate Part Creation and Modification* category.

The first question is in a multiple-choice single answer format. You should be within 1% of the multiple-choice answer before you move on to the modification single answer section (fill in the blank format).

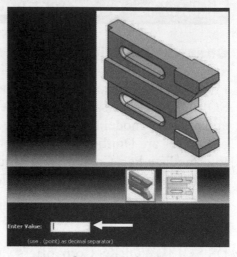

Each question is worth fifteen (15) points for a total of thirty (30) points. You are required to build a model with six or more features and to answer a question either on the overall mass, volume, or the location of the Center of mass for the created model relative to the default part Origin location. You are then requested to modify the part and answer a fill in the blank format question.

Screen shot from an exam

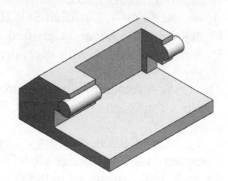

- *Basic Part Creation and Modification*: (Two questions - one multiple choice/one single answer - 15 points each).

  - Sketch Planes:

    - Front, Top, Right

  - 2D Sketching:

    - Geometric Relations and Dimensioning

  - Extruded Boss/Base Feature

  - Extruded Cut feature

  - Modification of Basic part

In the *Basic Part Creation and Modification* category there is a dimension modification question based on the first (multiple choice) question. You should be within 1% of the multiple-choice answer before you go on to the modification single answer section.

- *Intermediate Part Creation and Modification*: (Two questions - one multiple choice/one single answer - 15 points each).

  - Sketch Planes:

    - Front, Top, Right

  - 2D Sketching:

    Geometric Relations and Dimensioning

  - Extruded Boss/Base Feature

  - Extruded Cut Feature

  - Revolved Boss/Base Feature

  - Mirror and Fillet Feature

  - Circular and Linear Pattern Feature

  - Plane Feature

  - Modification of Intermediate part:

    - Sketch, Feature, Pattern, etc.

  - Modification of Intermediate part

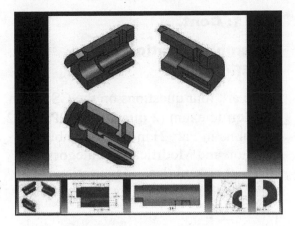

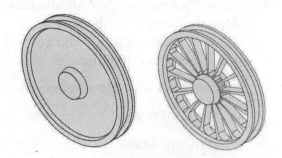

In the *Intermediate Part Creation and Modification* category, there are two dimension modification questions based on the first (multiple choice) question.

You should be within 1% of the multiple-choice answer before you go on to the modification single answer section.

☀ If you do not pass the certification exam (either segment), you will need to wait 30 days until you can retake that segment of the exam.

☀ To obtain additional CSWA exam information, visit the SOLIDWORKS VirtualTester Certification site at https://SOLIDWORKS.virtualtester.com/.

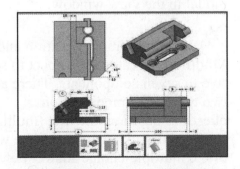

Hint: If you don't find an option within 1% of your answer please re-check your model(s).

Part 1: Cont.

Assembly Creation and Modification

There are four questions on the CSWA Academic exam (**2 questions** in Part 1, 2 questions in Part 2) in the Assembly Creation and Modification category: (2) different assemblies - (4) questions - (2) multiple choice/(2) single answer - 30 points each.

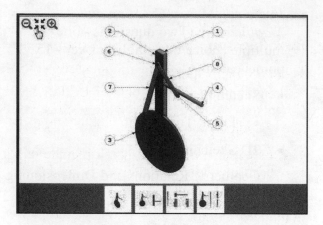

- *Assembly Creation and Modification*: (Two different assemblies - four questions - two multiple choice/two single answers - 30 points each).

  - Insert the first (fixed) component.

  - Insert all needed components.

  - Standard Mates.

  - Modification of key parameters in the assembly.

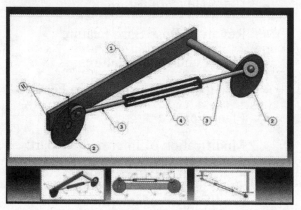

Download the needed components in a zip folder during the exam to create the assembly.

Use the new view indicator to increase or decrease the active model in the view window.

View indicator

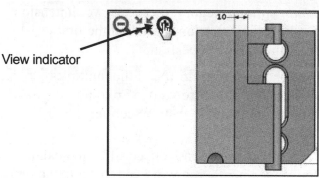

☀ In the Assembly Creation and Modification category, expect to see five to seven components. There are two dimension modification questions based on the first (multiple choice) question. You should be within 1% of the multiple-choice answer before you go on to the modification single answer section.

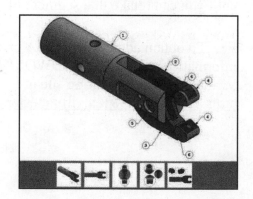

Part 2:

Introduction and Drafting Competencies

There are **three questions** on the CSWA Academic exam (Part 2) in the *Drafting Competencies* category. Each question is worth five (5) points.

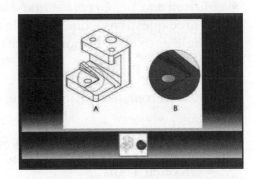

Screen shot from the exam

The three questions are in a multiple-choice single answer format. You are allowed to answer the questions in any order you prefer.

In the *Drafting Competencies* category of the exam, you are **not required** to create or perform an analysis on a part, assembly, or drawing but you are required to have general drafting/drawing knowledge and understanding of various drawing view methods.

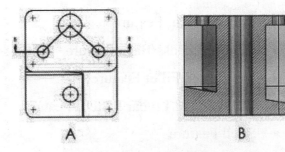

The questions are on general drawing views: Projected, Section, Break, Crop, Detail, Alternate Position, etc.

Drafting Competencies - To create drawing view 'B' it is necessary to select drawing view 'A' and insert which SolidWorks view type?

Advanced Part Creation and Modification

There are **three questions** on the CSWA Academic exam (Part 2) in the Advanced Part Creation and Modification category.

The first question is in a multiple-choice single answer format.

The other two questions (Modification of the model) are in the fill in the blank format.

The main difference between the Advanced Part Creation and Modification and the Basic Part Creation and Modification or the Intermediate Part Creation and Modification category is the complexity of the sketches and the number of dimensions and geometric relations along with an increased number of features.

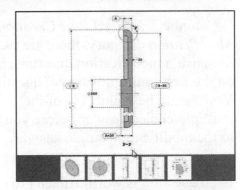

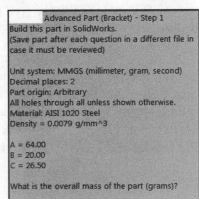

Advanced Part (Bracket) - Step 1
Build this part in SolidWorks.
(Save part after each question in a different file in case it must be reviewed)

Unit system: MMGS (millimeter, gram, second)
Decimal places: 2
Part origin: Arbitrary
All holes through all unless shown otherwise.
Material: AISI 1020 Steel
Density = 0.0079 g/mm^3

A = 64.00
B = 20.00
C = 26.50

What is the overall mass of the part (grams)?

- *Advanced Part Creation and Modification:*
  (Three questions - one multiple choice/two single answers - 15 points each).

  - Sketch Planes:

    - Front, Top, Right, Face, Created Plane, etc.

    - 2D Sketching or 3D Sketching

  - Sketch Tools:

    - Offset Entities, Convert Entities, etc.

  - Extruded Boss/Base Feature

  - Extruded Cut Feature

  - Revolved Boss/Base Feature

  - Mirror and Fillet Feature

  - Circular and Linear Pattern Feature

  - Shell Feature

  - Plane Feature

  - More Difficult Geometry Modifications

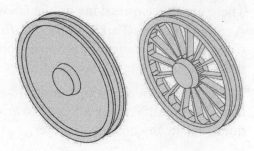

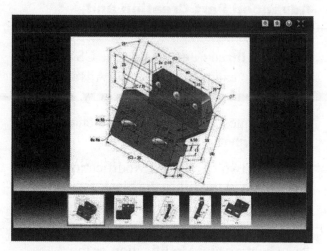

Advanced Part - Step 1 - (Connector)
Build this part in SOLIDWORKS.
(Save part after each question in a different file in case it must be reviewed)

Unit system: MMGS (millimeter, gram, second)
Decimal places: 2
Part origin: Arbitrary
All holes through all unless shown otherwise.
Material: Cast Stainless Steel
Density = 0.0077 g/mm^3

A = 104.00
B = 100.00
C = 20.00

What is the overall mass of the part (grams)?

Hint: If you don't find an option within 1% of your answer please re-check your model(s).

🔆 In the *Advanced Part Creation and Modification* category, there are two dimension modification questions based on the first (multiple choice) question. You should be within 1% of the multiple-choice answer before you go on to the modification single answer section.

Each question is worth fifteen (15) points for a total of forty five (45) points.

Screen shots from the exam

You are required to build a model, with six or more features and to answer a question either on the overall mass, volume, or the location of the Center of mass for the created model relative to the default part Origin location. You are then requested to modify the model and answer fill in the blank format questions.

Assembly Creation and Modification

There are four questions on the CSWA Academic exam (2 questions in part 1, **2 questions** in part 2) in the Assembly Creation and Modification category: (2) different assemblies - (4) questions - (2) multiple choice/ (2) single answer - 30 points each.

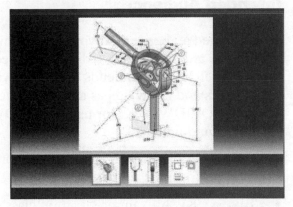

Screen shots from the exam

The first question is in a multiple-choice single answer format. You should be within 1% of the multiple-choice answer before you move on to the modification single answer section (fill in the blank format).

You are required to download the needed components from a provided zip file and insert them correctly to create the assembly.

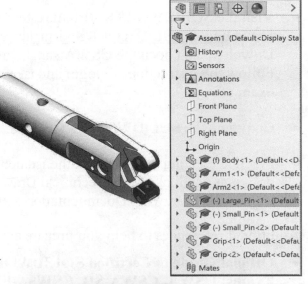

☀ In the Assembly Creation and Modification category, expect to see five to seven components. There are two dimension modification questions based on the first (multiple choice) question. You should be within 1% of the multiple-choice answer before you go on to the modification single answer section.

☀ No Surfacing questions are on the CSWA exam at this time.

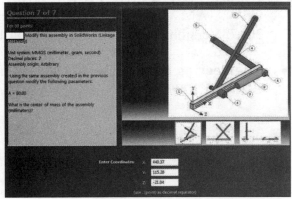

☀ At this time, Mechanical and Advanced mates are not required for the CSWA exam. You will need to know how to create a Limit Angle and Limit Distance mate.

Intended Audience

The intended audience is anyone with a minimum of 3 - 6 months of SOLIDWORKS experience and basic knowledge of engineering fundamentals and practices. SOLIDWORKS recommends that you review their SOLIDWORKS Tutorials on Parts, Assemblies and Drawings as a prerequisite and have at least 45 hours of classroom time learning SOLIDWORKS or using SOLIDWORKS with basic engineering design principles and practices.

To prepare for the CSWA exam, it is recommended that you first perform the following:

- Take a CSWA exam preparation class or review a text book written for the CSWA exam.

- Visit the SOLIDWORKS VirtualTester Certification site at https://SOLIDWORKS.virtualtester.com/. Download and open the CSWA sample exam folder. Follow the instructions to login and take a sample exam.

- Complete the SOLIDWORKS Tutorials.

- Practice creating models from the isometric working drawings sections of any Technical Drawing or Engineering Drawing Documentation text books.

Additional references to help you prepare are as follows:

- **Official Guide to Certified SOLIDWORKS Associate Exams: CSWA, CSWA-SD, CSWSA-FEA, CSWA-AM Version 4; 2017-2019**, Version 3; 2015 - 2017, Version 2; 2015 - 2012, Version 1; 2012, 2013.

- **Engineering Drawing and Design**, Jensen & Helsel, Glencoe, 1990.

During the Exam

During the exam, SOLIDWORKS provides the ability to click on a detail view below (as illustrated) to obtain additional details and dimensions during the exam.

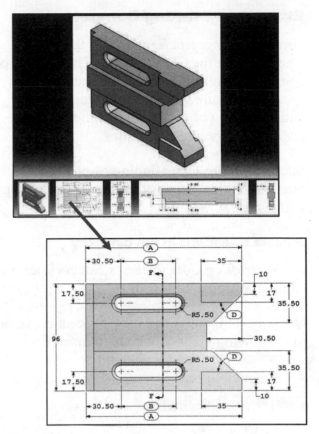

☀ No Simulation questions are on the CSWA exam at this time.

☀ No Sheetmetal questions are on the CSWA exam at this time.

FeatureManager names were changed through various revisions of SOLIDWORKS. Example: Extrude1 vs. Boss-Extrude1. These changes do not affect the models or answers in this book.

During the exam, use the control keys at the bottom of the screen to:

- *Show the Previous Question.*
- *Reset the Question.*
- *Show the Summary Screen.*
- *Move to the Next Question.*

When you are finished, press the End Examination button. The tester will ask you if you want to end the test. Click Yes.

If there are any unanswered questions, the tester will provide a warning message as illustrated.

☀ If you do not pass the certification exam (either segment), you will need to wait 30 days until you can retake that segment of the exam.

Use the clock in the tester to view the amount of time that you used and the amount of time that is left in the exam.

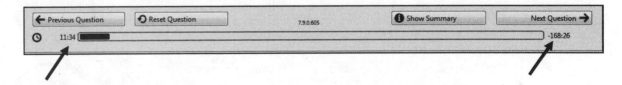

Examples: Drafting Competencies

Drafting Competencies is one of the five categories on the CSWA exam. There are three questions – multiple-choice format - 5 points each that require general knowledge and understanding of drawing view methods and basic 3D modeling techniques.

Spend no more than 10 minutes on each question in this category for the exam. Manage your time.

Sample Questions in the category

In the *Drafting Competencies* category, an exam question could read:

Question 1: Identify the view procedure. To create the following view, you need to insert a:

- A: Open Spline
- B: Closed Spline
- C: 3 Point Arc
- D: None of the above

The correct answer is B.

Question 2: Identify the illustrated view type.

- A: Crop view
- B: Section view
- C: Projected view
- D: None of the above

The correct answer is A.

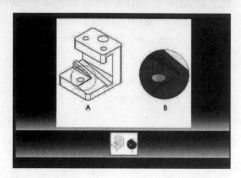

Drafting Competencies - To create drawing view 'B' it is necessary to select drawing view 'A' and insert which SolidWorks view type?

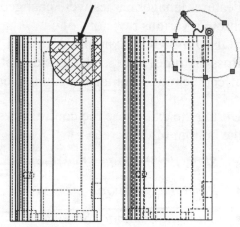

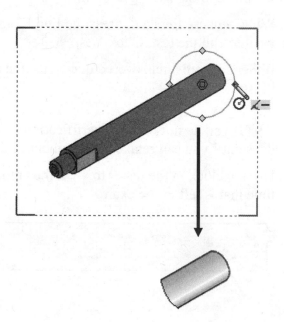

Question 3: Identify the illustrated Drawing view.

- A: Projected View

- B: Alternative Position View

- C: Extended View

- D: Aligned Section View

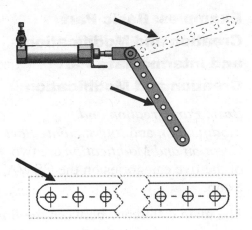

The correct answer is B.

Question 4: Identify the illustrated Drawing view.

- A: Crop View

- B: Break View

- C: Broken-out Section View

- D: Aligned Section View

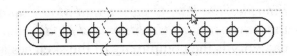

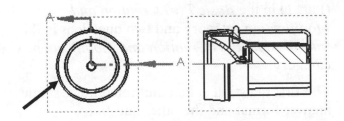

The correct answer is B.

Question 5: Identify the illustrated Drawing view.

- A: Section View

- B: Crop View

- C: Broken-out Section View

- D: Aligned Section View

The correct answer is D.

Question 6: Identify the view procedure. To create the following view, you need to insert a:

- A: Rectangle Sketch tool

- B: Closed Profile: Spline

- C: Open Profile: Circle

- D: None of the above

The correct answer is B.

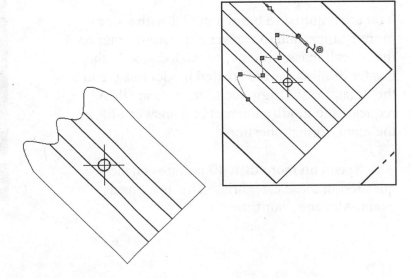

Examples: Basic Part Creation and Modification and Intermediate Part Creation and Modification

Basic Part Creation and Modification and *Intermediate Part Creation and Modification* are two of the five categories on the CSWA exam.

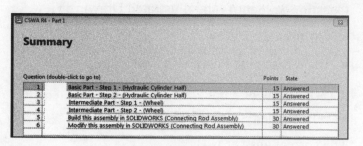

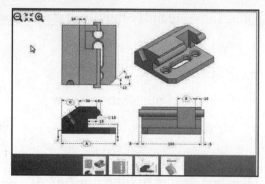

The main difference between the *Basic Part Creation and Modification* category and the *Intermediate Part Creation and Modification* or the *Advance Part Creation and Modification* category is the complexity of the sketches and the number of dimensions and geometric relations along with an increase in the number of features

There are two questions on the CSWA exam (Part 1) in the *Basic Part Creation and Modification* category and two questions in the *Intermediate Part Creation and Modification* category.

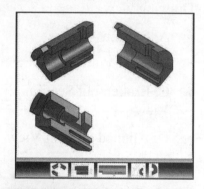

The first question is in a multiple-choice single answer format and the other question (Modification of the model) is in the fill in the blank format.

Each question is worth fifteen (15) points for a total of thirty (30) points.

You are required to build a model with six or more features and to answer a question either on the overall mass, volume, or the location of the Center of mass for the created model relative to the default part Origin location. You are then requested to modify the part and answer a fill in the blank format question.

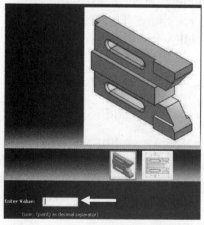

🔆 Spend no more than 40 minutes on the question in these categories. This is a timed exam. Manage your time.

Screen shots from an exam

Sample Questions in this category

Question 1:

Build the illustrated model from the provided information. Locate the Center of mass relative to the default coordinate system, Origin.

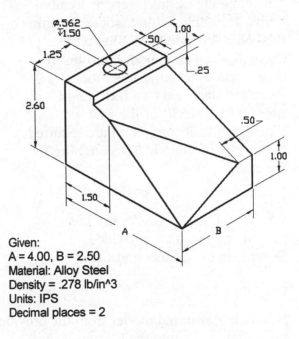

Given:
A = 4.00, B = 2.50
Material: Alloy Steel
Density = .278 lb/in^3
Units: IPS
Decimal places = 2

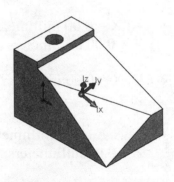

- A: X = -1.63 inches, Y = 1.48 inches, Z = -1.09 inches

- B: X = 1.63 inches, Y = 1.01 inches, Z = -0.04 inches

- C: X = 43.49 inches, Y = -0.86 inches, Z = -0.02 inches

- D: X = 1.63 inches, Y = 1.01 inches, Z = -0.04 inches

The correct answer is B.

☀ In the *Basic Part Creation and Modification* and *Intermediate Part Creation and Modification* category of the exam, you are required to read and understand an engineering document, set document properties, identify the correct Sketch planes, apply the correct Sketch and Feature tools and apply material to build a part.

Mass = 4.97 pounds

Volume = 17.86 cubic inches

Surface area = 46.77 square inches

Center of mass: (inches)
  X = 1.63
  Y = 1.01
  Z = -0.04

💡 Note the Depth/Deep ⊤ symbol with a 1.50 dimension associated with the hole. The hole Ø.562 has a three decimal place precision. Hint: Insert three features to build this model: Extruded Base and two Extruded Cuts. Insert a 3D sketch for the first Extruded Cut feature. You are required to have knowledge in 3D sketching for the exam.

💡 All SOLIDWORKS models (initial and final) are provided. Download the folders and files.

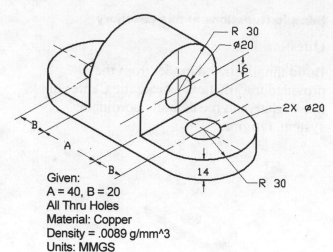

Given:
A = 40, B = 20
All Thru Holes
Material: Copper
Density = .0089 g/mm^3
Units: MMGS

Question 2:

Build the illustrated model from the provided information. Locate the Center of mass of the part.

- A: X = 0.00 millimeters, Y = 19.79 millimeters, Z = 0.00 millimeters

- B: X = 0.00 inches, Y = 19.79 inches, Z = 0.04 inches

- C: X = 19.79 millimeters, Y = 0.00 millimeters, Z = 0.00 millimeters

- D: X = 0.00 millimeters, Y = 19.49 millimeters, Z = 0.00 millimeters

- The correct answer is A.

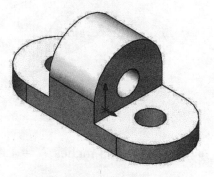

Mass properties of Part Modeling-Question2
  Configuration: Default
  Coordinate system: -- default --

Density = 0.01 grams per cubic millimeter

Mass = 1605.29 grams

Volume = 180369.91 cubic millimeters

Surface area = 29918.76 square millimeters

Center of mass: (millimeters)
  X = 0.00
  Y = 19.79
  Z = 0.00

Question 3:

Build the illustrated model from the provided information. Locate the Center of mass of the part.

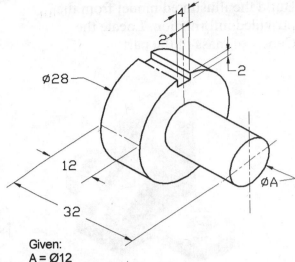

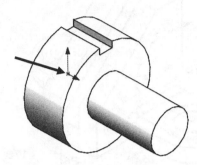

Given:
A = Ø12
Material: Cast Alloy Steel
Density = .0073 g/mm^3
Units: MMGS

- A: X = 10.00 millimeters, Y = -79.79 millimeters, Z: = 0.00 millimeters

- B: X = 9.79 millimeters, Y = -0.13 millimeters, Z = 0.00 millimeters

- C: X = 9.77 millimeters, Y = -0.10 millimeters, Z = -0.02 millimeters

- D: X = 10.00 millimeters, Y = 19.49 millimeters, Z = 0.00 millimeters

- The correct answer is B.

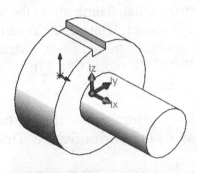

Mass properties of Part Modeling-Question3
    Configuration: Default
    Coordinate system: -- default --

Density = 0.01 grams per cubic millimeter

Mass = 69.77 grams

Volume = 9557.27 cubic millimeters

Surface area = 3069.83 square millimeters

Center of mass: (millimeters)
    X = 9.79
    Y = -0.13
    Z = 0.00

Question 4:

Build the illustrated model from the provided information. Locate the Center of mass of the part.

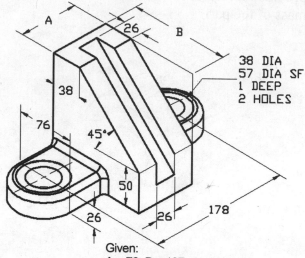

38 DIA
57 DIA SF
1 DEEP
2 HOLES

Given:
A = 76, B = 127
Material: 2014 Alloy
Density: .0028 g/mm^3
Units: MMGS
ALL ROUNDS EQUAL 6MM

There are numerous ways to build this model. Think about the various features that create the model. Hint: Insert seven features to build this model: Extruded Base, Extruded Cut, Extruded Boss, Fillet, Extruded Cut, Mirror and a second Fillet. Apply symmetry.

In the exam, create the left half of the model first, and then apply the Mirror feature. This is a timed exam.

- A: X = 49.00 millimeters, Y = 45.79 millimeters, Z = 0.00 millimeters

- B: X = 0.00 millimeters, Y = 19.79 millimeters, Z = 0.04 millimeters

- C: X = 49.21 millimeters, Y = 46.88 millimeters, Z = 0.00 millimeters

- D: X = 48.00 millimeters, Y = 46.49 millimeters, Z = 0.00 millimeters

The correct answer is C.

Mass = 3437.29 grams

Volume = 1227602.22 cubic millimeters

Surface area = 101091.11 square millimeters

Center of mass: (millimeters)
    X = 49.21
    Y = 46.88
    Z = 0.00

Question 5:

Build the illustrated model from the provided information. Locate the Center of mass of the part. All Thru Holes.

Think about the various features that create this model. Hint: Insert five features to build this part: Extruded Base, two Extruded Bosses, Extruded Cut, and Rib. Insert a Reference plane to create the Extruded Boss feature.

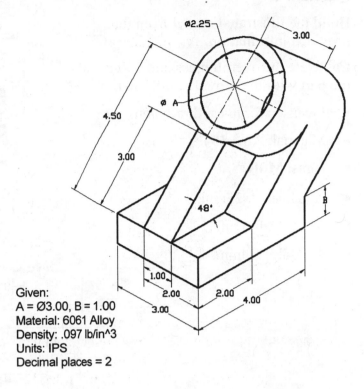

Given:
A = Ø3.00, B = 1.00
Material: 6061 Alloy
Density: .097 lb/in^3
Units: IPS
Decimal places = 2

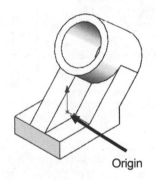

Origin

- A: X = 49.00 inches, Y = 45.79 inches, Z = 0.00 inches

- B: X = 0.00 inches, Y = 19.79 inches, Z = 0.04 inches

- C: X = 49.21 inches, Y = 46.88 inches, Z = 0.00 inches

- D: X = 0.00 inches, Y = 0.73 inches, Z = -0.86 inches

The correct answer is D.

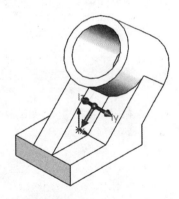

Mass properties of Part Modeling-Question5
  Configuration: Default
  Coordinate system: -- default --

Density = 0.10 pounds per cubic inch

Mass = 2.99 pounds

Volume = 30.65 cubic inches

Surface area = 100.96 square inches

Center of mass: (inches)
   X = 0.00
   Y = 0.73
   Z = -0.86

Question 6:

Build the illustrated model from the provided information.

Calculate the overall mass and volume of the part with the provided information.

- Precision for linear dimensions = **2**.

- Material: **AISI 304**.

- Units: **MMGS**.

- All Holes ⬇ **25**mm.

- All Rounds **5**mm.

- All Holes Ø**4**mm.

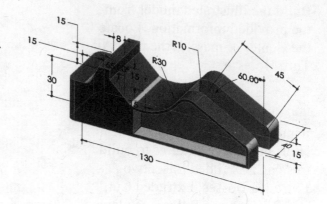

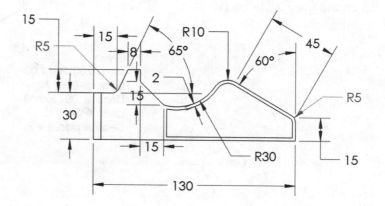

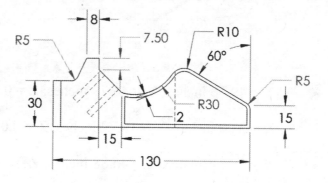

Front views

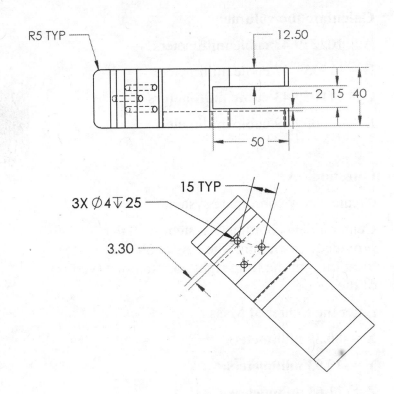

Top and Auxiliary

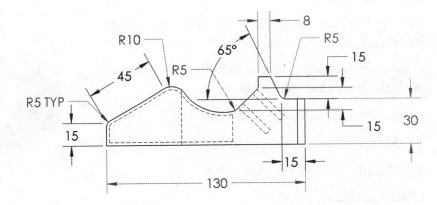

Back view

Calculate the mass:

A = **888.48grams**

B = 990.50grams

C = 788.48grams

D = 820.57grams

If you don't find your answer (within 1%) in the multiple-choice single answer format section, recheck your solid model for precision and accuracy. It could be as simple as missing a few fillets.

```
Mass properties of Part Modeling-Question 6
    Configuration: Default
    Coordinate system: -- default --

Density = 0.01 grams per cubic millimeter

Mass = 888.48 grams

Volume = 111059.43 cubic millimeters

Surface area = 25814.97  square millimeters

Center of mass: ( millimeters )
    X = -15.39
    Y = 15.93
    Z = -2.65
```

Calculate the volume:

A = 102259.43 cubic millimeters

B = 133359.47 cubic millimeters

C = 111059.43 cubic millimeters

D = 125059.49 cubic millimeters

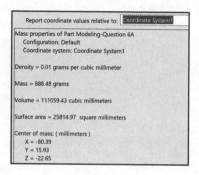

Question 6A:

Create a new coordinate system.

Center a new coordinate system with the provided illustration. The new coordinate system location is at the front right bottom point (vertex) of the model.

Enter the Center of Mass:

X = -80.39 millimeters

Y = -15.93 millimeters

Z = -22.65 millimeters

Report coordinate values relative to: Coordinate System1

Mass properties of Part Modeling-Question 6A
  Configuration: Default
  Coordinate system: Coordinate System1

Density = 0.01 grams per cubic millimeter

Mass = 888.48 grams

Volume = 111059.43 cubic millimeters

Surface area = 25814.97 square millimeters

Center of mass: (millimeters)
  X = -80.39
  Y = 15.93
  Z = -22.65

Question 6B:

Modify the illustrated model from the provided information.

Calculate the overall mass and volume of the part with the provided information.

- Modify all fillets (rounds) to 7mm.

- Modify the overall length to 140mm.

- Modify material to 1060 alloy.

Enter the mass:

309.75

Enter the volume:

114721.22

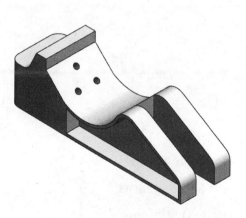

☀ If you don't find your answer (within 1%) in the multiple-choice single answer format section, recheck your solid model for precision and accuracy. It could be as simple as missing a few fillets.

Mass properties of Part Modeling-Question 6B
  Configuration: Default
  Coordinate system: -- default --

Density = 0.00 grams per cubic millimeter

Mass = 309.75 grams

Volume = 114721.22 cubic millimeters

Surface area = 27094.41 square millimeters

Center of mass: (millimeters)
  X = -16.25
  Y = 15.51
  Z = -2.67

Examples: Advanced Part Creation and Modification

Advanced Part Creation and Modification is one of the five categories on the CSWA exam. The main difference between the *Advanced Part Creation and Modification* and the *Basic Part Creation and Modification* category and the *Intermediate Part Creation and Modification* is the complexity of the sketches and the number of dimensions and geometric relations along with an increased number of features.

There are three questions - one multiple choice/two single answers - 15 points each. The question is either on the location of the Center of mass relative to the default part Origin or to a new created coordinate system and all of the mass properties located in the Mass Properties dialog box: total overall mass, volume, etc.

Sample Questions in the Category

In the *Advanced Part Creation and Modification* category, an exam question could read:

Advanced Part - Step 1 - (Connector)
Build this part in SOLIDWORKS.
(Save part after each question in a different file in case it must be reviewed)

Unit system: MMGS (millimeter, gram, second)
Decimal places: 2
Part origin: Arbitrary
All holes through all unless shown otherwise.
Material: Cast Stainless Steel
Density = 0.0077 g/mm^3

A = 104.00
B = 100.00
C = 20.00

What is the overall mass of the part (grams)?

Hint: If you don't find an option within 1% of your answer please re-check your model(s).

Screen shots from the exam

Question 1:

Build the illustrated model from the provided information. Locate the Center of mass of the part.

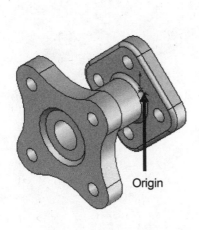

Origin

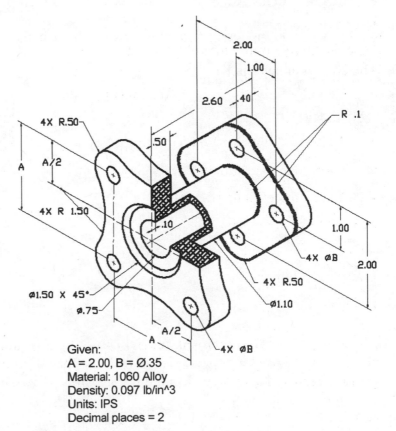

2.00
1.00
2.60 | .40
4X R.50
A
A/2
.50
R .1
4X R 1.50
.10
1.00
4X ØB
2.00
4X R.50
Ø1.50 X 45°
Ø.75
Ø1.10
A/2
A
4X ØB

Given:
A = 2.00, B = Ø.35
Material: 1060 Alloy
Density: 0.097 lb/in^3
Units: IPS
Decimal places = 2

Think about the steps that you would take to build the illustrated part. Identify the location of the part Origin.

Start with the back base flange. Review the provided dimensions and annotations in the part illustration.

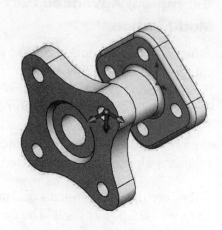

🔆 The key difference between the *Advanced Part Creation and Modification* and the *Basic Part Creation and Modification* category and the *Intermediate Part Creation and Modification* is the complexity of the sketches and the number of features, dimensions, and geometric relations. You may also need to locate the Center of mass relative to a created coordinate system location.

- A: X = 1.00 inches, Y = 0.79 inches, Z = 0.00 inches

- B: X = 0.00 inches, Y = 0.00 inches, Z = 1.04 inches

- C: X = 0.00 inches, Y = 1.18 inches, Z = 0.00 inches

- D: X = 0.00 inches, Y = 0.00 inches, Z = 1.51 inches

The correct answer is D.

Mass properties of Advanced Part Modeling-Question1
        Configuration: Default
        Coordinate system: -- default --

Density = 0.10 pounds per cubic inch

Mass = 0.59 pounds

Volume = 6.01 cubic inches

Surface area = 46.61 square inches

Center of mass: (inches)
    X = 0.00
    Y = 0.00
    Z = 1.51

Question 2:

Build the illustrated model from the provided information. Locate the Center of mass of the part.

Hint: Create the part with eleven features and a Reference plane: Extruded Base, Plane1, two Extruded Bosses, two Extruded Cuts, Extruded Boss, Extruded Cut, Extruded-Thin, Mirror, Extruded Cut, and Extruded Boss.

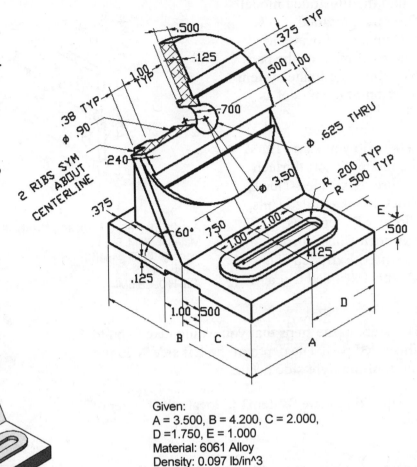

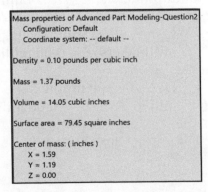

Given:
A = 3.500, B = 4.200, C = 2.000,
D =1.750, E = 1.000
Material: 6061 Alloy
Density: 0.097 lb/in^3
Units: IPS
Decimal places = 3

Think about the steps that you would take to build the illustrated part. Create the rectangular Base feature. Create Sketch2 for Plane1. Insert Plane1 to create Boss-Extrude2. Plane1 is the Sketch plane for Sketch3. Sketch3 is the sketch profile for Boss-Extrude2.

- A: X = 1.59 inches, Y = 1.19 inches, Z = 0.00 inches

- B: X = -1.59 inches, Y = 1.19 inches, Z = 0.04 inches

- C: X = 1.00 inches, Y = 1.18 inches, Z = 0.10 inches

- D: X = 0.00 inches, Y = 0.00 inches, Z = 1.61 inches

The correct answer is A.

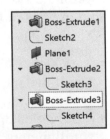

Mass properties of Advanced Part Modeling-Question2
  Configuration: Default
  Coordinate system: -- default --

Density = 0.10 pounds per cubic inch

Mass = 1.37 pounds

Volume = 14.05 cubic inches

Surface area = 79.45 square inches

Center of mass: (inches)
  X = 1.59
  Y = 1.19
  Z = 0.00

Question 3:

Build the illustrated model from the provided information. Locate the Center of mass of the part. Note the coordinate system location of the model as illustrated.

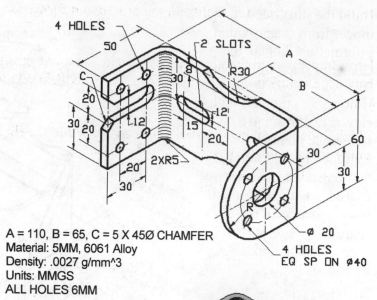

A = 110, B = 65, C = 5 X 45Ø CHAMFER
Material: 5MM, 6061 Alloy
Density: .0027 g/mm^3
Units: MMGS
ALL HOLES 6MM

Where do you start? Build the model. Insert thirteen features: Extruded-Thin1, Fillet, two Extruded Cuts, Circular Pattern, two Extruded Cuts, Mirror, Chamfer, Extruded Cut, Mirror, Extruded Cut and Mirror.

Think about the steps that you would take to build the illustrated part. The depth of the left side is 50mm. The depth of the right side is 60mm.

Create Coordinate System1 to locate the Center of mass.

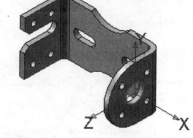

Coordinate system: +X, +Y. +Z

🔆 The SOLIDWORKS software displays positive values for (X, Y, Z) coordinates for a reference coordinate system. The CSWA exam displays either a positive or negative sign in front of the (X, Y, Z) coordinates to indicate direction as illustrated (-X, +Y, -Z).

- A: X = -53.30 millimeters, Y = -0.27 millimeters, Z = -15.54 millimeters

- B: X = 53.30 millimeters, Y = 0.27 millimeters, Z = 15.54 millimeters

- C: X = 49.21 millimeters, Y = 46.88 millimeters, Z = 0.00 millimeters

- D: X = 45.00 millimeters, Y = -46.49 millimeters, Z = 10.00 millimeters

The correct answer is A.

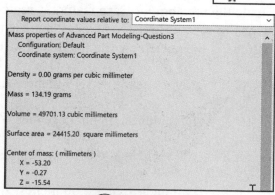

Report coordinate values relative to: Coordinate System1

Mass properties of Advanced Part Modeling-Question3
Configuration: Default
Coordinate system: Coordinate System1

Density = 0.00 grams per cubic millimeter

Mass = 134.19 grams

Volume = 49701.13 cubic millimeters

Surface area = 24415.20 square millimeters

Center of mass: (millimeters)
X = -53.20
Y = -0.27
Z = -15.54

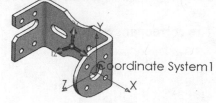

Coordinate System1

Question 4:

Build the illustrated model from the provided information. Locate the Center of mass of the part.

Hint: Insert twelve features and a Reference plane: Extruded-Thin1, two Extruded Bosses, Extruded Cut, Extruded Boss, Extruded Cut, Plane1, Mirror, and five Extruded Cuts.

Think about the steps that you would take to build the illustrated part. Create an Extrude-Thin1 feature as the Base feature.

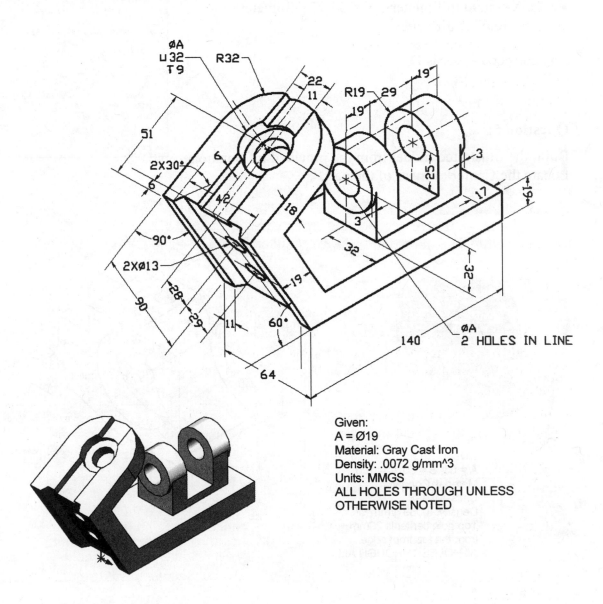

Given:
A = Ø19
Material: Gray Cast Iron
Density: .0072 g/mm^3
Units: MMGS
ALL HOLES THROUGH UNLESS
OTHERWISE NOTED

- A: X = -53.30 millimeters, Y = -0.27 millimeters, Z = -15.54 millimeters

- B: X = 53.30 millimeters, Y = 1.27 millimeters, Z = -15.54 millimeters

- C: X = 0.00 millimeters, Y = 34.97 millimeters, Z = 46.67 millimeters

- D: X = 0.00 millimeters, Y = 34.97 millimeters, Z = -46.67 millimeters

The correct answer is D.

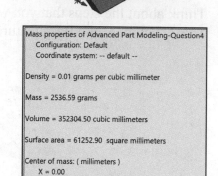

```
Mass properties of Advanced Part Modeling-Question4
    Configuration: Default
    Coordinate system: -- default --

Density = 0.01 grams per cubic millimeter

Mass = 2536.59 grams

Volume = 352304.50 cubic millimeters

Surface area = 61252.90  square millimeters

Center of mass: ( millimeters )
    X = 0.00
    Y = 34.97
    Z = -46.67
```

Question 5:

Build the illustrated model from the provided information. Locate the Center of mass of the part.

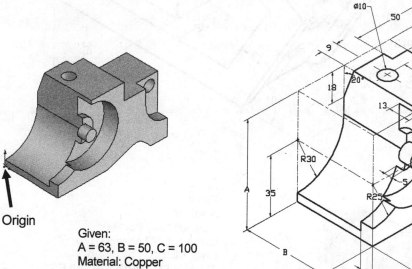

Origin

Given:
A = 63, B = 50, C = 100
Material: Copper
Units: MMGS
Density: .0089 g/mm^3
Top hole center is 20mm from the top front edge.
All HOLES THROUGH ALL

The center point of the top hole is located 30mm from the top right edge.

Think about the steps that you would take to build the illustrated part.

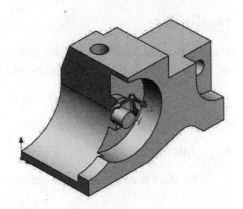

- A: X = 26.81 millimeters, Y = 25.80 millimeters, Z = -56.06 millimeters

- B: X = 43.30 millimeters, Y = 25.27 millimeters, Z = -15.54 millimeters

- C: X = 26.81 millimeters, Y = -25.75 millimeters, Z = 0.00 millimeters

- D: X = 46.00 millimeters, Y = -46.49 millimeters, Z = 10.00 millimeters

The correct answer is A.

This model has thirteen features and twelve sketches.

There are numerous ways to create the models in this chapter.

Mass properties of Advanced Part Modeling-Question5
Configuration: Default
Coordinate system: -- default --

Density = 0.01 grams per cubic millimeter

Mass = 1280.33 grams

Volume = 143857.58 cubic millimeters

Surface area = 26112.48 square millimeters

Center of mass: (millimeters)
X = 26.81
Y = 25.80
Z = -56.06

Question 6:

Build the illustrated model from the provided information. Calculate the overall mass and volume of the part with the provided information.

- Precision for linear dimensions = **2**.

- Material: **Plain Carbon Steel**.

- Units: **MMGS**.

- The part is **symmetrical** about the Front Plane.

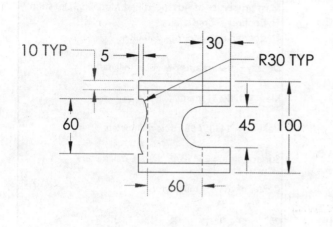

Top view

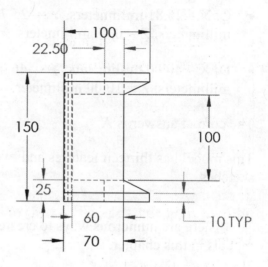

Front view

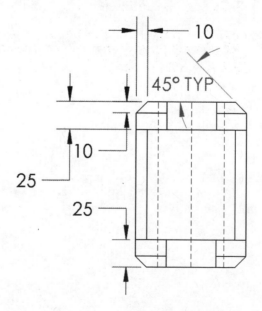

Right view

Calculate the mass:

A = 4411.5 grams

B = 4079.32 grams

C = 4234.30 grams

D = 5322.00 grams

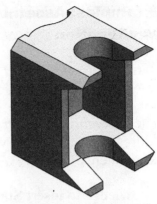

Calculate the volume:

A = 522989.22 cubic millimeters

B = 555655.11 cubic millimeters

C = 511233.34 cubic millimeters

D = 655444.00 cubic millimeters

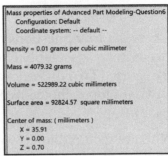

Question 6A:

Create a new coordinate system.

Center a new coordinate system with the provided illustration. The new coordinate system location is at the back right bottom point (vertex) of the model.

Enter the Center of Mass:

X = -64.09 millimeters

Y = 75.00 millimeters

Z = 40.70 millimeters

All models for this chapter are located in the CSWA model folder.

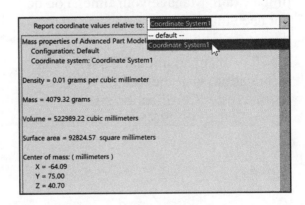

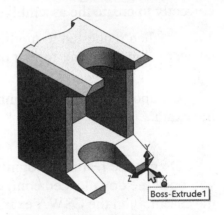

Examples: Assembly Creation and Modification

Assembly Creation and Modification is one of the five categories on the CSWA exam.

The *Assembly Creation and Modification* category addresses an assembly with numerous sub-components.

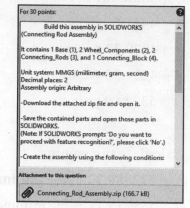

Knowledge to insert Standard mates is required in this category.

There are four questions on the CSWA Academic exam in the Assembly Creation and Modification category: (Two different assemblies - four questions - two multiple choice/two single answers - 30 points each).

You are required to download the needed components from a provided zip file and insert them correctly to create the assembly.

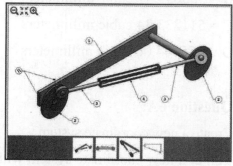

You are then requested to modify the assembly and answer fill in the blank format questions.

Components for the assembly are supplied in the exam.

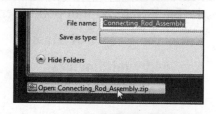

Do not use feature recognition when you open the downloaded components for the assembly in the CSWA exam. This is a timed exam. Manage your time. You do not need this information.

To obtain additional CSWA exam information, visit the SOLIDWORKS VirtualTester Certification site at https://SOLIDWORKS.virtualtester.com/.

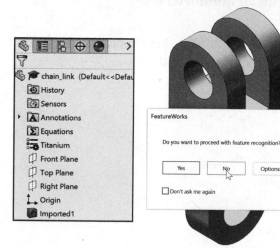

Sample Questions in this Category

In the *Assembly Creation and Modification* Assembly Modeling category, an exam question could read:

Build this assembly in SOLIDWORKS (Chain Link Assembly). It contains 2 long_pins (1), 3 short_pins (2), and 4 chain_links (3).

- Unit system: MMGS (millimeter, gram, second).

- Decimal places: 2.

- Assembly origin: Arbitrary.

IMPORTANT: Create the Assembly with respect to the Origin as shown in the Isometric view. This is important for calculating the proper Center of Mass. Create the assembly using the following conditions:

1. Pins are mated concentric to chain link holes (no clearance).

2. Pin end faces are coincident to chain link side faces.

A = 25 degrees, B = 125 degrees, C = 130 degrees

What is the center of mass of the assembly (millimeters)?

Hint: If you don't find an option within 1% of your answer please re-check your assembly.

A) X = 348.66, Y = -88.48, Z = -91.40

B) X = 308.53, Y = -109.89, Z = -61.40

C) X = 298.66, Y = -17.48, Z = -89.22

D) X = 448.66, Y = -208.48, Z = -34.64

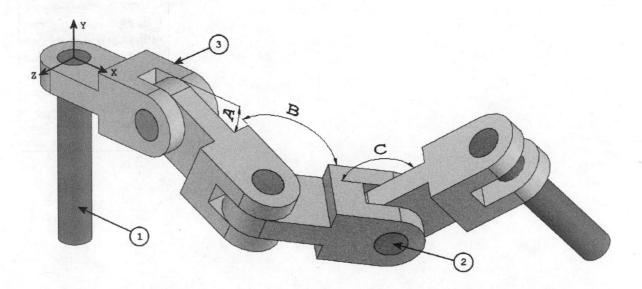

There are no step by step procedures in this section.

Download the needed components from the Chapter 6 CSWA Models folder.

The correct answer is:

A) X = 348.66, Y = -88.48, Z = -91.40

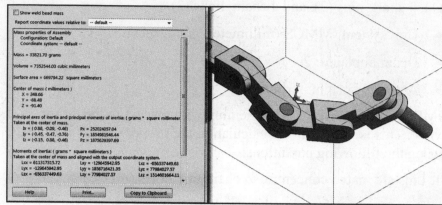

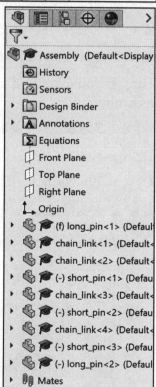

Chapter 11

Additive Manufacturing - 3D Printing

Below are the desired outcomes and usage competencies based on the completion of Chapter 11.

Desired Outcomes:	Usage Competencies:
• Knowledge of Additive Manufacturing.	• Examinc the advantages and disadvantages of Additive Manufacturing.
• Comprehend various 3D printer technologies and terminologies.	• Discuss 3D printer technology: Fused Filament Fabrication (FFF), STereoLithography (SLA), and Selective Laser Sintering (SLS).
• Familiarity with filament materials: PLA (Polylactic acid), FPLA (Flexible Polylactic acid), ABS (Acrylonitrile butadiene styrene), PVA (Polyvinyl alcohol), Nylon 618, and Nylon 645.	• Select the correct filament material. • Create an STL (*.stl) file, Additive Manufacturing file (*.amf), and a 3D Manufacturing Format file (*.3mf).
• Understand the following files types: STL (*.stl), Additive Manufacturing (*.amf), and 3D Manufacturing format (*.3mf).	• Choose optimum build orientation and slicer parameters. • Discuss 3D printer terminology: Raft, Skirt, Brim, Support, Touching, Build plate, Hot end, etc.
• Awareness of slicer and 3D printer parameters.	• Address fit tolerance for interlocking parts. • Define 3D Printing tips.
• Identify 3D Printer Add-ins.	• Print directly from SOLIDWORKS.

Notes:

Chapter 11 - Additive Manufacturing - 3D Printing

Chapter Objective

Provide a basic understanding between Additive vs. Subtractive manufacturing. Discuss Fused Filament Fabrication (FFF), STereoLithography (SLA), and Selective Laser Sintering (SLS) printer technology. Select suitable filament material. Comprehend 3D printer terminology. Knowledge of preparing, saving, and printing a model on a Fused Filament Fabrication 3D printer. Information on the Certified SOLIDWORKS Associate Additive Manufacturing (CSWA-AM) exam.

On the completion of this chapter, you will be able to:

- Discuss Additive vs Subtractive manufacturing.

- Review 3D printer technology: Fused Filament Fabrication (FFF), STereoLithography (SLA), and Selective Laser Sintering (SLS).

- Select the correct filament material:
  o PLA (Polylactic acid)
  o FPLA (Flexible Polylactic acid)
  o ABS (Acrylonitrile butadiene styrene)
  o PVA (Polyvinyl alcohol)
  o Nylon 618
  o Nylon 645

- Create an STL (*.stl) file, an Additive Manufacturing (*.amf) file and a 3D Manufacturing format (*.3mf) file.

- Prepare G-code.

- Comprehend general 3D printer terminology.

- Understand optimum build orientation.

- Enter slicer parameters:
  o Raft, brim, skirt, layer height, percent infill, infill pattern, wall thickness, fan speed, print speed, bed temperature, and extruder (hot end) temperature.

- Address fit tolerance for interlocking parts.

- Define general 3D Printing tips.

- Print directly from SOLIDWORKS.

- Knowledge of the Certified SOLIDWORKS Associate Additive Manufacturing exam.

Additive vs. Subtractive Manufacturing

In April 2012, *The Economist* published an article on 3D printing. In the article they stated that this was the "beginning of a third industrial revolution, offering the potential to revolutionize how the world makes just about everything."

Avi Reichental, who was President and CEO, 3D Systems stated, "With 3D printing, complexity is free. The printer doesn't care if it makes the most rudimentary shape or the most complex shape, and that is completely turning design and manufacturing on its head as we know it."

Over the past five years, companies are now using 3D printing to evaluate more concepts in less time to improve decisions early in product development. As the design process moves forward, technical decisions are iteratively tested at every step to guide decisions big and small, to achieve improved performance, lower manufacturing costs, delivering higher quality and more successful product introductions. In pre-production, 3D printing is enabling faster first article production to support marketing and sales functions, and early adopter customers. And in final production processes, 3D printing is enabling higher productivity, increased flexibility, reduced warehouse and other logistics costs, economical customization, improved quality, reduced product weight, and greater efficiency in a growing number of industries.

Technology for 3D printing continues to advance in three key areas: **printers** and **printing methods**, **design software**, and **materials** used in printing.

Already, 3D printing is being used in the medical industry to help save lives and in some space exploration efforts. But how will 3D printing affect the average, middle-class person in the future? Low cost 3D printers are addressing this consumer market.

Additive manufacturing is the process of joining materials to create an object from a 3D model, usually adding layer over layer.

Subtractive manufacturing relies upon the removal of material to create something. The blacksmith hammered away at heated metal to create a product. Today, a Computer Numerical Control CNC machine cuts and drills and otherwise removes material from a larger initial block to create a product.

🔅 Additive manufacturing, sometimes known as ***rapid prototyping***, can be slower than Subtractive manufacturing. Both take skill in creating the G-code and understanding the machine limitations.

🔅 Fused Filament Fabrication (FFF) and Fused Deposition Modeling (FDM) are used interchangeably in the book.

3D Printer Technology

Fused Filament Fabrication (FFF)

Fused filament fabrication (FFF) is a relatively new method of Additive manufacturing (also known as FDM) technology used for building three-dimensional prototypes or models layer by layer with a range of thermoplastics.

FFF technology is the most widely used form of 3D desktop printing at the consumer level, fueled by students, hobbyists, and office professionals.

The technology uses a continuous filament of a thermoplastic material. The thermoplastic filament is provided in a spool (open filament area) or in a refillable auto loading cartridge. Thermoplastic filament comes in two standard diameters: 1.75mm and 3mm (true size of 2.85mm).

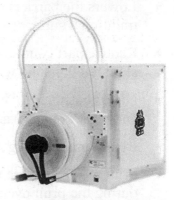

💡 Ultimaker 3 uses Near Field Communication technology with their 3mm filament. NFC technology informs the printer of the filament type and color. NFC spools tend to be more expensive than generic filament and you need to manually feed the filament correctly through the Bowden-tubes into the extruder (hot end). The spools are located in the back of the printer. This increases the footprint of the printer.

Ultimaker 3 3D printer

💡 Sindoh 3DWOX printers uses a replacement filament spool (1.75mm) inside their cartridge with a smart chip. The benefit is the automatic loading and unloading feature along with informing the printer of the filament type and properties. Replacement spools are available. The Sindoh 3DWOX 1 printer provides the ability to either use their filament or open source filament.

Sindoh 3DWOX
replacement spool and
cartridge

The thermoplastic material is heated in the nozzle (hot end) (160C - 250ºC) to form liquid, which solidifies immediately when it's deposited onto the build plate or platform. The nozzle travels at a controlled rate and moves in the X and Y direction. The build plate moves in the Z direction. This creates mechanical adhesion (not chemical).

Fused Filament Fabrication (FFF) technology can be described as the inverse process of a computer numerical cutting (CNC) machine. This technology only uses the amount of material required for the part, as opposed to a CNC machine which requires significant amounts of scrap material.

Sindoh 3DWOX
DP200 printer

Fused Filament Fabrication printers are available with multiple-head extruders. The most common usage for multiple-head extruders is to print in different colors or a different material for support or increase bed adhesion (raft).

A few advantages of Fused Filament Fabrication (FFF) Additive manufacturing:

- Lower cost (different entry levels) into the manufacturing environment.

Sindoh 3DWOX 2X

- Lowers the barriers (space, power, safety, and training) to traditional subtractive manufacturing.

- Reduce part count in an assembly from traditional subtractive manufacturing (complex parts vs. assemblies).

- Numerous thermoplastic filament types and colors (open and closed source) are available and affordable.

- Remote monitoring ability with a camera and LED lighting.

- Print more than one thermoplastic filament type and color during the print cycle (multiple-head extruders).

- Change or load new filament material during a print cycle.

- Customize percent infill and infill pattern type in the print.

- Only uses the amount of material required for the part. One exception, when using a raft and or support during the print.

- No post-curing process required.

- Reduce prototyping time.

- Faster development cycle.

- Quicker customer feedback.

- Quicker product customization and configuration.

- Parallel verticals: develop and prototype at the same time.

- Most work on either the Window or Mac OSX platform.

- Open source slicing engines. To name a few; (Slic3r, Skeinforge, Netfabb, KISSkice, and Cura).

- Open source filament.

A slicer takes a 3D model, most often in STereoLithography (*.stl) file format and translates the model into individual layers. It then generates the machine code that the 3D printer uses. You can also use Additive Manufacturing file (*.amf) or a 3D Manufacturing Format (*.3mf) file if supported.

Slicer software allows the user to calibrate 3D printer settings: filament type, part orientation, extruder speed, extruder temperature, bed temperature, cooling fan rate, raft type, support type, percent infill, infill pattern type, etc.

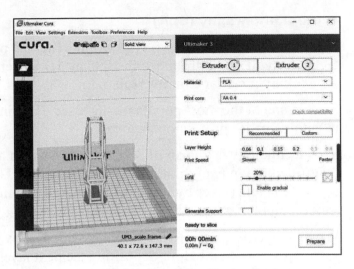

Ultimaker Cura slicer - Version: 3.4.1

A few disadvantages of Fused Filament Fabrication (FFF) Additive manufacturing:

- Slow build rates. Many printers lay down material at a speed of one to five cubic inches per hour. Depending on the part needed, other manufacturing processes may be significantly faster.

- May require post-processing. The surface finish and dimensional accuracy may be lower quality than other manufacturing methods.

- Poor mechanical properties. Layering and multiple interfaces can cause defects in the product.

- Frequent calibration is required. Without frequent calibration, prints may not be the correct dimensions, they may not stick to the build plate, and a variety of other not-so-wanted effects can occur.

- Limited by the accuracy of the stepper motors, extruder nozzle diameter, user calibration as well as print speed.

- Print time increases linearly as part tolerances become tighter. In general, FFF print tolerances range from 0.05mm to 0.5mm.

- To print something, you require a CAD model. You either need to know how to design using CAD software (SOLIDWORKS) or download a CAD model (native format or .stl) from a website (3D ContentCentral, Thingiverse, etc.).

StereoLithography (SLA)

Stereolithography (SLA) was the world's first 3D printing technology. It was introduced in 1988 by 3D Systems, Inc., based on work by inventor Charles Hull.

SLA is one of the most popular resin-based 3D printing technologies for professionals today.

SLA technology is a form of 3D printing in which a computer-controlled low-power, highly focused ultraviolet (UV) laser is used to create layers (layer by layer) from a liquid polymer (photopolymerization) that hardens on contact.

Markforged X7
SLS printer

When a layer is completed, a leveling blade is moved across the surface to smooth it before depositing the next layer. The platform moves by a distance equal to the layer thickness, and a subsequent layer is formed on top of the previously completed layers.

The process of tracing and smoothing is repeated until the part is complete. Once complete, the part is removed from the resin tank and drained of any excess polymer.

The part is in a "green" state. This green state differs from the completely cured state in one very important way: there are still polymerizable groups on the surface that subsequent layers can covalently bond to.

Through the application of heat and light, the strength and stability of printed parts improves beyond their original "green" state. However, each resin behaves slightly differently when post-cured, and requires different amounts of time and temperature to arrive at the material's optimum properties.

After the post-cure, you need to address the post-process. Post-processing is the removal of the supports and any needed polishing or sanding.

SLA prints are watertight and fully dense. No Infill is required.

Resolution of SLA technology varies from 0.05mm to 0.15mm with the industry average tolerance around 0.1mm. On average, this is significantly more precise than FFF technology and is the preferred rapid prototyping solution when extremely tight tolerances are required.

As the ultraviolet (UV) laser traces the layer, the liquid polymer solidifies and the excess areas are left as liquid in the tank.

(1micron = 1µm = 0.001mm).

One example of a popular SLA 3D desktop printer is the Formlabs Form 2. Formlabs was founded in September 2011 by three MIT Media Lab students.

The build area is 45mm × 145mm × 175mm with a layer thickness (Axis Resolution) of 25, 50, and 100 microns.

🔆 Desktop area increases significantly if you include their Form Wash and Form Cure products.

The resin (liquid polymer) is provided in a cartridge. There are different cartridges for various colors and materials. Not all resin cartridges support the 25, 50, and 100 micron resolutions.

Once the print is finished, remove it from the resin tank. Resin tanks are consumables and require replacement. Expect to replace a standard resin tank after 1,000-3,000 layers of printing (1-1.5 liters) of resin.

Wash the print with isopropyl alcohol (IPA). You should wait approximately 30 minutes for the IPA to fully evaporate after washing.

To ensure proper washing, Formlabs recommends their Form Wash. Form Wash automatically cleans uncured liquid resin printed part surfaces.

IPA dissolves uncured resin. The part is covered in uncured resin when it's removed from the resin tank. Use IPA in a well-ventilated area. Always wear protective gloves and eyewear. Find a safe way to dispose of the used IPA.

After the IPA wash and dry, perform a post-cure. At a basic level, exposure to sun light triggers the formation of additional chemical bonds within a printed part, making the material stronger and stiffer.

To ensure proper post-curing, Formlabs recommends their post-curing unit, Form Cure. Form Cure precisely combines temperature and 405 nm light to post-cure parts for peak material performance.

🔆 You cannot use a camera or webcam inside the build area. Formlabs Dashboard feature provides the ability to keep track of what layer the printer is on, resin consumptions, and various other stats.

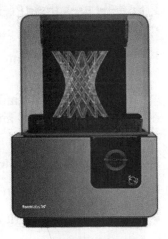

Formlabs Form 2
SLA 3D printer

Formlabs Form 2
resin cartridges

Formlabs
Form Wash

Formlabs
Form Cure

Formlabs uses their PreForm software. PreForm provides the ability to select the One-Click Print option, to automatically orient, support and layout the part.

Formlabs Form 2 printer works with the Window or Mac OSX platform.

🔅 Clean the resin tank after a print fails. Remove any cured resin on the elastic layer, discard print failures, and filter out debris. Clean any contamination on the clear acrylic tank window.

A few advantages of Stereolithography (SLA) Additive manufacturing:

- Final parts are stronger than using FFF technology.

- Parts are watertight and fully dense. No infill is required.

- Higher resolution than FFF technology.

A few disadvantages of Stereolithography (SLA) Additive manufacturing:

- Higher cost (different entry levels) into the manufacturing environment than FFF.

- Material costs are significantly higher than FFF due to the proprietary nature and limited availability of the photopolymers.

- Significantly slower fabrication speed than FFF.

- Suitable for low volume production runs of small, precise parts.

- Print only a single material type (color) at a time.

- Requires an isopropyl alcohol (IPA) wash.

- Ability to safety dispose the used isopropyl alcohol (IPA).

- Requires a post-cure.

- Parts are sensitive to long exposure to UV light.

- May need drain hole in the part to remove excess liquid polymer.

Selective Laser Sintering (SLS)

Selective Laser Sintering (SLS) technology is the most common additive manufacturing technology for industrial powder base applications.

SLS fuses particles together layer by layer through a high energy pulse laser. Similar to SLA, this process starts with a tank full of bulk material but is in a powder form vs. liquid. As the print continues, the bed lowers itself for each new layer as done in the SLS process.

3D Systems
ProX DMP 300

Both plastics and metals can be fused in this manner, creating much stronger and more durable prototypes.

Although the quality of the powders is dependent on the supplier's proprietary processes, the base materials used are typically more abundant than photopolymers, and therefore cheaper. However, there are additional costs in energy used for fabricating with this method which may reverse any savings realized in the material cost.

🔆 Speed and resolution of SLS printers typically match that of SLA, with industry averages at around 0.1mm tolerances. Only suitable for low volume production runs of small, precise parts.

🔆 SLS is the preferred rapid-prototyping method of metals and exotic materials.

Select the Correct Filament Material for FFF

There are many materials that are being explored for 3D printing; however, the two most dominant plastics for FFF technology are PLA (Polylactic acid) and ABS (Acrylonitrile-Butadiene-Styrene).

Both PLA and ABS are known as thermoplastics; that is, they become soft and moldable when heated and return to a solid when cooled.

There are three key printing stages for both thermoplastics.

1. Cold to warm: The thermoplastic starts in a hard state. It stays this way until heated to its glass transition temperature.

2. Warm to hot: The thermoplastic is now in a viscous state. It stays this way until heated to its melting temperature.

3. Hot to melting: The thermoplastic is now in a liquid state.

For PLA, the glass transition temperature is approximately 60°C. The melting temperature is approximately 155°C with a printing temperature range of 190C - 220°C.

For ABS, the glass transition temperature is approximately 100°C with a printing temperature range between 210C - 250°C. This is why a heated print bed is needed for ABS and is optional for PLA. The print bed must be kept well below the glass transition temperature and well above the melting temperature.

PLA and ABS are hygroscopic, meaning it attracts and absorbs moisture from the air. The effects of attracting water may result in one or more of the following problems: increased brittleness, diameter augmentation (potential problems with Bowden-tube printers), or filament bubbling once reaching the extruder (hot-end).

It is recommended that you store used filament in an air tight plastic bag (container) with a few silica gel packs. Place the bag in a dry, dark, controlled environment.

Silica gel pack

PLA

PLA (Polylactic Acid) is a biodegradable thermoplastic, made from renewable resources like corn starch or sugarcane. Relevant information:

- **Strength**: High | Flexibility: Low | Durability: Medium
- **Difficulty to use**: Low
- **Print temperature**: 190C - 220°C
- **Print bed temperature**: 20C - 40°C (not required)
- **Shrinkage/warping**: Minimal
- **Soluble**: No
- **Hot Head**: Standard Polytetrafluoroethylene, PTFE (Teflon)
- **Filament size**: 1.75mm and 2.85mm
- **Food safety**: Refer to manufacturer guidelines

PLA has a lower printing temperature (20C - 30°C) than ABS, and it doesn't warp as easily. PLA does not require a heated bed. Most non-heated beds are made from glass or metal.

 A removable flexible bed makes it ideal to retrieve printed parts.

PLA is normally used for its nice finish, easy and fast printing characteristics and for the large amounts of colors and varieties available.

Avoid using PLA if the print is exposed to temperatures of 60°C or higher or might be bent, twisted or dropped repeatedly.

Outside of 3D printing, PLA is typically used in medical implants, food packaging, and disposable tableware.

Darker colors and glow in the dark materials, often requires higher extruder temperatures (5C - 10°C).

PLA in general is more forgiving to temperature fluctuations and moisture than ABS and Nylon during a build cycle.

Flex/Soft PLA - Common flexible filaments are polyester-based (non-toxic). Recommended print temperature range is 220C - 250°C. It is highly recommended to drastically lower your printing speed to around 10-20mm/s. To take advantage of the filament's properties, print it with 10% infill or less. Most flexible filament adheres well to a heated bed.

PLA - Storage

PLA is mildly hygroscopic, meaning it attracts and absorbs moisture from the air.

PLA responds somewhat differently to moisture than ABS. Over time, it can become very brittle. In addition to bubbles or spurting at the nozzle (hot end), you may see discoloration and a reduction in 3D printed part properties.

Store the filament in an air tight plastic bag (container) with a few silica gel packs. Place the bag (container) in a dry, dark, temperature controlled environment. As an extra precaution, filament manufacturers often recommend using up rolls as soon as possible.

PLA can be dried using something as simple as a food dehydrator. It is important to note that this can alter the crystallinity ratio in the PLA and will lead to changes in extrusion temperature and other extrusion characteristics.

PLA - Part Accuracy

Compared to ABS, PLA demonstrates much less part warping. PLA is less sensitive to changes in temperature than ABS.

PLA undergoes more of a phase-change when heated and becomes much more liquid. If actively cooled, sharper details can be seen on printed corners without the risk of cracking or warping. The increased flow can also lead to stronger binding between layers, improving the strength of the printed part.

☀ In a small enclosed space, it is recommended to have your printer enclosed with a HEPA filtration filter.

ABS

ABS (Acrylonitrile-Butadiene-Styrene) is an oil-based thermoplastic, commonly found in (DWV) pipe systems, automotive trim, bike helmets, and toys (LEGO). Relevant information:

- **Strength**: High | Flexibility: Medium | Durability: High
- **Difficulty to use**: Medium
- **Print temperature**: 210C - 250°C
- **Print bed temperature**: 80C - 110°C (required)
- **Shrinkage/warping**: Medium
- **Soluble**: In esters, ketones, and acetone
- **Hot Head**: Standard Polytetrafluoroethylene, PTFE (Teflon)
- **Filament size**: 1.75mm and 2.85mm
- **Food safety**: Not food safe

ABS boasts slightly higher strength, flexibility, and durability. ABS is more sensitive to changes in temperature than PLA, which can result in cracking and warping if the print cools too quickly.

A heated bed plate is required for ABS. Most heated beds are made from glass or metal.

HEPA FILTER

It is recommended to have a ventilated printing area when using ABS material (oil-based thermoplastic). It is also recommended to have the printer enclosed with a HEPA filtration filter.

ABS is better suited for items that are frequently handled, dropped, or heated. It can be used for mechanical parts, especially if they are subjected to stress or must interlock with other parts.

Sindoh 3DWOX 1, 2X HEPA filter is built into the printer.

☀ For high temperature applications, ABS (glass transition temperature of 105°C) is more suitable than PLA (glass transition temperature of 60°C). PLA can rapidly lose its structural integrity as it approaches 60°C.

ABS - Storage

ABS is mildly hygroscopic. Diameter augmentation (potential problems with Bowden-tube printers) can be an issue.

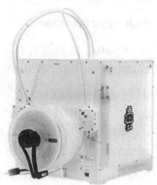

Store the filament in an air tight plastic bag (container) with a few silica gel packs. Place the bag in a dry, dark, temperature controlled environment. As an extra precaution, filament manufacturers often recommend using up rolls as soon as possible. ABS can be easily dried using a source of hot (preferably dry) air such as a food dehydrator.

Ultimaker 3 using Bowden-tubes

ABS - Part Accuracy

For most, the single greatest hurdle for accurate parts is good bed (platform) adhesion. Start with a clean, level, heated bed. Check the bed and extruder (hot end) are set to the correct temperature.

Eliminate all build area drafts (open windows, air conditioning vents, etc.). Use dry filament. Wet filament during printing prevents good layer adhesion and greatly weakens the part.

When printing on a glass plate, a (Polyvinyl Acetate) PVA based glue stick applied to the bed helps with bed adhesion. Elmer's or Scotch permanent glue sticks are inexpensive and easily found. Remember, less is more when applying the glue stick to the build plate.

You may need to add a raft (a horizontal latticework of filament located underneath the part) to the build. Over time, a heated metal bed can warp. Check for flatness.

☀ ABS provides a more matte appearance than PLA, but it can become very shiny after acetone vapor smoothing.

Nylon

Nylon (618, 645) is a popular family of synthetic polymers used in many industrial applications. Compared to most other filaments, it ranks as the number one contender when together considering strength, flexibility, and durability.

Relevant information for Nylon 618:

- **Strength**: High | Flexibility: High | Durability: High
- **Difficulty to use**: Medium
- **Print temperature**: 240C - 255°C
- **Print bed temperature**: 50C - 60°C (required)
- **Shrinkage/warping**: Medium
- **Soluble**: No
- **Hot Head**: Metal
- **Filament size**: 1.75mm and 2.85mm
- **Food safety**: Refer to manufacturer guidelines

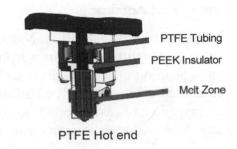

PTFE Tubing
PEEK Insulator
Melt Zone

PTFE Hot end

Relevant information for Nylon 645:

- **Strength**: High | Flexibility: High | Durability: High
- **Difficulty to use**: Medium
- **Print temperature**: 255C - 265°C
- **Print bed temperature**: 85C - 95°C (required)
- **Shrinkage/warping**: Considerable
- **Soluble**: No
- **Hot Head**: Metal
- **Filament size**: 1.75mm and 2.85mm
- **Food safety**: Refer to manufacturer guidelines

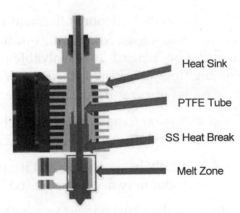

Heat Sink
PTFE Tube
SS Heat Break
Melt Zone

All metal Hot end

Nylon filament requires temperatures above 240°C to extrude. Most low-end 3D printers come standard with hot ends that use Polyether ether Ketone (PEEK) and Polytetrafluoroethylene (Teflon). Both PEEK and PTFE begin to breakdown above 240°C and will burn and emit noxious fumes. You should only use an all metal hot end.

💡 PLA and ABS is less likely to get stuck in the inner wall PTFE (Teflon) than an all metal hot end.

When should I use Nylon? Taking advantage of nylon's strength, flexibility, and durability use this 3D printer filament to create tools, functional prototypes, or mechanical parts (like hinges, buckles, or gears). Dry nylon prints buttery smooth and has a glossy finish.

Nylon - Storage

Nylon is very hygroscopic, more so than PLA or ABS. Nylon can absorb more than 10% of its weight in water in less than 24 hours. Successful 3D printing with nylon requires dry filament. When you print with nylon that isn't dry, the water in the filament explodes causing air bubbles during printing that prevents good layer adhesion and greatly weakens the part. It also ruins the surface finish. To dry nylon, place it in an oven at 50C - 60°C for 6-8 hours. After drying, store in an airtight container (Vacuum bag), preferably with dry silica gel packets. Place the container in a dry, dark, temperature controlled environment.

Nylon - Part Accuracy

Compared to ABS and PLA, Nylon and ABS warp approximately the same. PLA demonstrates much less part warping. A heated plate (50C - 90°C) is required for Nylon. When printing on a glass plate, a (Polyvinyl Acetate) PVA based glue stick applied to the bed is the best method of bed adhesion. Elmer's or Scotch permanent glue sticks are inexpensive and easily found. Remember, less is more when applying the glue stick to the plate. You will also have to clean it up after your build.

PVA (Polyvinyl Alcohol)

PVA (Polyvinyl Alcohol) filament is a water-soluble synthetic polymer. PVA filament dissolves in water. Many multi-head extruder users find PVA to be a useful support material because of its dissolvable properties. In general, PVA filament is used in conjunction with PLA not ABS. PVA adheres well to PLA and not ABS. Moreover, the extrusion temperature difference between PVA and ABS can be problematic.

PVA extrusion temperatures range between 160C - 190°C. A heated bed is recommended. Bed temperatures range between 40C - 50°C.

PVA is highly hygroscopic and is costly. PVA should be stored in an air tight box or container and may need to be dried before use.

Submerge the finished part in a bath of cold circulating water, until the PVA support structure is completely dissolved. This can be time consuming and messy.

Do not expose PVA filament to temperatures higher than 200°C for an extended period of time. An irreversible degradation of the material will occur, known as pyrolysis. It will jam the extruder nozzle (hot end). Unlike PLA and ABS, you cannot remove the jam by increasing the temperature. Clearing the jam in the nozzle will often require it to be re-drilled or replaced altogether.

A few filament companies provide breakaway material to replace the high cost of PVA. Breakaway is a support material used with multi head 3D printers. It is quick to remove and does not need further post-processing.

STereoLithography (*.stl) file

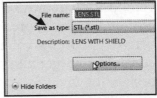

STereoLithography (*.stl) is a file format native to the Stereolithography CAD software created by 3D Systems. STL has several after-the-fact backronyms such as "Standard Triangle Language" and "Standard Tessellation Language."

An STL file describes only the surface geometry of a three-dimensional object without any representation of color, texture, or other common CAD model attributes. The STL format specifies both ASCII and Binary representations.

Binary files are more common, since they are more compact. An STL file describes a raw unstructured triangulated (point cloud) surface by the unit normal and vertices (ordered by the right-hand rule) of the triangles using a three-dimensional Cartesian coordinate system.

STL Save options allow you to control the number and size of the triangles by setting the various parameters in the CAD software.

Save an STL (*.stl) file in SOLIDWORKS

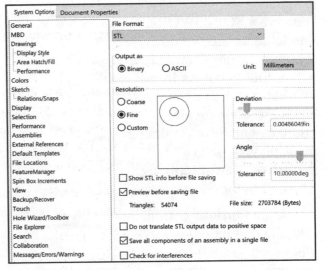

To save a SOLIDWORKS model as a STL file, click **File**, **Save As** from the Main menu or **Save As** from the Main menu toolbar. The Save As dialog is displayed. Select **STL(*.stl)** as the Save as type.

Click the **Options** button. The dialog box is displayed. View your options and the three types of Resolution: **Coarse**, **Fine** and **Custom**. The resolution options provide the ability to control the number and size of the triangles by setting various parameters in the CAD software. The Custom setting provides the ability to control the deviation and angle for the triangles. Click **OK**. View the generate point cloud of the part. Click **Yes**. The STL file is now ready to be imported into your 3D printer software.

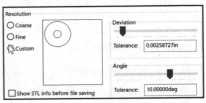

In SOLIDWORKS for a smoother STL file, change the Resolution to Custom. Change the deviation to 0.0005in (0.01mm). Change the angle to 5. Smaller deviations and angles produce a smoother file but increase the file size and print time.

Additive Manufacturing (*.amf) file

Additive Manufacturing (*.amf) is a file format that includes the materials that have been applied to the parts or bodies in the 3D model.

Additive Manufacturing (*.amf) is an open standard for describing objects for additive manufacturing processes such as 3D printing. The official ISO/ASTM 52915:2013 standard is an XML-based format designed to allow any computer-aided design software to describe the shape and composition of any 3D object to be fabricated on any 3D printer. Unlike its predecessor STL format, AMF has native support for color, materials, lattices, and constellation (groups). Therefore, it requires less post-processing to define data such as the position of your model relative to the selected 3D printer, orientation, color, materials, etc.

Save an Additive Manufacturing (*.amf) file in SOLIDWORKS

To save a SOLIDWORKS model as an Additive Manufacturing (*.amf) file, Click **File**, **Save As** from the Main menu or **Save As** from the Main menu toolbar. The Save As dialog is displayed. Select **Additive Manufacturing (*.amf)** as the Save as type.

Click the **Options** button. The dialog box is displayed. View your options. For most parts, utilize the default setting.

Close the dialog box.

Click **OK**. View the generate point cloud of the part.

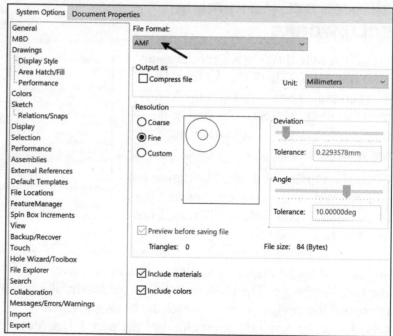

Click **Yes**. The file is now ready to be imported into your 3D printer software or print directly from SOLIDWORKS.

3D Manufacturing Format (*.3mf) file

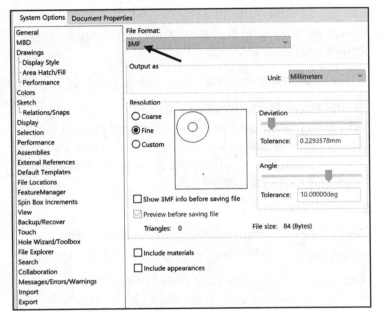

3D Manufacturing Format (*.3mf) is a file format developed and published by the 3MF Consortium. This format became natively supported in all Windows operating systems since Windows 8.1. 3MF has since garnered considerable support from large companies such as HP, 3D Systems, Stratasys, GE, Siemens, Autodesk and Dassault Systems although it is unknown how many actively use this file format. 3D Manufacturing Format file has similar native support for color, materials, lattices, and constellation (groups).

Save a 3D Manufacturing Format (*.3mf) file in SOLIDWORKS

To save a SOLIDWORKS model as a 3D Manufacturing Format (*.3mf) file, Click **File**, **Save As** from the Main menu or **Save As** from the Main menu toolbar. The Save As dialog is displayed. Select **3D Manufacturing Format (*.3mf)** as the Save as type.

Click the **Options** button. The dialog box is displayed. View your options. For most parts, utilize the default setting.

Close the dialog box. Click **OK**. View the generate point cloud of the part. Click **Yes**. The file is now ready to be imported into your 3D printer software or print directly from SOLIDWORKS.

The include materials and include appearances option is not selected by default.

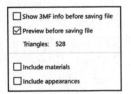

In SOLIDWORKS for a smoother STL file, change the Resolution to Custom. Change the deviation to 0.0005in (0.01mm). Change the angle to 5. Smaller deviations and angles produce a smoother file but increase the file size and print time.

What is a Slicer? How does a Slicer Work?

Creating the model is only the first step of 3D printing. You can 3D print directly from SOLIDWORKS. In addition, you can print directly from slicer software, if the slicer software supports this feature.

The slicer is software which converts a variety of file formats. The Additive Manufacturing file formats are STL (*.stl), Additive Manufacturing (*.amf), and 3D Manufacturing format (*.3mf) file. The slicer turns these file formats into printing instructions (G-code) for your 3D printer.

Slicing is the process of turning the 3D model into a toolpath for the 3D printer. Most people call it slicing because the first thing the slicing engine does is cut the 3D model into thin horizontal layers. Open source slicing engines include Slic3r, Skeinforge, Netfabb, KISSkice, Cura, etc. to name a few.

The G-code file contains the instructions based on settings you choose and calculates how much material the printer will need and how long it will take to print.

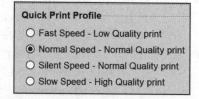

Slicer Parameters

Proper slicer parameters can mean the difference between a successful print, and a failed print. That's why it's important to know how slicers work and how various settings will affect the final print. There are open and close source slicers.

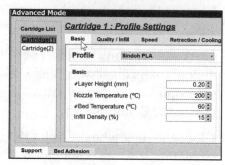

Sindoh 3DWOX slicer - Version: 1.4.2102.0

Layer Height

Layer height controls the resolution of the print. The setting specifies the height of each filament layer.

The higher values produce faster prints in lower resolution. Lower values produce slower prints in higher resolution.

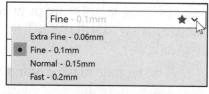

The default value for (PLA) using the Sindoh 3DWOX 2X - Version: 1.4.2102.0 is .20mm. There are four default settings: Fast, Normal, Silent, and Slow. Settings range from .05mm - .4mm. Sindoh uses a .4mm nozzle with 1.75mm filament.

The default value for (PLA) using in the Ultimaker 3 Cura slicer - Version: 3.4.1 is .1mm. There are four default settings: Fast, Normal, Fine, and Extra Fine. Settings range from .06mm - .2mm.

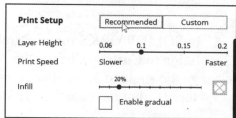

Ultimaker Cura slicer - Version: 3.4.1

Shell (Wall) Thickness

Wall thickness refers to the distance between one surface of the model and the opposite sheer surface. Set the wall thickness of the outside shell in the horizontal direction. Use in combination with the nozzle size to define the number of perimeter lines and the thickness of these perimeter lines.

The default value for (PLA) using the Sindoh 3DWOX 2X - Version: 1.4.2102.0 is .80mm. The minimum is .40mm.

The default value for (PLA) using the Ultimaker 3 Cura - Version: 3.4.1 is 1mm.

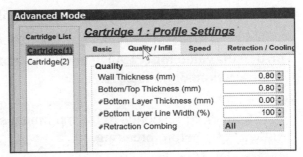

Sindoh 3DWOX slicer - Version: 1.4.2102.0

Shell		
Wall Thickness	1	mm
Wall Line Count	3	
Top/Bottom Thickness	1	mm
Top Thickness	1	mm
Top Layers	10	
Bottom Thickness	1	mm
Bottom Layers	10	

Ultimaker Cura slicer - Version: 3.4.1

Infill Density/Overlap

Infill is the internal structure of your object, which can be as sparse or as substantial as you would like it to be. A higher percentage will result in a more solid object, so 100% (not recommended) infill will make your object completely solid, while 0% infill will give you something completely hollow.

The higher the infill percentage, the more material and longer the print time. It will also increase weight and material cost.

When using any infill percentage, a pattern is used to create a strong and durable structure inside the print. A few standard patterns are Rectilinear, Honeycomb, Circular, Tri-hexagon, Cubic, Octet, and Triangular. In general, use a 10% - 15% infill with a maximum infill of 60%. 100% infill is not recommended. Part warping can be a concern.

Shells are the outer layers of a print which make the walls of an object, prior to the various infill levels being printed within. The number of shells affect stability and translucency of the model.

View the different infill and number of shells between the two models.

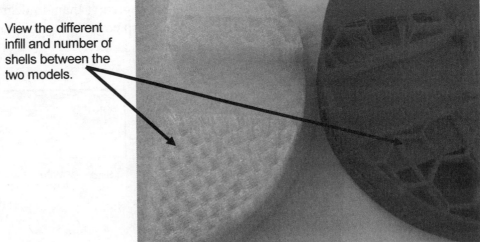

The default value for (PLA) using the Sindoh 3DWOX 2X - Version: 1.4.2102.0 is 15%.

The default value for (PLA) using the Ultimaker 3 Cura - Version: 3.4.1 is 20%.

💡 Strength corresponds to the maximum stress the print can take before breaking.

Infill	
Infill Pattern	Automatic ▾
Infill Overlap (%)	15 ⬍
Solid Infill Top	☑
Solid Infill Bottom	☑
Infill Before Wall	☐
Skin Outline	☐
Skin Type	Line ▾

Sindoh 3DWOX slicer - Version: 1.4.2102.0

Print Speed

Print speed refers to the speed at which the extruder travels while it lays down filament.

Print speed affects the following areas: Infill speed, Wall speed, Top/Bottom speed, Initial layer travel speed, Raft print speed, and Maximum travel resolution.

Print Speed		70	mm/s
🌡 Travel			<
❄ Cooling			<
🛆 Support			<

The default value for (PLA) using the Sindoh 3DWOX 2X - Version: 1.4.2102.0 is 40mm/s.

The default value for (PLA) using the Ultimaker 3 Cura - Version: 3.4.1 is 70mm/s.

Supports

Generate structures to support parts (features) of the model which have overhangs to the build plate.

Without these structures, such parts would collapse during the print process.

Support material

In order to create an overhang at any angle less than vertical, your printer offsets each successive layer. The lower the angle gets to horizontal, or 90°, the more each successive layer is offset.

General 45 degree rule. If your model has overhangs greater than 45 degrees, you need support material. If the part has numerous holes, sharp edges, long run (bridge), or thin bodies, support material may also be required.

Support Types

Touching Buildplate

The Touching Buildplate option provides supports only where the part touches the build plate. This reduces build time, clean up and support material.

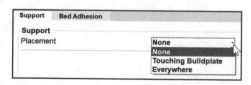

Support	Bed Adhesion	
Support		
Placement		None ▾
		None
		Touching Buildplate
		Everywhere

Sindoh 3DWOX slicer - Version: 1.4.2102.0

Use the touching Buildplate option when you have overhangs and tricky angles toward the bottom of a design, but do not wish to plug up holes, hollow spaces, or arches in the rest of the design.

Everywhere

The Everywhere option provides support material everywhere on the part, not only where the part touches the build plate.

Bed (Platform) Adhesion

Bed adhesion is one of the most important elements for getting a good 3D print. Set various options to ensure good bed adhesion and to prevent part warping. If needed, use blue painter's tape, hair spray or a glue stick.

Bed Adhesion Type- Raft

A Raft is a horizontal latticework of filament located underneath the part. A Raft is used to help the part stick to the build plate (heated or non-heated).

Rafts are also used to help stabilize thin tall parts with small build plate footprints.

When the print is complete, remove the part from the build plate. Peel the raft away from the part. If needed, use a scraper or spatula.

Bed Adhesion Type- Skirt

A Skirt is a layer of filament that surrounds the part with a 3mm - 4mm offset. The layer does not connect the part directly to the build plate. The Skirt primes the extruder and establishes a smooth flow of filament. In some slicers, the skirt is added automatically when you select the None option, for bed adhesion type.

Bed Adhesion Type- Brim

A Brim is basically like a Skirt for the part. A Brim has a zero offset from the part. It is a layer of filament laid down around the base of the part to increase its surface area. A Brim, however, does not extend underneath the part, which is the key difference between a brim and a raft.

Touching Build plate

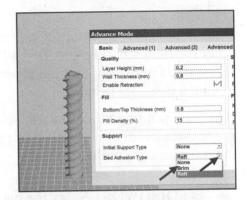

Part with a Raft on the Build plate

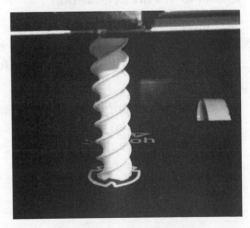

Part with a Skirt on the Build plate

Part Orientation

Insert the file into your printer's slicer. The model is displayed in the build plate area.

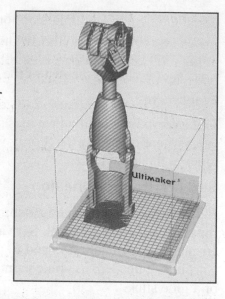

Depending on your printer's slicer software, you may or may not receive a message indicating the object is too large for the current build plate.

If the object is too large, you will need to scale it down or redesign it into separate parts.

Use caution when scaling critical features if you require fasteners or a minimum wall thickness.

You should always center the part and have it lay flat on the build plate. Bed adhesion is one of the most important elements for getting a good 3D print.

If you are printing more than one part, space them evenly on the plate or position them for a single build.

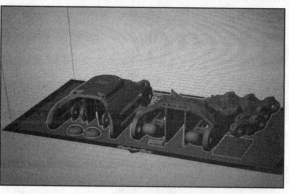

🔅 In SOLIDWORKS, lay the parts out in an assembly. Save the assembly as a part. Save the part file as an STL (*.stl), or Additive Manufacturing (*.amf), or 3D Manufacturing format (*.3mf).

| STL (*.stl) |
| 3D Manufacturing Format (*.3mf) |
| Additive Manufacturing File (*.amf) |
| eDrawings (*.eprt) |
| 3D XML (*.3dxml) |
| Microsoft XAML (*.xaml) |

Consideration should be used when printing an assembly. If the print takes 20 hours, and a failure happens after 19 hours, you just wasted a lot of time versus printing each part individually.

Example 1: Part Orientation

Part orientation is very important on build strength and the amount of raft and support material required for the build. Incorrect part orientations can lead to warping, curling, and delamination.

If maximizing strength is an issue, select the part orientation on the build plate so that the "grain" of the print is oriented to maximize the strength of the part.

Example 1: First Orientation - Vertical

In the first orientation (vertical), due to the number of holes and slots, additional support material is required (with minimum raft material) to print the model.

Removing the material in these geometrics can be very time consuming.

Example 1: Second Orientation - Horizontal

In the second orientation (horizontal), additional raft material is used, and the support material is reduced.

The raft material can be easily removed with a pair of needle-nose pliers and no support material clean-up is required for the holes and slots. Note: In some cases, raft material is not needed.

Example 2: Part Orientation

The lens part is orientated in a vertical position with the large face flat on the build plate. This reduces the required support material and ensures proper contact (maximum surface area) with the build plate.

Optimize Print Direction

Some slicers (Sindoh 3DWOX desktop) provide the ability to run an optimization print direction analysis of the part under their Advanced Mode. The areas in evaluations are **Thin Region, Area of Overhang Surface** and the **Amount of needed support** material. This can be very useful when you are unsure of the part orientation.

Suppress mates in an assmbly to have the model lay flat on the build plate. If the model does not lay flat, you will require a raft and additional supports. This will increase build time and material cost.

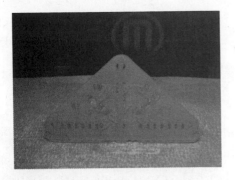

First orientation - vertical

Second orientation - horizontal

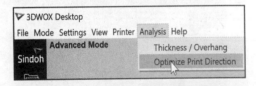

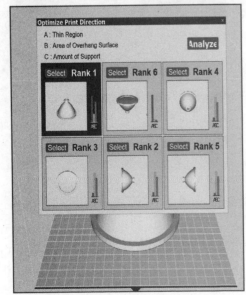

Sindoh 3DWOX slicer - Version: 1.4.2102.0

The needed support material is created mainly internal to the part to print the CBORE feature. Note: There is some outside support material on the top section of the part.

Raft material   Internal support material for the CBORE

💡 Proper part orientation for thin parts will make the removal of the raft easier.

Remove the Model from the Build Plate

Non-heated Build Plate

Most Non-heated build plates are made from glass or metal. If needed, use blue painter's tape, glue stick or hair spray to assure good model adhesion. After the build, remove the plate from the printer.

Ultimaker 3 glass build plate

Utilize a flat edge tool (thin steel spatula). Gently work under the part, and lift the part directly away from yourself. Clean the build plate. Return the plate. Re-level the plate after every build.

The Sindoh 3DWOX DP201 printer (non-heated) has a flexible magnet removable plate that does not require blue painter's tape, glue stick or hair spray to assure good model adhesion. This eliminates the need for scrapers or any sharp tools to remove the print. After the build, remove the plate and bend.

Sindoh 3DWOX DP201
flexible magnet build plate

🔅 Bed adhesion is one of the most important elements for getting a good 3D print.

Heated Build Plate

Most heated build plates are made from glass or metal. If needed, use a glue stick or hair spray to assure good model adhesion. If you have a heated build plate your temperatures can range between 30C - 90ºC depending on the material, so be careful. After the build, remove the plate from the printer. Utilize a flat edge tool (thin steel spatula). Gently work under the part, and lift the part directly away from yourself. Clean the build plate. Return the plate. Re-level the plate after every build.

Prusa i3 MK3 flexible metal
magnet build plate

A few manufacturers (Sindoh 3DWOX 1 and 2X and Prusa i3 MK3) provide a flexible heated magnet metal build plate.

This eliminates the need for scrapers or any sharp tools most of the time. After the build, remove the flexible magnet plate and bend.

🔅 Most heated metal bed plates are coated. Over time, the coating wears and the plate needs to be replaced.

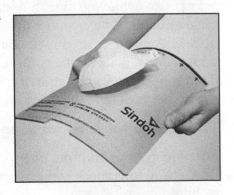

Sindoh 3DWOX 1 and 2X
flexible metal magnet build plate

Know your Printer's Limitations

Overall part size can be an issue. Most affordable 3D printers typically are small enough to fit on your desktop. Typical build volumes range between 200 x 200 x 185 mm and 228 x 200 x 300 mm. The Sindoh 3DWOX 2X has one of the largest build volumes for an affordable desktop FFF 3D printer.

Sindoh 3DWOX 2X

There are features that are too small to be printed on a desktop 3D printer. An important, but often overlooked variable in what the printer can achieve is thread width. Thread width is determined by the diameter of the extruder nozzle. Most printers have a 0.4mm nozzle. A circle created by the printer is approximately two thread widths deep: 0.8mm thick with a 0.4mm nozzle to 1mm thick for a 0.5mm nozzle. A good rule of thumb is "The smallest feature you can create is double the thread width."

Tolerance for Interlocking Parts

For objects with multiple interlocking parts, design for a tolerance fit. Getting tolerances correct, can be difficult using FFF technology.

In general, use the below suggested guidelines.

- Use ±0.1mm (±.004 in.) tolerance for a tight fit (press fit parts, connectors).

- Use ±0.2mm (±.008 in.) tolerance for a print in place (hinge).

- Use ±0.3mm (±.012 in.) tolerance for loose fit (pin in hole).

Test the fit yourself with the particular model to determine the right tolerance for the items you are creating and material you are using.

💡 Tolerance may vary depending on filament type, manufacturer, color, humidity, build plate flatness, bed temperature, extruder temperature, etc.

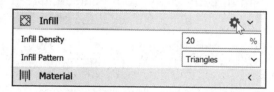

Ultimaker Cura slicer - Version: 3.4.1

General Printing Tips

Reduce Infill (Density/Overlap)

Infill is a settable variable in most Slicers. The amount of infill can affect the top layers, bottom layers, infill line distance and infill overlap.

Reduced infill can have a negative effect on part strength. There are always trade-offs.

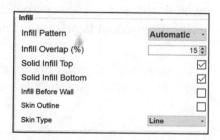

Sindoh 3DWOX slicer - Version: 1.4.2102.0

The material inside the part (infill) exerts a force on the entire printed part as it cools. More material increases cost of the part and build time.

Parts with a lower percentage of infill should have a lower internal force between layers and can reduce the chance of curling, cracking, and layer delamination along with a low build cost and time.

Control Build Area Temperature

For a consistent quality build, control the build area and environment temperature. Eliminate all drafts and control air flow that may cause a temperature gradient within the build area.

Sindoh 3DWOX - Standard enclosed build area

Changes in temperature during a build cycle can cause curling, cracking and layer delamination, especially on long thin parts. From 1000s of hours in 3D printing experience, we have found that having a top cover and sides along with a consistent room temperature provides the best and repeatable builds.

Some printer companies like Ultimaker and FlashForge, sell additional enclosure kits (doors, sides, and tops). This helps regulate the temperature and reduce drafts for improved print quality.

The kits also lower the sound of the printer and provide a more secure print area.

FlashForge 3D Printer Creator Pro

When troubleshooting issues with your printer, it is always best to know if the nozzle and heated bed are achieving the desired temperatures. A thermocouple and a thermometer come in handy. I prefer a Type-K Thermocouple connected to a multi-meter. A non-contact IR or Laser-based thermometer also works well.

One important thing to remember is that IR and Laser units are not 100% accurate when it comes to shiny reflective surfaces. A Type-K thermocouple can be taped to the nozzle or heated bed using Kapton tape, for a very accurate temperature measurement.

Cover/Door kit for Ultimaker 3

Add Pads

Sometimes, when you are printing a large flat object, you may view warping at the corners or extremities. One way to address this is to create small pads to your part during the modeling process. Create the model for the print. Think before you print and know your printer limitations. The pads can be any size and shape, but generally, diameter 10mm cylinders that are 1-2 layers thick work well. After the part is printed, remove them.

Makerbot Image

Unique Shape or a Large Part

If you need to make parts larger than your build area or create parts that have intricate projections, here are a few suggestions:

- Fuse smaller sections together using acetone (if using ABS). Glue if using PLA.

- Design smaller parts to be attached together (without hardware).

- Design smaller parts to be screwed together (with hardware).

Safe Zone Rule

Parts may have a safe zone. The safe zone is called "self-supporting" and no support material is required to build the part.

The safe zone can range between 30° to 150°. If the part's features are below 30° or greater than 150°, it should have support material during the build cycle. This is only a rule. Are there other factors to consider? Yes. They are layer thickness, extrusion speed, material type, length of the overhang along with the general model design of features.

Design your part for your printer. Use various modeling techniques (ribs, fillets, pads, etc.) during the design process to eliminate or to minimize the need for supports and clean up.

First Layer Not Sticking

One of the toughest aspects of 3D printing is to get your prints to stick to the build surface or bed platform. Investigate the following:

- Clean the bed. Remove any residue of tape, glue, hair spray, etc.

- Apply new blue painter's tape (non-heated), hair spray or glue to the build plate if needed.

- Level the bed (build platform). Perform an automatic (Assisted Bed Leveling) or manual leveling.

- Check extruder (hot end) temperature. Different filaments require different hot end temperatures.

- Check heated build plate temperature. Different filaments require different bed temperatures.

- Control the build area temperature around the printer. Eliminate all drafts.

- Layer height. Min layer height = 1/4 nozzle diameter. Max layer height = 1/2 nozzle diameter. Layer height too low, might cause the filament to be pushed back into the nozzle (plugging). If layer height is too high, the layers won't stick to the build plate.

Level Build Platform

An unleveled build platform will cause many headaches during a print. You can quickly check the platform by performing the business card test: use a single business card to judge the height of the extruder nozzle over the build platform. Achieve a consistent slight resistance when you position the business card between the tip of the extruder and the bed platform for all leveling positions.

Most 3D printers have an automatic Assisted Bed Leveling feature as illustrated.

Sindoh 3DWOX Printers

Minimize Internal Support

Design the part for the printer. Use various modeling techniques in SOLIDWORKS (ribs, fillets, pads, etc.) during the design process to eliminate or to minimize the need for support and final part clean up.

Design a Water Tight Mesh

A water-tight mesh is achieved by having closed edges creating a solid volume. If you were to fill your geometry with water, would you see a leak? You may have to clean up any internal geometry that could have been left behind accidentally from Booleans.

Clearance

If you are creating separate or interlocking parts, make sure there is a large enough distance between tight areas. 3D printing production makes moving parts without assembly a possibility. Take advantage of this strength by creating enough clearance that the model's pieces do not fuse together or trap support material inside.

In General (FFF Printers)

- Keep your software and firmware up to date.

- Think before you print. Design the model for your printer.

- Understand the printer's limitations. Adjust one thing at a time between prints and keep notes about the settings and effect on the print. Label test prints and take photographs.

- Control the build area and environment temperature. Eliminate all drafts and control air flow that may cause a temperature gradient within the build area.

- Level and clean the build plate before a build.

- Select the correct filament (material) for the application. Materials are still an area of active exploration.

- Set the correct extruder (hot end) and bed plate temperature.

- Most low-end 3D printers use an extruder (hot end) with Polyether ether Ketone (PEEK) or Polytetrafluoroethylene (Teflon). Both PEEK and PTFE begin to breakdown above 240°C and will burn and emit noxious fumes. Use an all metal hot end above 240°C.

- Select the correct part orientation.

- Suppress mates in an assmbly to have the model lay flat on the build plate.

- Most parts can be printed successfully with 15 - 20% infill.

- Select the correct settings for your Slicer. If in doubt, use the factory default settings.

- Control the filament storage environment (temperature, humidity, etc.).

- General 45 degree rule. If your model doesn't have any overhangs greater than 45 degrees, you should not need support material. If the part has numerous holes, sharp edges, long run (bridge), or thin bodies, support material may be needed.

- If in doubt, create your first build with a raft and support.

- If needed, orientate the printed part on the bed plate for maximum strength (lines perpendicular to the force being applied).

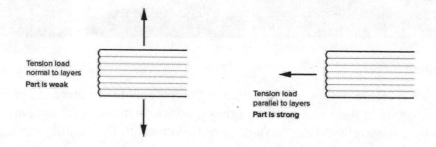

Print Directly from SOLIDWORKS

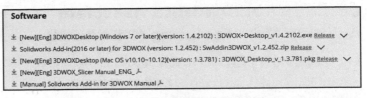

Download and install the printer drivers.

Download and install the SOLIDWORKS slicer Add-in. In this case, the SOLIDWORKS 3DWOX.

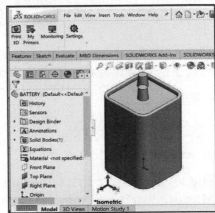

🔅 Sceen shots and procedure will vary depending on the 3D printer manufacture's slicer Add-in.

When the 3DWOX Add-in installation is complete, a 3DWOX tab is displayed in the CommandManager.

Open a SOLIDWORKS model.

Click the 3DWOX tab. View your options:

- **Print 3D**.

- **My Printers**.

- **Monitoring**.

- **Settings**.

The **Print 3D** button provides the ability to access the slicer within SOLIDWORKS. The SOLIDWORKS part model is automatically converted into an STL file and appears on the selected (Settings button) printer bed. The Sindoh 3DWOX slicer is displayed.

The **My Printers** button provides the ability to manage your printers (printer name, IP and availability) over a network.

The **Monitoring** button provides the ability to connect to the network and view your printing real time with an internal camera.

The **Settings** button displays the printers that you are connected to.

Certified SOLIDWORKS Associate Additive Manufacturing (CSWA-AM)

The Certified SOLIDWORKS Associate Additive Manufacturing (CSWA-AM) exam indicates a foundation in and apprentice knowledge of today's 3D printing technology and market.

The CSWA-AM exam is meant to be taken after the completion of the 10-part learning path located on MySOLIDWORKS.com.

The learning content is free, but the creation of a MySOLIDWORKS account is needed in order to access the content.

Lessons	Status	Languages
Introduction To Additive Manufacturing	✓ Completed	ENG
Machine Types	✓ Completed	ENG
Materials	✓ Completed	ENG
Model Preparation	✓ Completed	ENG
File Export Settings	✓ Completed	ENG
Machine Preparation	✓ Completed	ENG
Printing the Part	✓ Completed	ENG
Post Printing	✓ Completed	ENG
Part Finishing	✓ Completed	ENG
Software Options	✓ Completed	ENG

The lessons cover **Introduction to Additive Manufacturing** (7 minutes), **Machine Types** (8 minutes), **Materials** (7 minutes), **Model Preparation** (9 minutes), **File Export Settings** (8 minutes), **Machine Preparation** (7 minutes), **Printing the Part** (6 minutes), **Post Printing** (7 minutes), **Part Finishing** (9 minutes) and **Software Options** (9 minutes).

After each lesson, there is a short online quiz covering the topic area. The lessons are focused on two types of 3D printer technology: Fused Filament Fabrication (FFF) and STereoLithography (SLA). There are a few questions on Selective Laser Sintering (SLS) technology and available software-based printing aids.

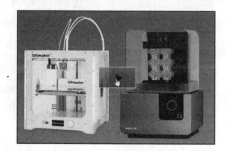

The exam covers the Ultimaker 3 (FFF) and the FormLabs Form 2 (SLA) machine as examples.

Fused Filament Fabrication (FFF) and Fused Deposition Modeling (FDM) are used interchangeably in the certification exam.

The CSWA-AM (Additive Manufacturing) exam covers numerous areas: material types, printing technologies, machine types and processes, part design and orientation for 3D printing, printer preparation, post printing finishing STereoLithography (SLA), Slicer software features and functionality and available software-based printing aids.

The exam is 50 questions. Each question is worth 2 points. You are allowed to answer the questions in any order. Total exam time is 60 minutes. You need a minimum passing score of 80 or higher. The exam is out of 100 points.

Summary

Stereolithography (SLA) was the world's first 3D printing technology. It was introduced in 1988 by 3D Systems, Inc., based on work by inventor Charles Hull.

SLA is one of the most popular resin-based 3D printing technologies for professionals today.

Selective Laser Sintering (SLS) technology is the most common additive manufacturing technology for industrial powder base applications. SLS fuses particles together layer by layer through a high energy pulse laser. Similar to SLA, this process starts with a tank full of bulk material but is in a powder form vs. liquid.

Sindoh 3DWOX 1

Fused filament fabrication (FFF) is an additive manufacturing technology used for building three-dimensional prototypes or models layer by layer with a range of thermoplastics.

FFF technology is the most widely used form of 3D desktop printing at the consumer level, fueled by students, hobbyists and office professionals.

The two most dominant plastics for FFF technology are PLA (Polylactic acid) and ABS (Acrylonitrile-Butadiene-Styrene).

With 1000s of hours using multiple low cost FFF 3D printers, we have found that learning about additive print technology is a great experience. Students face a multitude of obstacles with their first 3D prints. Understanding what went wrong and knowing the capabilities of the 3D printer produces positive results.

Never focus too much on one single issue. These machines are complex, and trouble often arises from multiple reasons. A slipping filament may not only be caused by a bad gear drive, but also by an obstructed nozzle, a wrong feed value, a too low (or too high) temperature or a combination of all these.

Always store filament in an air tight plastic bag (container) with a few silica gel packs. Place the bag in a dry, dark, temperature controlled environment. As an extra precaution, filament manufacturers often recommend using up rolls as soon as possible.

It's best to monitor the first 3 - 5 layers when starting a print. Never assume you will have perfect bed adhesion. If there is a bed adhesion issue, check to see if the bed is level. Check the bed and extruder (hot end) temperatures and then either apply blue painter's tape (non-heated), a glue stick or hair spray. You can also add a raft to the part.

Always clean and level the build plate before every print.

As printed parts cool (PLA, ABS and Nylon), various areas of the object cool at different rates. Depending on the model being printed and the filament material, this effect can lead to warping, curling and or layer delamination.

Acetone is often used in post processing to smooth ABS, also giving the part a glossy finish. ABS can be sanded and is often machined (for example, drilled) after printing. PLA can also be sanded and machined; however, greater care is required.

For high temperature applications, ABS (glass transition temperature of 105°C) is more suitable than PLA (glass transition temperature of 60°C). PLA can rapidly lose its structural integrity as it approaches 55C - 60°C.

Design the part for your printer. Try to use the 45 degree rule. If your model doesn't have any overhangs greater than 45 degrees, you should not need support. There are exceptions to the 45 degree rule. The most common ones are straight overhangs, and fully suspended islands.

Use various CAD modeling techniques (ribs, fillets, pads, etc.) during the design process to eliminate or to minimize the need for supports and clean up.

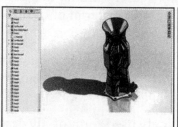

Native Data Transfer

Maintain Design Integrity

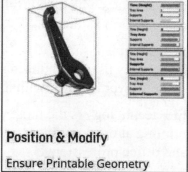

Position & Modify

Ensure Printable Geometry

From a structural situation, select the correct % infill, infill pattern and orientate the part on the build plate to print the layers perpendicular in relation to the movement of the build platform.

3DXpert is a SOLIDWORKS Add-in. 3DXpert for SOLIDWORKS provides an extensive toolset to analyze, prepare and optimize your design for additive manufacturing. It also provides the ability to print an assembly file as a single part.

Design Supports

Ensure Quality Prints with Minimal Supports

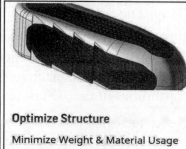

Optimize Structure

Minimize Weight & Material Usage and Apply Surface Textures

Key tools are:

- **Native Data Transfer**.

- **Position & Modification**.

- **Optimize Structure**.

- **Arrange Build Plate**.

Arrange Build Plate

Best Utilization of Tray Area and Printer Time

Appendix

SOLIDWORKS Keyboard Shortcuts

Below are some of the pre-defined keyboard shortcuts in SOLIDWORKS:

Action:	Key Combination:
Model Views	
Rotate the model horizontally or vertically	**Arrow** keys
Rotate the model horizontally or vertically 90 degrees	**Shift** + **Arrow** keys
Rotate the model clockwise or counterclockwise	**Alt** + left of right **Arrow** keys
Pan the model	**Ctrl** + **Arrow** keys
Magnifying glass	**g**
Zoom in	**Shift** + **z**
Zoom out	**z**
Zoom to fit	**f**
Previous view	**Ctrl** + **Shift** + **z**
View Orientation	
View Orientation menu	**Spacebar**
Front view	**Ctrl** + **1**
Back view	**Ctrl** + **2**
Left view	**Ctrl** + **3**
Right view	**Ctrl** + **4**
Top view	**Ctrl** + **5**
Bottom view	**Ctrl** + **6**
Isometric view	**Ctrl** + **7**
NormalTo view	**Ctrl** + **8**
Selection Filters	
Filter edges	**e**
Filter vertices	**v**
Filter faces	**x**
Toggle Selection Filter toolbar	**F5**
Toggle selection filters on/off	**F6**
File menu items	
New SOLIDWORKS document	**Ctrl** + **n**
Open document	**Ctrl** + **o**
Open From Web Folder	**Ctrl** + **w**
Make Drawing from Part	**Ctrl** + **d**
Make Assembly from Part	**Ctrl** + **a**
Save	**Ctrl** + **s**
Print	**Ctrl** + **p**
Additional items	
Access online help inside of PropertyManager or dialog box	**F1**
Rename an item in the FeatureManager design tree	**F2**

Action:	Key Combination:
Rebuild the model	**Ctrl + b**
Force rebuild - Rebuild the model and all its features	**Ctrl + q**
Redraw the screen	**Ctrl + r**
Cycle between open SOLIDWORKS document	**Ctrl + Tab**
Line to arc/arc to line in the Sketch	**a**
Undo	**Ctrl + z**
Redo	**Ctrl + y**
Cut	**Ctrl + x**
Copy	**Ctrl + c**
Paste	**Ctrl + v**
Delete	**Delete**
Next window	**Ctrl + F6**
Close window	**Ctrl + F4**
View previous tools	**s**
Selects all text inside an Annotations text box	**Ctrl + a**

In a sketch, the **Esc** key un-selects geometry items currently selected in the Properties box and Add Relations box.

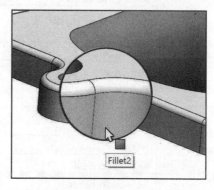

In the model, the **Esc** key closes the PropertyManager and cancels the selections.

Use the **g** key to activate the Magnifying glass tool. Use the Magnifying glass tool to inspect a model and make selections without changing the overall view.

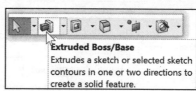

Use the **s** key to view/access previous command tools in the Graphics window.

Modeling - Best Practices

Best practices are simply ways of bringing about better results in easier, more reliable ways. The Modeling - Best Practice list is a set of rules helpful for new users and users who are trying to experiment with the limits of the software.

These rules are not inflexible, but conservative starting places; they are concepts that you can default to, but that can be broken if you have good reason. The following is a list of suggested best practices:

- Create a folder structure (parts, drawings, assemblies, simulations, etc.). Organize into project or file folders.

- Construct sound document templates. The document template provides the foundation that all models are built on. This is especially important if working with other SOLIDWORKS users on the same project; it will ensure consistency across the project.

- Generate unique part filenames. SOLIDWORKS assemblies and drawings may pick up incorrect references if you use parts with identical names.

- Apply Custom Properties. Custom Properties is a great way to enter text-based information into the SOLIDWORKS parts. Users can view this information from outside the file by using applications such as Windows Explorer, SOLIDWORKS Explorer, and Product Data Management (PDM) applications.

- Understand part orientation. When you create a new part or assembly, the three default Planes (Front, Right and Top) are aligned with specific views. The plane you select for the Base sketch determines the orientation.

- Learn to sketch using automatic relations.

- Limit your usage of the Fixed constraint.

- Add geometric relations, then dimensions in a 2D sketch. This keeps the part from having too many unnecessary dimensions. This also helps to show the design intent of the model. Dimension what geometry you intend to modify or adjust.

- Fully define all sketches in the model. However, there are times when this is not practical, generally when using the Spline tool to create a freeform shape.

- When possible, make relations to sketches or stable reference geometry, such as the Origin or standard planes, instead of edges or faces. Sketches are far more stable than faces, edges, or model vertices, which change their internal ID at the slightest change and may disappear entirely with fillets, chamfers, split lines, and so on.

- Do not dimension to edges created by fillets or other cosmetic or temporary features.

- Apply names to sketches, features, dimensions, and mates that help to make their function clear.

- When possible, use feature fillets and feature patterns rather than sketch fillets and sketch patterns.

- Apply the Shell feature before the Fillet feature, and the inside corners remain perpendicular.

- Apply cosmetic fillets and chamfers last in the modeling procedure.

- Combine fillets into as few fillet features as possible. This enables you to control fillets that need to be controlled separately, such as fillets to be removed and simplified configurations.

- Create a simplified configuration when building very complex parts or working with large assemblies.

- Use symmetry during the modeling process. Utilize feature patterns and mirroring when possible. Think End Conditions.

- Use global variables and equations to control commonly applied dimensions (design intent).

- Add comments to equations to document your design intent. Place a single quote (') at the end of the equation, then enter the comment. Anything after the single quote is ignored when the equation is evaluated.

- Avoid redundant mates. Although SOLIDWORKS allows some redundant mates (all except distance and angle), these mates take longer to solve and make the mating scheme harder to understand and diagnose if problems occur.

- Fix modeling errors in the part or assembly when they occur. Errors cause rebuild time to increase, and if you wait until additional errors exist, troubleshooting will be more difficult.

- Create a Library of Standardize notes and parts.

- Utilize the Rollback bar. Troubleshoot feature and sketch errors from the top of the design tree.

- Determine the static and dynamic behavior of mates in each sub-assembly before creating the top-level assembly.

- Plan the assembly and sub-assemblies in an assembly layout diagram. Group components together to form smaller sub-assemblies.

- When you create an assembly document, the base component should be fixed, fully defined or mated to an axis about the assembly origin.

- In an assembly, group fasteners into a folder at the bottom of the FeatureManager. Suppress fasteners and their assembly patterns to save rebuild time and file size.

- When comparing mass, volume and other properties with assembly visualization, utilize similar units.

- Use limit mates sparingly because they take longer to solve and whenever possible, mate all components to one or two fixed components or references. Long chains of components take longer to solve and are more prone to mate errors.

Helpful On-line Information

The SOLIDWORKS URL:
http://www.SOLIDWORKS.com
contains information on Local
Resellers, Solution Partners,
Certifications, SOLIDWORKS users
groups and more.

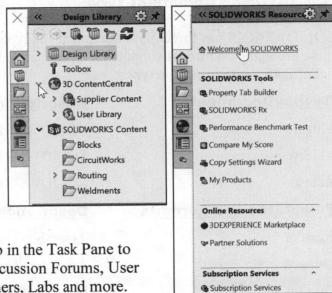

Access 3D ContentCentral using the
Task Pane to obtain engineering
electronic catalog model and part
information.

Use the SOLIDWORKS Resources tab in the Task Pane to
obtain access to Customer Portals, Discussion Forums, User
Groups, Manufacturers, Solution Partners, Labs and more.

Helpful on-line SOLIDWORKS information is available from
the following URLs:

- http://www.swugn.org/

List of all SOLIDWORKS User groups.

- https://www.solidworks.com/sw/education/certification-programs-cad-students.htm

The SOLIDWORKS Academic
Certification Programs.

- http://www.solidworks.com/sw/industries/education/engineering-education-software.htm

The SOLIDWORKS Education
Program.

- https://solidworks.virtualtester.com/#home_button

The SOLIDWORKS Certification Center - Virtual tester site.

*On-line tutorials are for educational purposes only. Tutorials are
copyrighted by their respective owners.

SOLIDWORKS Document Types

SOLIDWORKS has three main document file types: Part, Assembly and Drawing, but there are many additional supporting types that you may want to know. Below is a brief list of these supporting file types:

Design Documents	Description
.sldprt	SOLIDWORKS Part document
.slddrw	SOLIDWORKS Drawing document
.sldasm	SOLIDWORKS Assembly document

Templates and Formats	Description
.asmdot	Assembly Template
.asmprp	Assembly Template Custom Properties tab
.drwdot	Drawing Template
.drwprp	Drawing Template Custom Properties tab
.prtdot	Part Template
.prtprp	Part Template Custom Properties tab
.sldtbt	General Table Template
.slddrt	Drawing Sheet Template
.sldbombt	Bill of Materials Template (Table-based)
.sldholtbt	Hole Table Template
.sldrevbt	Revision Table Template
.sldwldbt	Weldment Cutlist Template
.xls	Bill of Materials Template (Excel-based)

Library Files	Description
.sldlfp	Library Part file
.sldblk	Blocks

Other	Description
.sldstd	Drafting standard
.sldmat	Material Database
.sldclr	Color Palette File
.xls	Sheet metal gauge table

Engineering Change Order (ECO)

D&M	Engineering Change Order	ECO # _____
		Page 1 of __

| Product Line | ☐ Hardware
☐ Software
☐ Quality
☐ Tech Pubs | Author
Date
Authorized Mgr.
Date |

Change Tested By

Reason for ECO(Describe the existing problem, symptom and impact on field)

D&M Part No.	Rev From/To	Part Description	Description	Owner

ECO Implementation/Class		Departments	Approvals	Date
All in Field	☐	Engineering		
All in Test	☐	Manufacturing		
All in Assembly	☐	Technical Support		
All in Stock	☐	Marketing		
All on Order	☐	DOC Control		
All Future				
Material Disposition		ECO Cost		
Rework	☐	DO NOT WRITE BELOW THIS LINE (ECO BOARD ONLY)		
Scrap	☐	Effective Date		
Use as is	☐	Incorporated Date		
None	☐	Board Approval		
See Attached	☐	Board Date		

This text follows the ASME Y14 2009 Engineering Drawing and Related Documentation Practices for drawings. Display of dimensions and tolerances are as follows:

TYPES of DECIMAL DIMENSIONS (ASME Y14.5 2009)			
Description:	**UNITS:** **MM**	**Description:**	**UNITS:** **INCH**
Dimension is less than 1mm. Zero precedes the decimal point.	0.9 0.95	Dimension is less than 1 inch. Zero is not used before the decimal point.	.5 .56
Dimension is a whole number. Display no decimal point. Display no zero after decimal point.	19	Express dimension to the same number of decimal places as its tolerance. Add zeros to the right of the decimal point.	1.750
Dimension exceeds a whole number by a decimal fraction of a millimeter. Display no zero to the right of the decimal.	11.5 11.51	If the tolerance is expressed to 3 places, then the dimension contains 3 places to the right of the decimal point.	

TABLE 1 TOLERANCE DISPLAY FOR INCH AND METRIC DIMENSIONS (ASME Y14.5 2009)		
DISPLAY:	**UNITS:** **INCH:**	**UNITS:** **METRIC:**
Dimensions less than 1	.5	0.5
Unilateral Tolerance	$1.417^{+.005}_{-.000}$	$36^{0}_{-0.5}$
Bilateral Tolerance	$1.417^{+.010}_{-.020}$	$36^{+0.25}_{-0.50}$
Limit Tolerance	.571 .463	14.50 11.50

GLOSSARY

Alphabet of Lines: Each line on a technical drawing has a definite meaning and is drawn in a certain way. The line conventions recommended by the American National Standards Institute (ANSI) are presented in this text.

Alternate Position View: A drawing view superimposed in phantom lines on the original view. Utilized to show range of motion of an assembly.

Anchor Point: The origin of the Bill of Material in a sheet format.

Annotation: An annotation is a text note or a symbol that adds specific information and design intent to a part, assembly, or drawing. Annotations in a drawing include specific note, hole callout, surface finish symbol, datum feature symbol, datum target, geometric tolerance symbol, weld symbol, balloon, and stacked balloon, center mark, centerline marks, area hatch and block.

ANSI: American National Standards Institute.

Area Hatch: Apply a crosshatch pattern or solid fill to a model face, to a closed sketch profile, or to a region bounded by a combination of model edges and sketch entities. Area hatch can be applied only in drawings.

ASME: American Society of Mechanical Engineering, publisher of ASME Y14 Engineering Drawing and Documentation Practices that controls drawing, dimensioning and tolerancing.

Assembly: An assembly is a document in which parts, features and other assemblies (sub-assemblies) are put together. A part in an assembly is called a component. Adding a component to an assembly creates a link between the assembly and the component. When SOLIDWORKS opens the assembly, it finds the component file to show it in the assembly. Changes in the component are automatically reflected in the assembly. The filename extension for a SOLIDWORKS assembly file name is *.sldasm.

Attachment Point: An attachment point is the end of a leader that attaches to an edge, vertex, or face in a drawing sheet.

AutoDimension: The Autodimension tool provides the ability to insert reference dimensions into drawing views such as baseline, chain, and ordinate dimensions.

Auxiliary View: An Auxiliary View is similar to a Projected View, but it is unfolded normal to a reference edge in an existing view.

AWS: American Welding Society, publisher of AWS A2.4, Standard Location of Elements of a Welding Symbol.

Axonometric Projection: A type of parallel projection, more specifically a type of orthographic projection, used to create a pictorial drawing of an object, where the object is rotated along one or more of its axes relative to the plane of projection.

Balloon: A balloon labels the parts in the assembly and relates them to item numbers on the bill of materials (BOM) added in the drawing. The balloon item number corresponds to the order in the Feature Tree. The order controls the initial BOM Item Number.

Baseline Dimensions: Dimensions referenced from the same edge or vertex in a drawing view.

Bill of Materials: A table inserted into a drawing to keep a record of the parts and materials used in an assembly.

Block: A symbol in the drawing that combines geometry into a single entity.

BOM: Abbreviation for Bill of Materials.

Broken-out Section: A broken-out section exposes inner details of a drawing view by removing material from a closed profile. In an assembly, the Broken-out Section displays multiple components.

CAD: The use of computer technology for the design of objects, real or virtual. CAD often involves more than just shapes.

Cartesian Coordinate System: Specifies each point uniquely in a plane by a pair of numerical coordinates, which are the signed distances from the point to two fixed perpendicular directed lines, measured in the same unit of length. Each reference line is called a coordinate axis or just axis of the system, and the point where they meet is its origin.

Cell: Area to enter a value in an EXCEL spreadsheet, identified by a Row and Column.

Center Mark: A cross that marks the center of a circle or arc.

Centerline: An axis of symmetry in a sketch or drawing displayed in a phantom font.

CommandManager: The CommandManager is a Context-sensitive toolbar that dynamically updates based on the toolbar you want to access. By default, it has toolbars embedded in it based on the document type. When you click a tab below the Command Manager, it updates to display that toolbar. For example, if you click the Sketch tab, the Sketch toolbar is displayed.

Component: A part or sub-assembly within an assembly.

ConfigurationManager: The ConfigurationManager is located on the left side of the SOLIDWORKS window and provides the means to create, select and view multiple configurations of parts and assemblies in an active document. You can split the

ConfigurationManager and either display two ConfigurationManager instances, or combine the ConfigurationManager with the FeatureManager design tree, PropertyManager or third party applications that use the panel.

Configurations: Variations of a part or assembly that control dimensions, display and state of a model.

Coordinate System: SOLIDWORKS uses a coordinate system with origins. A part document contains an original origin. Whenever you select a plane or face and open a sketch, an origin is created in alignment with the plane or face. An origin can be used as an anchor for the sketch entities, and it helps orient perspective of the axes. A three-dimensional reference triad orients you to the X, Y, and Z directions in part and assembly documents.

Copy and Paste: Utilize copy/paste to copy views from one sheet to another sheet in a drawing or between different drawings.

Cosmetic Thread: An annotation that represents threads.

Crosshatch: A pattern (or fill) applied to drawing views such as section views and broken-out sections.

Cursor Feedback: The system feedback symbol indicates what you are selecting or what the system is expecting you to select. As you move the mouse pointer across your model, system feedback is provided.

Datum Feature: An annotation that represents the primary, secondary and other reference planes of a model utilized in manufacturing.

Depth: The horizontal (front to back) distance between two features in frontal planes. Depth is often identified in the shop as the thickness of a part or feature.

Design Table: An Excel spreadsheet that is used to create multiple configurations in a part or assembly document.

Detail View: A portion of a larger view, usually at a larger scale than the original view. Create a detail view in a drawing to display a portion of a view, usually at an enlarged scale. This detail may be of an orthographic view, a non-planar (isometric) view, a section view, a crop view, an exploded assembly view or another detail view.

Detailing: Detailing refers to the SOLIDWORKS module used to insert, add and modify dimensions and notes in an engineering drawing.

Dimension Line: A line that references dimension text to extension lines indicating the feature being measured.

Dimension Tolerance: Controls the dimension tolerance values and the display of non-integer dimensions. The tolerance types are *None, Basic, Bilateral, Limit, Symmetric, MIN, MAX, Fit, Fit with tolerance* or *Fit (tolerance only)*.

Dimension: A value indicating the size of the 2D sketch entity or 3D feature. Dimensions in a SOLIDWORKS drawing are associated with the model, and changes in the model are reflected in the drawing, if you DO NOT USE DimXpert.

Dimensioning Standard - Metric: - ASME standards for the use of metric dimensioning required all the dimensions to be expressed in millimeters (mm). The (mm) is not needed on each dimension, but it is used when a dimension is used in a notation. No trailing zeroes are used. The Metric or International System of Units (S.I.) unit system in drafting is also known as the Millimeter, Gram Second (MMGS) unit system.

Dimensioning Standard - U.S: - ASME standard for U.S. dimensioning use the decimal inch value. When the decimal inch system is used, a zero is not used to the left of the decimal point for values less than one inch, and trailing zeroes are used. The U.S. unit system is also known as the Inch, Pound, Second (IPS) unit system.

DimXpert for Parts: A set of tools that applies dimensions and tolerances to parts according to the requirements of the ASME Y.14.41-2009 standard.

DimXpertManager: The DimXpertManager lists the tolerance features defined by DimXpert for a part. It also displays DimXpert tools that you use to insert dimensions and tolerances into a part. You can import these dimensions and tolerances into drawings. DimXpert is not associative.

Document: In SOLIDWORKS, each part, assembly, and drawing is referred to as a document, and each document is displayed in a separate window.

Drawing Sheet: A page in a drawing document.

Drawing Template: A document that is the foundation of a new drawing. The drawing template contains document properties and user-defined parameters such as sheet format. The extension for the drawing template filename is .DRWDOT.

Drawing: A 2D representation of a 3D part or assembly. The extension for a SOLIDWORKS drawing file name is .SLDDRW. Drawing refers to the SOLIDWORKS module used to insert, add, and modify views in an engineering drawing.

Edit Sheet Format: The drawing sheet contains two modes. Utilize the Edit Sheet Format command to add or modify notes and Title block information. Edit in the Edit Sheet Format mode.

Edit Sheet: The drawing sheet contains two modes. Utilize the Edit Sheet command to insert views and dimensions.

eDrawing: A compressed document that does not require the referenced part or assembly. eDrawings are animated to display multiple views in a drawing.

Empty View: An Empty View creates a blank view not tied to a part or assembly document.

Engineering Graphics: Translates ideas from design layouts, specifications, rough sketches, and calculations of engineers & architects into working drawings, maps, plans and illustrations which are used in making products.

Equation: Creates a mathematical relation between sketch dimensions, using dimension names as variables, or between feature parameters, such as the depth of an extruded feature or the instance count in a pattern.

Exploded view: A configuration in an assembly that displays its components separated from one another.

Export: The process to save a SOLIDWORKS document in another format for use in other CAD/CAM, rapid prototyping, web or graphics software applications.

Extension Line: The line extending from the profile line indicating the point from which a dimension is measured.

Extruded Cut Feature: Projects a sketch perpendicular to a Sketch plane to remove material from a part.

Face: A selectable area (planar or otherwise) of a model or surface with boundaries that help define the shape of the model or surface. For example, a rectangular solid has six faces.

Family Cell: A named empty cell in a Design Table that indicates the start of the evaluated parameters and configuration names. Locate Comments in a Design Table to the left or above the Family Cell.

Fasteners: Includes Bolts and nuts (threaded), Set screws (threaded), Washers, Keys, and Pins to name a few. Fasteners are not a permanent means of assembly such as welding or adhesives.

Feature: Features are geometry building blocks. Features add or remove material. Features are created from 2D or 3D sketched profiles or from edges and faces of existing geometry.

FeatureManager: The FeatureManager design tree located on the left side of the SOLIDWORKS window provides an outline view of the active part, assembly, or drawing. This makes it easy to see how the model or assembly was constructed or to examine the various sheets and views in a drawing. The FeatureManager and the Graphics window are dynamically linked. You can select features, sketches, drawing views and construction geometry in either pane.

First Angle Projection: In First Angle Projection the Top view is looking at the bottom of the part. First Angle Projection is used in Europe and most of the world. However, America and Australia use a method known as Third Angle Projection.

Fully defined: A sketch where all lines and curves in the sketch, and their positions, are described by dimensions or relations, or both, and cannot be moved. Fully defined sketch entities are shown in black.

Foreshortened radius: Helpful when the centerpoint of a radius is outside of the drawing or interferes with another drawing view: Broken Leader.

Foreshortening: The way things appear to get smaller in both height and depth as they recede into the distance.

French curve: A template made out of plastic, metal or wood composed of many different curves. It is used in manual drafting to draw smooth curves of varying radii.

Fully Defined: A sketch where all lines and curves in the sketch, and their positions, are described by dimensions or relations, or both, and cannot be moved. Fully defined sketch entities are displayed in black.

Geometric Tolerance: A set of standard symbols that specify the geometric characteristics and dimensional requirements of a feature.

Glass Box method: A traditional method of placing an object in an *imaginary glass box* to view the six principal views.

Global Coordinate System: Directional input refers by default to the Global coordinate system (X-, Y- and Z-), which is based on Plane1 with its origin located at the origin of the part or assembly.

Graphics Window: The area in the SOLIDWORKS window where the part, assembly, or drawing is displayed.

Grid: A system of fixed horizontal and vertical divisions.

Handle: An arrow, square or circle that you drag to adjust the size or position of an entity such as a view or dimension.

Heads-up View Toolbar: A transparent toolbar located at the top of the Graphic window.

Height: The vertical distance between two or more lines or surfaces (features) which are in horizontal planes.

Hidden Lines Removed (HLR): A view mode. All edges of the model that are not visible from the current view angle are removed from the display.

Hidden Lines Visible (HLV): A view mode. All edges of the model that are not visible from the current view angle are shown gray or dashed.

Hole Callouts: Hole callouts are available in drawings. If you modify a hole dimension in the model, the callout updates automatically in the drawing if you did not use DimXpert.

Hole Table: A table in a drawing document that displays the positions of selected holes from a specified origin datum. The tool labels each hole with a tag. The tag corresponds to a row in the table.

Import: The ability to open files from other software applications into a SOLIDWORKS document. The A-size sheet format was created as an AutoCAD file and imported into SOLIDWORKS.

Isometric Projection: A form of graphical projection, more specifically, a form of axonometric projection. It is a method of visually representing three-dimensional objects in two dimensions, in which the three coordinate axes appear equally foreshortened and the angles between any two of them are 120°.

Layers: Simplifies a drawing by combining dimensions, annotations, geometry and components. Properties such as display, line style and thickness are assigned to a named layer.

Leader: A solid line created from an annotation to the referenced feature.

Line Format: A series of tools that controls Line Thickness, Line Style, Color, Layer and other properties.

Local (Reference) Coordinate System: Coordinate system other than the Global coordinate system. You can specify restraints and loads in any desired direction.

Lock Sheet Focus: Adds sketch entities and annotations to the selected sheet. Double-click the sheet to activate Lock Sheet Focus. To unlock a sheet, right-click and select Unlock Sheet Focus or double click inside the sheet boundary.

Lock View Position: Secures the view at its current position in the sheet. Right-click in the drawing view to Lock View Position. To unlock a view position, right-click and select Unlock View Position.

Mass Properties: The physical properties of a model based upon geometry and material.

Menus: Menus provide access to the commands that the SOLIDWORKS software offers. Menus are Context-sensitive and can be customized through a dialog box.

Model Item: Provides the ability to insert dimensions, annotations, and reference geometry from a model document (part or assembly) into a drawing.

Model View: A specific view of a part or assembly. Standard named views are listed in the view orientation dialog box such as isometric or front. Named views can be user-defined names for a specific view.

Model: 3D solid geometry in a part or assembly document. If a part or assembly document contains multiple configurations, each configuration is a separate model.

Motion Studies: Graphical simulations of motion and visual properties with assembly models. Analogous to a configuration, they do not actually change the original assembly model or its properties. They display the model as it changes based on simulation elements you add.

Mouse Buttons: The left, middle, and right mouse buttons have distinct meanings in SOLIDWORKS. Use the middle mouse button to rotate and Zoom in/out on the part or assembly document.

Oblique Projection: A simple type of graphical projection used for producing pictorial, two-dimensional images of three-dimensional objects.

OLE (Object Linking and Embedding): A Windows file format. A company logo or EXCEL spreadsheet placed inside a SOLIDWORKS document are examples of OLE files.

Ordinate Dimensions: Chain of dimensions referenced from a zero ordinate in a drawing or sketch.

Origin: The model origin is displayed in blue and represents the (0,0,0) coordinate of the model. When a sketch is active, a sketch origin is displayed in red and represents the (0,0,0) coordinate of the sketch. Dimensions and relations can be added to the model origin but not to a sketch origin.

Orthographic Projection: A means of representing a three-dimensional object in two dimensions. It is a form of parallel projection, where the view direction is orthogonal to the projection plane, resulting in every plane of the scene appearing in affine transformation on the viewing surface.

Parametric Note: A Note annotation that links text to a feature dimension or property value.

Parent View: A Parent view is an existing view on which other views are dependent.

Part Dimension: Used in creating a part, they are sometimes called construction dimensions.

Part: A 3D object that consist of one or more features. A part inserted into an assembly is called a component. Insert part views, feature dimensions and annotations into 2D drawing. The extension for a SOLIDWORKS part filename is .SLDPRT.

Perspective Projection: The two most characteristic features of perspective are that objects are drawn smaller as their distance from the observer increases, and foreshortened: the size of an object's dimensions along the line of sight are relatively shorter than dimensions across the line of sight.

Plane: To create a sketch, choose a plane. Planes are flat and infinite. Planes are represented on the screen with visible edges.

Precedence of Line Types: When obtaining orthographic views, it is common for one type of line to overlap another type. When this occurs, drawing conventions have established an order of precedence.

Precision: Controls the number of decimal places displayed in a dimension.

Projected View: Projected views are created for Orthogonal views using one of the following tools: Standard 3 View, Model View or the Projected View tool from the View Layout toolbar.

Properties: Variables shared between documents through linked notes.

PropertyManager: Most sketch, feature, and drawing tools in SOLIDWORKS open a PropertyManager located on the left side of the SOLIDWORKS window. The PropertyManager displays the properties of the entity or feature so you specify the properties without a dialog box covering the Graphics window.

RealView: Provides a simplified way to display models in a photo-realistic setting using a library of appearances and scenes. RealView requires graphics card support and is memory intensive.

Rebuild: A tool that updates (or regenerates) the document with any changes made since the last time the model was rebuilt. Rebuild is typically used after changing a model dimension.

Reference Dimension: Dimensions added to a drawing document are called Reference dimensions and are driven; you cannot edit the value of reference dimensions to modify the model. However, the values of reference dimensions change when the model dimensions change.

Relation: A relation is a geometric constraint between sketch entities or between a sketch entity and a plane, axis, edge or vertex.

Relative view: The Relative View defines an Orthographic view based on two orthogonal faces or places in the model.

Revision Table: The Revision Table lists the Engineering Change Orders (ECO), in a table form, issued over the life of the model and the drawing. The current Revision letter or number is placed in the Title block of the Drawing.

Right-Hand Rule: Is a common mnemonic for understanding notation conventions for vectors in 3 dimensions.

Rollback: Suppresses all items below the rollback bar.

Scale: A relative term meaning "size" in relationship to some system of measurement.

Section Line: A line or centerline sketched in a drawing view to create a section view.

Section Scope: Specifies the components to be left uncut when you create an assembly drawing section view.

Section View: You create a section view in a drawing by cutting the parent view with a cutting, or section line. The section view can be a straight cut section or an offset section defined by a stepped section line. The section line can also include concentric arcs. Create a Section View in a drawing by cutting the Parent view with a section line.

Sheet Format: A document that contains the following: page size and orientation, standard text, borders, logos, and Title block information. Customize the Sheet format to save time. The extension for the Sheet format filename is .SLDDRT.

Sheet Properties: Sheet Properties display properties of the selected sheet. Sheet Properties define the following: Name of the Sheet, Sheet Scale, Type of Projection (First angle or Third angle), Sheet Format, Sheet Size, View label, and Datum label.

Sheet: A page in a drawing document.

Silhouette Edge: A curve representing the extent of a cylindrical or curved face when viewed from the side.

Sketch: The name to describe a 2D profile is called a sketch. 2D sketches are created on flat faces and planes within the model. Typical geometry types are lines, arcs, corner rectangles, circles, polygons, and ellipses.

Spline: A sketched 2D or 3D curve defined by a set of control points.

Stacked Balloon: A group of balloons with only one leader. The balloons can be stacked vertically (up or down) or horizontally (left or right).

Standard views: The three orthographic projection views, Front, Top and Right positioned on the drawing according to First angle or Third angle projection.

Suppress: Removes an entity from the display and from any calculations in which it is involved. You can suppress features, assembly components, and so on. Suppressing an entity does not delete the entity; you can unsuppress the entity to restore it.

Surface Finish: An annotation that represents the texture of a part.

System Feedback: Feedback is provided by a symbol attached to the cursor arrow indicating your selection. As the cursor floats across the model, feedback is provided in the form of symbols riding next to the cursor.

System Options: System Options are stored in the registry of the computer. System Options are not part of the document. Changes to the System Options affect all current and future documents. There are hundreds of Systems Options.

Tangent Edge: The transition edge between rounded or filleted faces in hidden lines visible or hidden lines removed modes in drawings.

Task Pane: The Task Pane is displayed when you open the SOLIDWORKS software. It contains the following tabs: SOLIDWORKS Resources, Design Library, File Explorer, Search, View Palette, Document Recovery and RealView/PhotoWorks.

Templates: Templates are part, drawing and assembly documents that include user-defined parameters and are the basis for new documents.

Third Angle Projection: In Third angle projection the Top View is looking at the Top of the part. First Angle Projection is used in Europe and most of the world. America and Australia use the Third Angle Projection method.

Thread Class or Fit: Classes of fit are tolerance standards; they set a plus or minus figure that is applied to the pitch diameter of bolts or nuts. The classes of fit used with almost all bolts sized in inches are specified by the ANSI/ASME Unified Screw Thread standards (which differ from the previous American National standards).

Thread Lead: The distance advanced parallel to the axis when the screw is turned one revolution. For a single thread, lead is equal to the pitch; for a double thread, lead is twice the pitch.

Tolerance: The permissible range of variation in a dimension of an object. Tolerance may be specified as a factor or percentage of the nominal value, a maximum deviation from a nominal value, an explicit range of allowed values, be specified by a note or published standard with this information, or be implied by the numeric accuracy of the nominal value.

Toolbars: The toolbar menus provide shortcuts enabling you to access the most frequently used commands. Toolbars are Context-sensitive and can be customized through a dialog box.

T-Square: A technical drawing instrument, primarily a guide for drawing horizontal lines on a drafting table. It is used to guide the triangle that draws vertical lines. Its name comes from the general shape of the instrument where the horizontal member of the T slides on the side of the drafting table. Common lengths are 18", 24", 30", 36" and 42".

Under-defined: A sketch is under defined when there are not enough dimensions and relations to prevent entities from moving or changing size.

Units: Used in the measurement of physical quantities. Decimal inch dimensioning and Millimeter dimensioning are the two types of common units specified for engineering parts and drawings.

Vertex: A point at which two or more lines or edges intersect. Vertices can be selected for sketching, dimensioning, and many other operations.

View Palette: Use the View Palette, located in the Task Pane, to insert drawing views. It contains images of standard views, annotation views, section views, and flat patterns (sheet metal parts) of the selected model. You can drag views onto the drawing sheet to create a drawing view.

Weld Bead: An assembly feature that represents a weld between multiple parts.

Weld Finish: A weld symbol representing the parameters you specify.

Weld Symbol: An annotation in the part or drawing that represents the parameters of the weld.

Width: The horizontal distance between surfaces in profile planes. In the machine shop, the terms length and width are used interchangeably.

Zebra Stripes: Simulate the reflection of long strips of light on a very shiny surface. They allow you to see small changes in a surface that may be hard to see with a standard display.

INDEX